5·22·95

D1075229

The African Leopard

Biology and Resource Management in the Tropics Series

Biology and Resource Management in the Tropics Series
Michael J. Balick, Anthony B. Anderson, and Kent H. Redford, Editors

Alternatives to Deforestation: Steps Toward Sustainable Use of the Amazon Rain Forest,
   edited by Anthony B. Anderson (1990)

Useful Palms of the World: A Synoptic Bibliography,
   edited by Michael J. Balick and Hans T. Beck (1990)

The Subsidy from Nature: Palm Forests, Peasantry, and Development on an Amazon Frontier,
   by Anthony B. Anderson, Peter H. May, and Michael J. Balick (1991)

Contested Frontiers in Amazonia,
   by Marianne Schmink and Charles Wood (1992)

Conservation of Neotropical Forests,
   edited by Kent H. Redford and Christine Padoch (1992)

Footprints of the Forest: Ka'apor Ethnobotany,
   by William Balée

The African Leopard

ECOLOGY AND BEHAVIOR
OF A SOLITARY FELID

Theodore N. Bailey

COLUMBIA UNIVERSITY PRESS

New York

Columbia University Press
New York   Chichester, West Sussex

Library of Congress Cataloging-in-Publication Data

Bailey, Theodore N.
    The African leopard : a study of the ecology and behavior of a
solitary felid / Theodore N. Bailey.
        p.   cm.—(Biology and resource management in the tropics
series)
    Includes bibliographical references (p.   ) and index.
    ISBN 0–231–07872–2
    1. Leopard—Africa—Behavior.   2. Leopard—Africa—Ecology.
I. Title.   II. Series.
QL737.C23B26   1993
599.74'428—dc20   93–828
                                                  CIP

Casebound editions of Columbia University Press books
are printed on permanent and durable acid-free paper.

Printed in the United States of America
c 10 9 8 7 6 5 4 3 2 1

To all my family and to the memory of my father
who first kindled my interest in Africa

Not so the leopard, the most catlike of all cats, the quintessential cat. Secretive, silent, smooth and supple as a piece of silk, he is an animal of darkness, and even in the dark he travels alone.

—MAITLAND EDEY,
The Cats of Africa

Contents

Foreword

THE cats are probably the least-studied—and most poorly understood—of all mammal species. Of the thirty-seven species of big and small cats inhabiting the world, we know little more about most of them than that they exist. The reasons are obvious. For the most part, cats are solitary, secretive, and nocturnal, and they inhabit remote, inhospitable regions of the earth. Cats make themselves unavailable for traditional methods of observation and study. Only lions and cheetahs, by virtue of their selection of open habitat, can be observed and studied directly. All the others must be observed indirectly. The most accepted means of indirect observation is radiotelemetry, a technique developed only in the last two decades. Because the methodology and equipment were simply not available before then, research on cats lagged behind that on other species. In the early 1970s very little objective research had been conducted on any of the "secretive" cats, big or small.

Our studies of Idaho mountain lions and bobcats in the 1960s were the first ever to utilize marked individuals and radiotelemetry in analyzing the dynamics and behavior of any felid population. In these two studies we were able to objectively document territorialism and its function in the regulation of population size. This finding alone was of great importance, both from the theoretical and practical standpoints.

Upon completing these two projects, we looked to the other big cats with the idea of doing comparable research. The mountain lion and bobcat research had been designed to seek evidence for or against unifying principles in two adaptable felids in North America—research on a similar cat in the Old World seemed appropriate.

The leopard was an ideal research candidate. It is perhaps the most adaptable, widely distributed mammal in the Eastern Hemisphere and is thus the mountain lion's Old World counterpart. Despite the fact it has historically ranged much of the Eastern Hemisphere, virtually no research had been conducted on its natural history or ecological role.

Worldwide conservation concerns in the 1960s for all spotted cats added to the leopard's attractiveness as a research subject. Various nations imposed voluntary bans on trade in leopard skins, and in 1972 the United States classified the leopard as an endangered species. Clearly the time was right for an in-depth ecological study of this species.

Encouraged by the late C. R. "Pink" Gutermuth of the World Wildlife Fund, I traveled to Africa in 1970 to investigate the possibility of such a project. I made subsequent trips over the next two years, seeking to sort out all the logistical, biological, and political considerations. After visiting a number of areas in East and South Africa, I decided Kruger National Park in the Republic of South Africa was the best place for the study to be conducted. The South African Parks Board agreed and supported my application to the National Science Foundation in the United States.

After receiving the NSF grant, I sought a qualified field biologist to undertake the research. The person chosen had to be not only academically qualified and adequately trained but able to endure the hardships of long periods of isolation in a tropical habitat and life in an unfamiliar culture far from home.

Ted Bailey filled the bill. Born and raised in rural southern Ohio, Ted had engaged in rigorous outdoor activity since childhood. While I was putting together the leopard project, he was winding down the bobcat study for his doctoral dissertation. His wife, Mary, was a partner in that work and was ready for the next challenge.

Ted eagerly accepted the opportunity, and this book is the result of his dedication and commitment. The delay in its publication does not detract from the significance and completeness of the work.

Because they are consummate predators, all cats sit at the apex of the food chain. As such, they act as bellwethers of the condition of the environment—healthy populations of cats mean healthy populations of prey and a healthy environment. Yet one-third of the world's species of cats are currently considered threatened, if not endangered. As the world's human population grows and irreversibly alters many environments, we may lose some species of cats before we learn anything about them. Dedicated research like Dr. Bailey's is essential for gaining the new knowledge that conservation efforts require. His work will become a landmark in guiding the conservation of leopards throughout much of their range.

Maurice Hornocker, Director
Hornocker Wildlife Research Institute

Acknowledgments

THE efforts and support of many people and organizations in the United States, Canada, and the Republic of South Africa made this study possible. First and foremost, I am grateful to Dr. Maurice G. Hornocker, former leader of the Idaho Cooperative Wildlife Research Unit, University of Idaho, who had responsibility for conduct of the project and who hired me to undertake the field research. He proposed and obtained financial support for the study while I was conducting doctoral research on another felid, the bobcat. His three exploratory trips throughout Africa, contacts with colleagues, and internationally recognized research on the North American mountain lion, or cougar, set the stage for a favorable reception when I first arrived in southern Africa in 1973 to make actual arrangements for the study and to select study areas. I deeply appreciate his advice and support before, during, and after the fieldwork and his constant encouragement during the extended writing period while I was in Alaska with entirely different responsibilities.

Dr. Hornocker also enlisted the aid of Dr. Ian McTaggart-Cowan, former dean of Graduate Studies at the University of British Columbia, Canada, who cooperated with, obtained financial support for, and provided encouragement during the study. E. Bizeau, former assistant unit leader, and E. Beymer of the Idaho Cooperative Wildlife Research Unit guaranteed that the project functioned smoothly from an administrative viewpoint. I am also grateful to the late A. Johnson, of Johnson Electronics, Moscow, Idaho, who personally manufactured and often modified my radio telemetry equipment and ensured that it arrived promptly and performed reliably throughout the study. The fact that I had no major radio equipment problems during the entire study, when such equipment was still in the early stages of its development, attested to Albert's skill and meticulous workmanship.

Many people in the Republic of South Africa supported the leopard study. R. Knobel, then chief director of the National Parks Board of the Republic of South Africa, permitted me to study leopards in Kruger National Park. Then deputy chief director of the Parks Board, A. M. Brynard, was especially helpful when I first visited the republic. His personal interest in the project, arranging travel, accommodations, and prompt communication with authorities regarding my radiotelemetry equipment, prevented potential delays in initiating the study.

I am grateful to Dr. U. de V. Pienaar, who at that time was head of the Department of Nature Conservation in Kruger National Park. He made available to me the advice of his staff and facilities at the Department of Nature Conservation, accommodations and staff privileges for me and my family, and provided the support of other park facilities. Dr. P. van Wyk, then chief of research in the Department of Nature Conservation, was most helpful in providing support for my many and often unexpected requests; his botanical expertise helped me become familiar with the park's flora.

I am especially indebted to Dr. G. L. ("Butch") Smuts, who was then a senior research officer in Kruger National Park. He enthusiastically supported the project from the day I arrived to the day I departed. Butch and his family were ideal neighbors and became special friends of our family during the project. His help in providing manpower, immobilizing drugs, and contacts with park and other wildlife biologists and rangers helped the project function smoothly. His help in safely releasing incidentally captured, and often furious lions, from leopard traps during the early phase of the study was much appreciated.

I acknowledge the support of others in the Department of Nature Conservation, including Dr. S. C. J. Joubert for his observations of leopards and for providing game census data; W. P. D. Gertenbach for help in identifying flora; A. and H. Braack for use of laboratory equipment and facilities; and G. L. van Rooyen and his staff for constantly repairing and modifying leopard traps and providing dart syringes and components. Other staff members always ready to provide information on leopards included I. J. Whyte, T. W. Dearlove, and Park Rangers D. Swart, T. Whitfield, "Flip" Nel, B. R. Bryden, L. Olivier, and B. Pretorious. Maude van Niekerk provided secretarial support.

I also gratefully acknowledge the support of Philemon Nkuna who accompanied me throughout the remote regions of the park's central and southern districts during my first visit and helped me interpret leopard signs to select potential leopard study areas. Lazarus Mangane accompanied me daily during the first five months of the study. His help in preparing baits, setting traps, and handling captured leopards—often physically demanding work in intense heat—ensured the rapid capture of leopards at the beginning of the study. His pleasant and good-natured attitude, humor, and resourcefulness made the working experience enjoyable and rewarding. Philemon Chauke, who replaced Lazarus on several occasions, also provided excellent field support.

Veterinarians V. de Vos, D. Gradwell, and E. Young, and their staff, who were then with the Transvaal Division of Veterinary Services provided support by examining live leopards and conducting postmortems on dead ones. I am also grateful to D. Argo and J. Segerman from the South African Institute for Medical Research for identifying mites and ticks collected from leopards. A. Verster, J. Boomker, and S. E. Thomas of the Veterinary Research Institute identified internal parasites, conducted histopathological examinations, and identified coccidial oocysts from captured leopards.

National Parks Board pilots J. Newman and H. van Niekerk provided excellent aircraft support when I needed to locate leopards that left the study areas. By-Products Plant supervisor J. Marias and his staff were always accommodating, providing bait for leopard traps whenever it was available. J. van Deventer, South African Automobile Association, kept my vehicle in good running condition and my many thorn-punctured tires repaired, and H. Prinsloo of the Stevenson-Hamilton Library at Skukuza obtained several needed publications on leopards.

I also thank P. van der Walt for permitting me to participate in an aerial game census over the Kalahari-Gemsbok National Park and see a leopard from the air. Dr. "Gus" Mills and his wife, Margy, provided excellent accommodations in the Kalahari Desert and gave me an opportunity to observe brown hyenas. Dr. J. L. Anderson provided me an opportunity to view leopard habitat and wildlife in the Umfolozi and Hluhluwe Game Reserves in Natal. Dr. P. M. Brooks shared with me his unpublished information on leopard kills and prey populations in those game reserves. Dr. S. M. Hirst, then chief of research, Nature Conservation Branch of the Transvaal Provincial Administration supplied information and advice in selecting a study area. I. Crabtree, Warden of the Sabi-Sand Game Reserve provided needed reports, observations of his captive leopard; he also permitted me to search for one of my leopards in his reserve.

Financial support for the study was provided by grants to Dr. Maurice Hornocker from the National Science Foundation and the World Wildlife Fund, and to Dr. Ian Met. Cowan from National Research Council of Canada. Project costs were reduced through the generous support of the National Parks Board of South Africa, which extended to me use of its staff, aircraft, laboratory and office facilities, immobilizing drugs, and capture equipment at Kruger National Park at no cost. I also thank the Parks Board for granting me and my family many of the privileges reserved for its permanent staff at Skukuza and for providing us housing accommodations at reduced costs.

Drs. M. G. Hornocker and G. B. Schaller reviewed initial drafts of chapters 1 to 10 of this manuscript and Drs. G. L. Smuts and U. de V. Pienaar and his staff at Skukuza reviewed chapters 1–3. I thank them for their helpful comments and criticisms. A draft of the manuscript was later reviewed by J. F. Eisenberg and M. E. Sunquist; I acknowledge them for their positive comments, support, and encouragement to complete it. I especially thank M. E. Sunquist for offering constructive recommendations and references after reviewing the complete manuscript. I also appreciate the support, encouragement, constructive editing, and helpful advice of E. Lugenbeel, A. McCoy, A. Gibbons, L. Wood, and T. Bonner at Columbia University Press.

Above all, I thank my wife, Mary, whose continuous support before, during, and after the fieldwork was invaluable. In addition to the responsibilities of raising our children, Becky, Kim, and Brian, in a foreign environment, she daily recorded data from my field notebooks, helped me radio-track leopards and census prey, and occasionally helped me weigh and handle captured leopards. Perhaps even more important, she was a source of constant encouragement as I attempted to write this book far from Africa and leopards, and she typed, several times, many of its chapters. As our children matured, they sustained their interest in Africa and its wildlife. When my daughter Kim was a student attending the University of Idaho, she obtained from the university library many references on leopards that I needed (but could not get in Alaska). We will never forget the time we lived in Africa and will always treasure our African friends and experiences. To all who made it possible, we thank you for what may well be the most exciting experience of our lives.

PART ONE

The Study

1

Introduction

THE worldwide demand for the furs of spotted cats in the 1960s reached such proportions it was estimated that up to fifty thousand African leopards[1] were being killed annually for their skins (Edey 1968; Myers 1976). In 1968 and 1969 fur brokers in the United States alone imported the skins of more than seventeen thousand leopards (Myers 1973). Alarmed at the rate at which leopards were being killed, the United States classified the leopard as endangered throughout its entire range[2] in the early 1970s, and the International Union for the Conservation of Nature (IUCN) listed five subspecies of leopards as rare and endangered. In view of the worldwide downward trend in leopard numbers the IUCN urged intensive research on leopards. For example, Simon (1969) stated "In spite of being one of the largest and most widely distributed members of the cat family, the leopard remains essentially unstudied in the wild. . . . It is important that the facts should be established by undertaking a study of the animal's biology, status, and population dynamics." Because little was known about leopards when I began my study in 1973, I hoped to obtain ecological and behavioral information that could be added to our knowledge of this magnificent and then little-known felid. This study was undertaken some time ago, and the delay in the appearance of this book resulted from my

1. See appendix A for a list of scientific and common names of mammals, bird, reptiles, and fish mentioned in the text.
2. See appendix B for a summary of the 1982 reclassification of the leopard by the United States Fish and Wildlife Service.

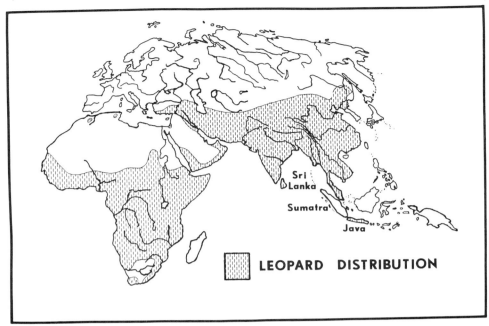

FIGURE 1.1 A general geographic distribution of the leopard. The actual distribution of the leopard is more restricted. Small, isolated, and endangered populations of leopards occur at the peripheries of this range.

involvement in other wildlife projects in Alaska, where I had little access to publications on either leopards or Africa.

Fossil evidence, some as old as 1.5 to 2.0 million years (Hemmer 1976; Brain 1981), suggests leopards were once more widely distributed than today. During the Pleistocene, leopards inhabited Africa, Asia, and Europe as far west as England—a vast area that included nearly the same area as that inhabited by the lion and tiger combined (Fisher et al. 1969). At the turn of this century, the range of the leopard included Africa and stretched north across Asia from the Caucasus Mountains in southern Russia eastward through southern Turkmenia and southwestern Tadzhikestan to the Amur River in eastern Siberia. There, leopards were once reported as far north as 52° N latitude near Khaborovsk (Stroganov 1969). In southern Asia leopards occurred throughout the forested regions of India, the Indo-China Peninsula, southern China, and the island of Sri Lanka (Turnbull-Kemp 1967; Guggisberg 1975) (see fig. 1.1). The leopard's historical presence on the island of Sumatra is doubtful, although fossils of leopards occurred there as well as further south in Java (Seidensticker 1986).

Despite this once widespread distribution, leopards have been greatly reduced in numbers over their vast former range. Of thirty-three countries surveyed south of the Sahara in the 1970s, leopard populations in Africa were greatly reduced in five; status was uncertain in fifteen, secure within some regions while at the same time uncertain or declining within other regions in three, and secure because of parks, preserves, or low human densities in ten countries (Myers 1976). Despite the reduction in international commercial trade of leopard skins beginning in the 1970s, the leopard's status in Africa continued to deteriorate because of habitat loss and use of poisons and baited traps to protect livestock (Myers 1986; Hamilton 1986). For example, between 1962 and 1981 leopards in Kenya may have declined in over 80% of the country (Hamilton 1986).

Little is still known about the population status of leopards in Asia. In Iran in the mid-1970s, leopards were apparently abundant only in the Alborz Mountains (Joslin 1986). They were present in the Karchat Hills of Pakistan (Schaller 1977) and were still found in the larger forests of India (Sankhala 1977) and Nepal (Seidensticker 1976a; Schaller 1977). Leopard populations have been greatly reduced throughout China (Zong-Yi and Sung 1986) and have declined outside parks and reserves in Sri Lanka (Santiapillai et al. 1982).

At the edges of their geographic range, leopards occur in small, isolated populations. The Barbery leopard (*Panthera pardus panthera*) of North Africa is not only isolated but separated from other leopards by hundreds, perhaps thousands, of kilometers. The last known stronghold of this rare leopard was in the Central Atlas Mountains and the forests of Oulmes in Morocco (Guggisberg 1975). The Anatolian leopard (*Panthera pardus tulliana*) apparently exists only as scattered individuals in southwestern Turkey and appears on its way to extinction (Borner 1977). Perhaps slightly more numerous is the subspecies *Panthera pardus jarvisi* in the desert of southern Israel (Ilany 1986). Little is known about *Panthera pardus nimr* of the Arabian Peninsula. The most northern race of leopard, *Panthera pardus orientalis,* the Amur leopard, is now largely confined to several reserves in North Korea and the Maritime Territory of the Soviet Far East (Russia) where in the 1970s it was estimated there were fewer than one hundred individuals (Prynn 1980).

The leopard is often considered to epitomize the features and behavior of the larger cats:

> Not so the leopard, the most catlike of all cats, the quintessential cat. Secretive, silent, smooth, and supple as a piece of silk, he is an animal of darkness, and even in darkness he travels alone. Leopards, because of their furtive habits, are almost impossible to see.
>
> (Edey 1968)

> The leopard can surely be described as the most perfect of the big cats, beautiful in appearance and graceful in its movements.   (Guggisberg 1975)

Man's admiration for the leopard's beauty and strength expressed itself as early as the Roman Empire, when leopards were sought for use in the highly popular circus games. In medieval Europe leopards were held in captivity as highly prized status symbols by royalty. Considerable confusion existed during this period, however, regarding the leopard's taxonomy. At one time some people believed that the male leopard, a "pardus," and the female cheetah, a "panthera," were the same species. Others believed lionesses commonly mated with male leopards, or "pards," to produce spotted maneless lions. Some early naturalists claimed that leopards and panthers were two separate species. The leopard was supposedly smaller, long-tailed, dark, and forest-dwelling, whereas the panther was larger and light in color and inhabited more open country. Part of the confusion was due to the leopard's tendency toward melanism in moist, densely forested areas. Dark or black leopards are apparently common in the tropical Asian countries of Burma (Myanmar) and the southern parts of the Malay Peninsula, southern India, Java, southwestern China, some parts of Nepal, and the province of Assam in northeastern India. The black form of the leopard is inherited as an autosomal recessive to the spotted form but they both have identical morphology and interbreed without difficulty; captive black female leopards do not appear to have as many young on the average as captive spotted females (Robinson 1969). Heterozygous spotted males and females can produce black offspring; the frequency of the black gene in offspring is estimated at about 37% among breeding captive spotted leopards. The matings of black parents produce only black offspring. The spotted and black forms coexist as a polymorphism in certain parts of Asia (especially Malaysia and Thailand); the survival advantage of the black form in the wild is unknown, but could be related to greater concealment (Robinson 1969). Black leopards are less common in Africa although they have been reported in Ethiopia, the Aberdares forest of Kenya, the Virunga volcanoes region, and in Cameroon (Guggisberg 1975).

The leopard has also been considered one of the most ferocious and cunning members of the large cat family. Its reputation for ferocity, however, stems largely from early accounts of hunters' encounters with injured leopards. Perhaps the most famous account was that of Carl Akeley. This American naturalist had foolishly wounded a leopard and returned the following day to track it down. In the process the injured leopard attacked Akeley who expended all his ammunition to no avail. He reloaded and attempted to shoot the leopard again, but it was too late:

> Immediately I was face to face with the leopard in mid-air. My trusty rifle was knocked flying from my hands and in my arms was the leopard—eighty pounds of furious bloodthirsty cat. I knew she intended to sink her teeth into my throat and hold me tight in the grip of her jaws and forepaws while with her hind paws she dug out my stomach—for this practice is the pleasant way of leopards. (Akeley and Akeley 1931)

He killed the leopard by choking it with his hands and crushing its ribs.

Although many of these early accounts provided little factual information about leopards, there were some exceptions. In his story of the famous man-eating leopard of Rudraprayag, Jim Corbett revealed his penchant for observing and recording details of the animals he hunted. Besides being keenly aware of the leopard's hunting habits and movements, he described their vocalizations, scratch marks, and relations with other leopards. For example, he noted that "male leopards are very resentful of intrusions of others of their kind in the area they consider to be their own." He also noted that fights between leopards were unusual because they invariably keep to their own areas (Corbett 1947).

Southern Africa's famous naturalist and former Kruger Park warden Colonel Stevenson-Hamilton also provided factual information on leopard habits (Stevenson-Hamilton 1947), and later Turnbull-Kemp (1967) summarized what was then known about leopards. His book *The Leopard* provided valuable insights into the leopard's life but emphasized the lack of scientific information.

The leopard, up to the 1970s, remained essentially unstudied in the wild. Then observations of leopards, often associated with other studies, were reported by Schaller (1972), Eisenberg and Lockhart (1972), and Muckenhirn and Eisenberg (1973). Hamilton's (1976) intensive study of leopard movements, using radiotelemetry in Tsavo National Park, Kenya, from 1972 to 1974, provided a valuable source of comparative data with my study. Since then only a few other intensive studies of leopards have been reported from Africa (Bothma and Le Riche 1984, 1986; Norton and Lawson 1985; Norton and Henley 1987).

Intensive studies of populations of solitary felids were extremely difficult until the development of reliable radiotelemetry equipment. One exception was a study of a population of cougars in the northwestern United States (Idaho) by Hornocker (1969, 1970). Later this same cougar population was studied with radiotelemetry and their social organization described in detail (Seidensticker et al. 1973). During the same period, I studied a population of bobcats, another solitary but smaller felid, describing their social organization and ecology (Bailey 1972, 1974, 1979). The experience and information gained from these two studies directed by Maurice Hornocker eventually led to this study of leopards.

My study was designed to obtain information on the ecology and behavior of leopards inhabiting two nearby, but different, environments. The information obtained during the study is presented in the first ten chapters, each of which can be read independently. In the final two chapters I synthesize the information, compare it to that reported for leopards elsewhere, and discuss several aspects of leopard conservation.

The years I spent in Africa were deeply rewarding and difficult to describe to those who have not tasted the beauty and richness of Africa's landscapes, wildlife, and people. As the study progressed, I began to recognize leopards and other animals as individuals and anticipated them in my daily travels: the bull elephant with the crooked tusk who often startled me when he suddenly appeared, seemingly out of nowhere, from the dense brush; the giraffe with the missing tail that watched me suspiciously, regardless of how many times I passed through his area; and the motionless, sunning crocodile on a small island in the Sabie River. Male leopard number 23 was more than a number; he was an unpredictable and cunning individual that defied my initial attempts to capture him and eventually died in the jaws of another predator. He and the other leopards became as familiar to me as the giant fig trees shading the Sabie River or the granite koppie at Sihehleni. But I knew a time would come when they would live only in my memory—that I would never meet them again as individuals, never experience the feeling of meeting old friends.

2

Kruger National Park and the Leopard Study Areas

THE Kruger National Park (KNP) was an ideal place to study leopards. Its dense vegetation and abundant prey provided an ideal leopard habitat representative of the woodland-savanna vegetation (Cloudsley-Thompson 1969) that characterizes much of the leopards' range throughout Africa. An immense area of deciduous savanna (Venter and Gertenbach 1986) in the Republic of South Africa's Transvaal Lowveld, the park is the oldest and one of the largest in Africa. Located between 22° 25' and 25° 32' S and longitudes 30° 50'and 32° E, the 19,485-km^2 park is 320 km from north to south and up to 65 km east to west (see fig. 2.1).

Despite the park's immense size, apparently it was never an ecological unit because game traditionally migrated from the present park into the water-abundant foothills to the west and then returned each wet season (Pienaar 1963). When the western boundary was completely fenced in 1961, the traditional east-west movement pattern of annual game migrations changed. The park's eastern boundary, fenced in 1975, stopped the last traditional movements of game in and out of the park. Although several private game reserves provide protected habitat adjacent to the park's western boundary, much of the park is now surrounded by agricultural, developed, or native-trust land. Kruger National Park is divided for administrative purposes into three districts, each with a characteristic flora and fauna (see fig. 2.1). The northern, and largest, district encompasses about 10,466 km^2, lies north of the Olifants River, and is covered by *Mopane* or mixed *Mopane-Combretum* woodland. The majority of the park's elephants, buffalo, eland, roan antelope, and tsessebe live within this

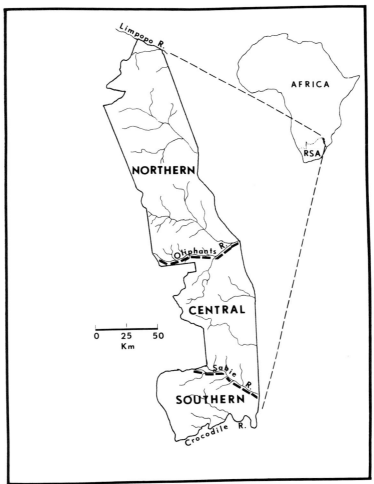

FIGURE 2.1 Kruger National Park, its administrative districts, and some of its major rivers. Upper right shows the park's location in the Republic of South Africa (RSA) and Africa.

district. The park's central district lies between the Olifants and Sabie Rivers, encompasses 5,517 km^2 of open or woodland savanna, and supports most of the park's zebra, wildebeest, giraffe, lion, and spotted hyena. The 3,522-km^2 southern district encompasses many diverse habitats, primarily *Acacia* woodland and thicket, and supports the park's highest impala population (Pienaar 1963).

The Study Areas

For ecological as well as practical reasons, I selected the southern district of the park for my study and confined most of my observations to an area within 20 km of park headquarters at Skukuza. Compared to the other districts, this region encompassed some of the best, if not the best, leopard habitat in the entire park. The area was also accessible, conducive to live-trapping and radio-tracking, and near a source of bait for leopard traps. I selected two different leopard study areas because I wanted to compare the ecology and behavior of leopards in different environments.

THE SABIE RIVER STUDY AREA

The Sabie River Study Area (SRSA) was situated along the perennial Sabie River. The river flows southeasterly between the central and southern regions of KNP and is one of the most densely wildlife-populated areas in the park (Pienaar et al. 1966a, 1966b). It supports large numbers of impala, kudu, and other smaller ungulates and considerable numbers of buffalo, hippopotamus, and elephant. In 1966 the estimated biomass of game for a strip 3.2 km wide on each side of the Sabie River was 9,070 kg/km^2 (Pienaar et al. 1966a). With high populations of smaller mammals such as baboons, vervet monkeys, porcupines, and cane rats, and game birds such as guinea fowl and francolin, this area provided excellent habitat for leopards. When I initially surveyed the area in early 1973, I found abundant prey and numerous tracks of leopards.

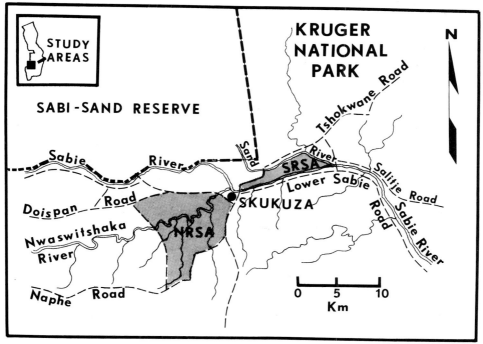

FIGURE 2.2 Location of the Sabie River (SRSA) and Nwaswitshaka River (NRSA) leopard study areas (northeast and southwest, respectively, of Skukuza) within the southern district of Kruger National Park.

2.1. The Sabie River during the dry season. This perennial river formed the southern boundary of the Sabie River study area. Large *Ficus sycomorus* and beds of *Phragmites* reeds were common, and the area supported a high density of impala and leopards.

In an attempt to use natural boundaries, I selected a 17-km^2 study area between the Sabie and Sand Rivers just northeast of Skukuza (see fig. 2.2). The Sabie River formed the study area's southern boundary and the Sand River, a smaller perennial river, formed much of the study area's northern boundary. The Sand River joined the Sabie River at the study area's eastern extremity. A paved road open to tourist traffic formed the study area's only unnatural boundary on the west. I established a trap line for leopards inside the study area along a firebreak road that ran adjacent to the Sabie River and from 0.1 km to 1.0 km south of the Sand River. I also used tourist roads opposite the rivers and outside the study area to monitor radio-collared leopards. I initially included similar habitat west of Skukuza within a study area, but eventually terminated studies there when it became apparent the area was not needed to increase sample size. Only two male leopards were captured and monitored in the area west of Skukuza. The Sabie River, and to a lesser extent the Sand River, was a seasonal barrier to the movements of leopards and other game. During wet seasons the larger Sabie River was seldom crossed by game, except elephants. During the dry season, however, when water levels dropped (see photo 2.1), the river became progressively easier to cross, especially at rocky areas. Monkeys, baboons, and leopards could easily jump from rock to rock, and other game could wade across. The shallow, narrow Sand River was regularly crossed by game, including leopards, except after periods of heavy rainfall.

THE NWASWITSHAKA RIVER STUDY AREA

The Nwaswitshaka River Study Area (NRSA) was situated along the lower reaches of this seasonal river, which originates in the southwestern hills of the park and flows northeasterly into the Sabie River at Skukuza. Because it flows for only brief periods after heavy rains, it provides only temporary water for game. However, instead of drinking at the Nwaswitshaka River, game usually drank at small natural water holes adjacent to the river during wet seasons. After these seasonal water holes evaporated, game used a permanent water hole, drank water where it seeped through the sand in the riverbed, or left the area to find water elsewhere. The only permanent water hole was at a windmill near the western boundary of the study area. About 81 km² of the lower watershed of the Nwaswitshaka River was selected as a study area (see fig. 2.2).

Paved tourist roads, which followed the ridges separating adjacent watersheds, served as study area boundaries on the north, east, and south sides of the study area. An unpaved tourist road that crossed the Nwaswitshaka River 13 km upstream from Skukuza formed the area's western boundary. Only one road, a firebreak road, passed through the center of the study area, and I used it regularly to establish a trapline and monitor radio-collared leopards.

The carrying capacity of game in the Nwaswitshaka River region is considerably less than that along the Sabie River. Pienaar et al. (1966b) estimated it supported about 575 kg/km², or about 6% of the biomass supported by the Sabie River region. Because seasonal lack of water was apparently a limiting factor, I assumed game was more plentiful adjacent to the Nwaswitshaka River than in areas far from the river.

Physiography

Kruger National Park lies just east of South Africa's most prominent topographical feature, the Great Escarpment. At the scarp's face, about 90 km west of the park, the elevation changes abruptly from about 1,824 to 1,155 m. Then, through a series of foothills, the land slopes gently downward into progressively flatter country within the park. Inside the park, altitudes vary from 823 m in the hilly southwestern region to 183 m in the eastern, low-lying Lebombo Flats. The highest point is along the Lebombo Range on the eastern boundary at 496 m.

The park generally appears flat, but actually comprises two physical regions: an undulating, elevated western half and a flatter, lower-lying eastern half. The country is extremely broken only in the northwestern and southwestern corners of the park, but prominent outcrops of ancient rocks, called koppies, provide welcome changes in relief throughout much of the park's otherwise repetitious landscape. Surface drainage is eastward with five of the park's six perennial rivers, the Crocodile, Sabie, Olifants, Letaba, and Levubu, originating at the escarpment. Each river flows directly across the park, through the Lebombo Range and into Mozambique, while the Limpopo River drains a vast area northeast and northwest of the park. A number of seasonal rivers, such as the Nwaswitshaka, originate within the park and also flow generally eastward. Numerous steep-banked but usually dry streambeds, or dongas, drain the more elevated, undulating western half of the park.

Immensely old rocks of the western Lowveld were exposed as surface erosion penetrated up the major rivers to the escarpment during the mid-Cretaceous (Cole 1961). These old weather-resistant rocks from the Archean era occur throughout the western half of the park as gniesses, schists, granites, and dolorites (Schutte 1986). In the eastern half of the park, these rocks are overlaid by more recent basalts of the Karoo system, and the Lebombo Range itself is rhyolitic in origin. A narrow layer of sandstone that separates these two major geologic formations also demarcates a change in vegetation.

Soils within the eastern Lowveld, which are generally shallow with a poor water-retaining capacity, are leached of their more soluble minerals and have a low humus content (Cole 1961). The shallowness of the sandy, granitic soils in the western half of the park (Venter and Gertenbach 1986) is apparent when one observes elephants pushing over large trees with apparent ease. In the eastern half of the park, the poorly drained basaltic soils are slightly more fertile because they have not been leached as long as the granitic soils.

No prominent topographic features were in the SRSA. From a low, broad ridge running roughly down the long east-west axis of the study area, the land merely sloped toward the nearest river and averaged 257 m in elevation. The slightly steeper, more undulating southern half of the study area was drained by a number of small dongas into the Sabie River, but on the more gently sloping northern half such dongas were scarce. In the northeastern region, where the Sand River turned southward, there was an extensive flat area of heavy clay soil, which was characterized by numerous small waterholes during the wet seasons. A hilly region north of the study area was drained by the Mutlumuvi, Malayathiaya, and the Nwanyikulu, all seasonal streams that flowed into the Sand River. In contrast, the terrain south of the study area was relatively flat but drained by several streams such as the Mhlambanyathi, which flowed into the Sabie River.

The NRSA was a hilly region compared to the SRSA. The average elevation along the northern and southern watershed ridges was 335 and 366 m, respectively, and the river itself dropped from 320 m at the western end of the study area to 274 m at Skukuza. Several small koppies occurred in the study area south of the Nwaswitshaka River. The largest of these, Sihehleni, rose to 385 m. Two large seasonal streams, the Msimuku and Nhlanganeni, and another smaller one, the Mathhekeyane, drained the elevated portions of the study area.

The Seasons

The climate of KNP is characterized by hot, rainy summers and warm, dry winters. During the dry season the skies are clear for days, and even during the wet season sunshine occurs on most days because rainfall is usually of short duration.

Water is the most critical factor influencing wildlife, and the whole tempo of life in the park is attuned to the seasons. During the wet seasons, when water is plentiful and forage abundant, the game is healthy and well dispersed. But after the rains cease and the waterholes evaporate, game abandon their former haunts and move to regions near permanent sources of water. In the wet seasons, one constantly hears insects, frogs, and birds, even during the hottest part of the day, but during the dry season the land becomes quiet, and often the only sound is the wind.

THE WET SEASON

The first rains typically fall in September or October. Average daily temperatures have already slowly begun to rise from the extreme lows in July, and one can feel the heat gradually building up, especially in the early mornings. Usually several days before the rains, clouds form in the previously cloudless sky, and one senses the animals are aware of the imminent change. Impala appear restless as they trudge back and forth to dwindling water supplies, and birds seem unusually quiet. Then comes a flash of lightning and the clap of thunder. All is now still except the wind. Suddenly raindrops begin to fall. The rains have arrived.

A record rainfall ushered in the first wet season of the leopard study. The rains began at 8:13 p.m. on September 27, 1973. By 9:00 a.m., three days later when it stopped, 192.1 mm had fallen, causing extensive flooding and killing game. After the rain I found dead impala at the bases of trees and under leafless shrubs where they had sought cover. In a nearby private game reserve, more than five hundred game animals died during the rainfall (Crabtree 1974). Fortunately such rains were unusual. Less spectacular rains ushered in the wet season of 1974 when only 36 mm fell between October 7 and 10.

Most of the park's annual precipitation falls between November and March, usually in the form of heavy downpours (see fig. 2.3). Annual precipitation varies from about 740 mm in the southwest to about 440 mm in the northeast regions of the park (Venter and Gertenbach 1986). At Skukuza, near the leopard study areas, average monthly rainfall varied from 5.7 to 95.8 mm, and annually

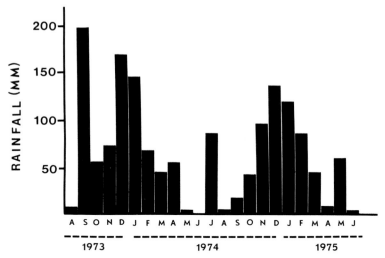

FIGURE 2.3 The distribution of monthly rainfall at Skukuza during the study period.

averaged 554 mm between 1926 and 1985 (Venter and Gertenbach 1986). The period from 1961 to 1970 was dry, but from 1971 to 1978, during my study, rainfall was 120% above average (Gertenbach 1980), which created ideal forage for certain species of ungulates.

Summers in the park are very hot. Although maximum monthly temperatures at Skukuza sometimes surpass 44°C, the diurnal variation in average temperature varies from 12°C to nearly 20°C in both seasons (Gertenbach 1983). Relative humidity in the summer thus drops from about 70% to 80% in the early morning to 45% to 60% in the afternoon. Winds generally prevail from the direction of the Indian Ocean, that is from the northeast to southwest, at about 2.5 kph to 4.5 kph.

The park's drab appearance changes rapidly after the rains. Within several weeks leaves burst from their buds, grass grows 15 to 20 cm, waterholes fill, and the whole park has a lush green aspect. This appearance is maintained until April or May when the rains gradually cease. By June or July the grass turns yellow and brown, leaves fall from the trees and bushes, and game concentrates near water. It is the beginning of the dry season.

THE DRY SEASON

The onset of the dry season is not as clearly defined as the wet season. Some precipitation may occur throughout the dry season, but in general, July through August have the least rainfall. Sometimes, as on July 1, 1974, violent thunderstorms cover the ground with hail and temporarily fill dried waterholes and

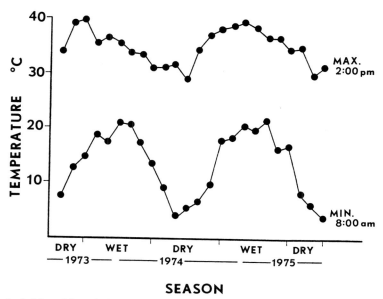

FIGURE 2.4 Monthly minimum and maximum temperatures recorded at Skukuza during the leopard study period.

seasonal streams with water. But the low humidity, prevalent winds, and prolonged periods of intense sunshine rapidly evaporate the water, and the park quickly resumes its winter appearance.

Nights can be cold in the park during the dry season, but temperatures often reach 25° to 30°C in the afternoons (see fig. 2.4). The lowest temperature recorded at Skukuza during the leopard study was −2.1°C in July 1975, but most winters are frost free (Venter and Gertenbach 1986). Early morning temperatures for the coolest months (July and August) averaged 10.7° and 13.8°C, respectively in 1974.

Brushfires are easily started in the late winter when the grass stratum is extensive, high, and often very dry. When lightning occurs with little or no rainfall, grass fires start easily and the air is filled with smoke and the odor of burning vegetation. Because of wildfires, the park is divided into more than four hundred fireblocks surrounded by 3,800 km of firebreak roads. Once a fire is detected, backfires are built along the perimeter to contain the fire inside the blocks. Before the establishment of fireblocks, uncontrolled fires burned rampantly across hundreds of square kilometers of veld. The impact of such fires on the vegetation of the park was significant, and the vegetation in the park may have been a pyrophilous climax (Van Wyk and Wager 1968; Van Wyk 1971).

The Flora

Kruger National Park lies within the subarid to arid woodland savanna vegetation zone (Phillips 1965). However, the commonly used term "bushveld" better describes the vegetation. Bushveld is a mixture of small to medium-size trees and shrubs with an understory of grasses and herbs. In some areas of the park lie open grassy plains with scattered large trees or small shrubs. In other areas dense stands of trees and shrubs, or shrubs and grasses occur (Gertenbach 1983).

Bushveld vegetation is composed of three distinct strata. An upper stratum of deciduous trees 5 m to 11 m in height dominates 70% of the park. A second substratum of shrubs and stunted trees 1 m to 3 m in height scattered throughout the park dominates the remaining 30%. Beneath the trees and shrubs is a stratum of more than two hundred species of grasses and one thousand species of forbs (Venter and Gertenbach 1986).

Brush encroachment is occurring in KNP (Van Wyk and Fairall 1969; Van Wyk 1971) and appears most rapid in the wetter southwestern district. Because brush encroachment is most evident in experimental plots protected from fires for fifteen to twenty-five years, lack of fire appears to be a contributing factor. However, the rate of encroachment is also related to rainfall and the nature of the intruding species. On some fire-protected plots, shrubs and trees have increased roughly 30% in five years. The most important encroaching species are sicklebush, *Dichrostachys cinerea*, which often forms dense stands, *Terminalia sericea, Cassia petersiana, Maytenus senegalensis*, and *Strychnos madagascariensis*. To control brush encroachment, controlled burning has occurred in the park since 1954.

From six to thirty-five major vegetation types (Van Wyk 1971, 1972), vegetation zones, and landscapes (Gertenbach 1983) have been identified in the park (see fig. 2.5). A *Terminalia-Dichrostachys* savanna woodland is found on the sandy loam soils of granitic origin in the southwestern region of the park. The shrub layer is of minor importance in this type, but the tall thatch grass *Hyperthelia dissoluta* is found over large areas. An *Acacia-Sclerocarya* savanna woodland covers most of the basaltic soils on the eastern half of the park south of the Olifants River, but it continues westward in the central district to the boundary. Here, large *Acacia nigrescens* and *Sclerocarya birrea caffra* trees are scattered over fairly open plains with grasses such as *Themeda, Panicum, Digitaria*, and *Eragrostis*. To the west, north of the Olifants River, is the *Combretum-Mopane* woodland. *Combretum apiculatum* dominates the tree stratum on high ground and *mopane*, in its shrub form, dominates in the lower areas. The Sandveld communities are complex floristic communities in the sandy northwestern region of the park.

The remaining vegetation type, *Combretum-Acacia* woodland, occurs throughout the undulating western half of the park, south of the Olifants River. Me-

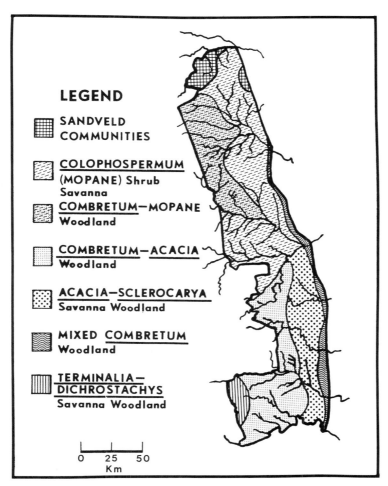

LEGEND

▦ SANDVELD COMMUNITIES

▨ COLOPHOSPERMUM (MOPANE) Shrub Savanna

▨ COMBRETUM—MOPANE Woodland

☐ COMBRETUM—ACACIA Woodland

⬚ ACACIA—SCLEROCARYA Savanna Woodland

〰 MIXED COMBRETUM Woodland

▥ TERMINALIA— DICHROSTACHYS Savanna Woodland

0   25   50
Km

FIGURE 2.5 Some of the major vegetation types of Kruger National Park.

dium-size *Combretum apiculatum* trees dominate the ridges and slopes but are replaced by larger (*Acacia nigrescens*) trees in the low-lying areas. Shrubs are conspicuous and grasses such as *Setaria flabellata*, *Panicum maximum*, *Brachiaria spp.*, *Eragrostis spp.*, *Themeda triandra*, and *Digitaria eriantha* are common. Both leopard study areas occurred within this vegetation type.

The vegetation in the SRSA was dominated by thorny acacias (Gertenbach 1983). "The country all around Sabie Bridge (i.e., Skukuza) was covered with dense acacia bush," said Col. James Stevenson-Hamilton (1937) when he first saw the region in 1902. On the undulating portions of the study area *Acacia nigrescens*, *Sclerocarya birrea caffra*, and *Spirostachys africana* were the dominant large trees; *Acacia grandicornuta* occurred almost exclusively on the flat brackish areas (see table 2.1). These brackish areas were fairly open because shrubs and

TABLE 2.1 *Trees and Shrubs in 100-Square-Meter Transects in the Sabie River and Nwaswitshaka River Study Areas*

| | Sabie River | | | | Nwaswitshaka River | | | |
| | Undulations | | Flats | | Undulations | | Flats | |
Vegetation	n	%	n	%	n	%	n	%
Large Trees								
Acacia nigrescens	3	43	—	—	5	13	1	20
Sclerocarya caffra	2	29	—	—	1	2	—	—
Spirotachys africana	1	14	—	—	1	2	—	—
Zizyphus mucronata	1	14	—	—	6	15	—	—
Acacia tortilis	—	—	—	—	2	5	—	—
Acacia nilotica	—	—	—	—	2	5	—	—
Combretum apiculatum	—	—	—	—	2	5	1	20
Acacia grandicornuta	—	—	6	100	19	48	3	60
Others	—	—	—	—	2	5	—	—
Subtotal	7	100	6	100	40	100	5	100
Small and Medium-Size Trees								
Zizyphus mucronata	30	13	—	—	15	10	4	7
Combretum apiculatum	19	8	—	—	4	3	—	—
Acacia grandicornuta	6	3	42	67	1	—	23	38
Dichrostachys cinerea	164	69	—	—	48	32	13	21
Euclea divinorum	2		16	25	26	17	1	1
Acacia nigrescens	1	2	—	—	24	16	3	5
Acacia nilotica	2		—	—	3	2	13	21
Others	12	5	5	8	31	20	4	7
Subtotal	236	100	63	100	152	100	61	100
Shrubs								
Grewia spp.	124	70	—	—	121	82	12	71
Acacia exuvialis	34	19	—	—	17	12	4	23
Others	20	11	—	—	9	6	1	6
Subtotal	178	100	—	—	147	100	17	100

grasses were sparse (see photo 2.2). Common small to medium-size trees such as *Dichrostachys cinerea, Zizyphus mucronata,* and *Combretum apiculatum,* and dense shrub communities dominated by *Grewia spp.* and *Acacia exuvialis* covered most of the undulating region. A narrow but important zone of vegetation occurred adjacent to the Sabie and Sand Rivers. The dominant trees were huge, white-barked *Ficus sycomorus,* the evergreen *Diospyros mespiliformis,* and *Syzygium cordatum.* All provided welcome shade during the hot dry season. Other common trees in the riparian zone were *Acacia albida, Acacia robusta, Trichilia emetica, Kigelia africana, Combretum imberbe, Adina microcephala,* and *Combretum erythrophyllum.* Extremely dense stands of the tall reed grass *Phragmites communis* occurred along the river banks and on the river's islands. These stands provided excellent cover for bushbuck, buffalo, lion, and leopards.

Vegetation in the NRSA was not as dense as that in the SRSA (see photo 2.3). The dense riparian zone adjacent to the Nwaswitshaka River had fewer shrubs and small to medium-size trees, but more large trees, compared to the

2.2. An open vegetative cover type in the Sabie River study area was the brackish flats, characterized by clay soils, short grasses, and short but widely spaced *Acacia grandicornuta* trees.

Sabie River area (see table 2.2). *Zizyphus mucronata* and several thorny *Acacias* dominated the large tree stratum; *Dichrostachys cinerea* and *Euclea divinorum* dominated the smaller tree substratum (see table 2.1). On the undulations, away from the river, trees and shrubs such as *Combretum apiculatum*, *Combretum zeyheri*, *Pterocarpus rotundifolius*, and *Terminalia sericea* were common.

Small open areas, 1 to 20 ha in size, were common on the brackish depressions adjacent to the Nwaswitshaka River. Within some, sparse, short grasses grew under scattered *Euclea divinorum* trees 3 m to 4 m tall. These areas were associated with seasonal waterholes and were favored by zebra, wildebeest, warthog, white rhinoceros, and impala. Other brackish depressions contained nearly homogenous stands of *Acacia grandicornuta* or *Acacia nilotica* with sparse understories of shrubs. These areas were intensively used by impala.

The riparian vegetation zone along the Nwaswitshaka River was less distinct than that along the Sabie and Sand Rivers. Fig trees, *Ficus sycomorus*, were absent and dense reed beds more scattered (see photo 2.4). The most conspicuous trees were huge *Diospyros mespiliformis*, but *Combretum imberbe* and *Schotia brachypetala* were also common. Narrower but similar zones of riparian vegetation also occurred along the tributaries of the Nwaswitshaka River.

2.3. More open vegetative cover dominated by *Acacia* and *Combretum* trees characterized the Nwaswitshaka River study area. Impala and leopard densities were lower here.

The Impact of Humans on Kruger National Park

Although primitive hunter-gatherers inhabited southern Africa at least ten thousand years ago (Lee 1979), humans apparently had little impact on what is now KNP until after the first white man, Louis Trigardt, passed through the area in 1837 (Stevenson-Hamilton 1937). After gold was discovered nearby in 1869, humans began to significantly impact wildlife. Although President Paul Kruger was already aware of the serious decline in game in 1884, it took fourteen years to establish the Sabi Game Reserve.

TABLE 2.2 *Densities (Numbers/Hectare) of Trees and Shrubs in the Sabie and Nwaswitshaka River Leopard Study Areas*

Size	Sabie River		Nwaswitshaka River	
	Undulations	Flats	Undulations	Flats
Trees				
Large[1]	70	120	120	100
Small to medium[2]	2,360	1,260	1,520	1,220
Shrubs	1,780	0	1,470	340

[1] > 23 cm diameter
[2] < 23 cm diameter

2.4. The seasonal Nwaswitshaka River was dry except during the height of the rainy seasons. Leopards traveled along its dry, sandy bed and favored its *Phragmites* and riparian vegetation zones.

HISTORICAL BACKGROUND

The KNP of today began as the Sabi Game Reserve in 1898. Little, however, was done with the reserve until Colonel Stevenson-Hamilton was appointed first warden in 1902. Through his efforts, the old Sabi Game Reserve and nearby Shingwidzi Game Reserve were combined and proclaimed the KNP in 1926. But, even before this period, human impact on wildlife was significant. The last original white rhinoceros was shot in the area in the 1890s, and the last original black rhinoceros was seen in 1936. A rinderpest epizootic introduced by domestic cattle in eastern Africa swept through the area in 1897–98 and killed most of the buffalo. Ivory hunters killed off most of the elephants by 1905 when fewer than ten were discovered in a remote area between the Letaba and Olifants rivers (Pienaar et al. 1966b). To protect herbivores, carnivore control programs were conducted between 1903 and 1960. All species of carnivores were systematically destroyed at every opportunity before 1933; thereafter, only large mammalian carnivores and crocodiles were destroyed (Pienaar 1969). For example, 1,272 lions and 660 leopards were killed between 1903 and 1927 (Smuts 1982), and between 1954 and 1960, 51 cheetahs were shot (Pienaar 1969). Carnivore control was finally discontinued as a policy in 1960.

Fires also impacted many species in the park. From 1880 to 1945 domestic sheep owners periodically started fires in the southwestern region of the park to provide grassland for grazing. They were removed and burning was discontinued from 1947 to 1954, and encroachment of grasslands by bush and reversion of open savannas to woodland-savanna, woodlands, and thickets occurred. Grazing species such as reedbuck, wildebeest, zebra, sable and roan antelope, waterbuck, eland, and tsessebe declined, as well as ostrich and cheetah, while browsing species such as kudu and impala increased.

WILDLIFE MANAGEMENT PROGRAMS

Park officials have attempted, in various ways, to reintroduce, enhance, protect, and provide habitat for, treat diseases of, and control wildlife populations in the park. Efforts to reestablish and enhance wildlife populations include reintroduction of the white rhinoceros since 1961, red duiker since 1963, oribi since 1964, cheetah since 1967, black rhinoceros since 1971, and eland since 1972. The provision of permanent drinking places for wildlife was top priority to enhance populations because severe droughts during 1910–13, 1926–28, 1933–36, 1944–48, 1962–65, and 1967–70 decimated wildlife throughout the park. A goal of three hundred artificial drinking sites was nearly attained when I began my studies in 1973. This included 268 boreholes at 230 sites in addition to 36 earthen and 33 concrete dams on streams and rivers. The consequences of dispersing wildlife into new areas and their impacts on the vegetation have not been without controversy (Smuts 1978, 1982).

Protection and enhancement of wildlife habitat by fire is another important aspect of park management. During my study the park was comprised of four hundred fireblocks separated by more than 3,800 km of firebreak roads. This network is used to control natural wildfires, such as those that burned more than 33% of the park in 1973, and to initiate prescribed fires to reduce brush encroachment of grasslands.

The spread of disease, primarily hoof-and-mouth disease, from the park is controlled by fences, which now completely surround the park. The first fence, along the southern boundary was built in 1959; the western and northern fences were completed by 1960; the eastern border was fenced in 1975. To protect other species, such as the rare roan antelope, from anthrax, individuals are annually inoculated. Although nearly two hundred elephants were poached for ivory in the early 1980s, when ivory prices were high, poaching in general has not been a major problem (Paynter 1986).

The park's culling programs are perhaps the most controversial. Species culled are those with few natural predators (elephant, hippopotamus, buffalo) and those reaching high population levels because of the fencing and artificial waterholes (impala, zebra, wildebeest). The population control programs began in 1964, when 104 hippopotamus were removed from the drought-stricken Letaba River. Zebra, wildebeest, and impala were later culled, and elephant and buffalo culling began in 1968 (Smuts 1982). In the mid-1970s lions and spotted hyenas were controlled to assist in easing population declines of zebra and wildebeest in the park's central district (Joubert 1986).

TOURISM

Tourism is the park's main source of income. In 1927 when the park was first opened to vehicles, only three cars entered the park (Stevenson-Hamilton 1937). By 1970 more than 360,000 visitors were viewing the park's wildlife each year, and by 1974 more than 2,000 km of roads and thirteen rest camps were available for overnight visitors. Initially, the park was open only during the dry season because of the threat of malaria. The paving of tourist roads began in 1964, and a year later the entire area south of the Letaba River was open to tourists year-round. About 33% of the park's visitors leave the park each day (day visitors); only those with accommodations at a rest camp are allowed to remain over-night.

Strict regulations protect the park's flora and fauna from tourists—and tourists from potentially dangerous wildlife. Vehicles are limited to selected roads; tourists may not drive off of roads nor leave their vehicles; speed limits are strictly enforced; and visitors must leave the park or be in a fenced rest camp before dark.

Summary

1. Leopards were studied in Kruger National Park, a 19,485 km^2 area 320 km long (N-S) and up to 65 km wide (E-W) in the northeastern corner of the Republic of South Africa.
2. Two leopard study areas were chosen in the densely vegetated and prey-abundant southern district of the park.
3. The 17-km^2 Sabie River leopard study area lay between two perennial rivers and supported high concentrations of wildlife during the dry seasons.
4. The nearby 81-km^2 Nwaswitshaka River study area, was bisected by the seasonal Nwaswitshaka River and supported less wildlife than the Sabie River study area during the dry seasons.
5. Although appearing flat, the topography of the park is actually undulating low hills. Exceptions include some rugged hills in the northwestern and southwestern regions and some low-lying flats in the eastern half of the park. Elevations range from 823 m to 183 m.
6. The climate in the park is characterized by hot rainy summers and cool dry winters. Average rainfall varies from 440 mm to 740 mm, and temperatures often exceed 30°C in the summers.
7. The woodland-savanna, or bushveld, vegetation of the park comprises three distinct strata: An upper stratum of trees 5 m to 11 m high; a

substratum of stunted trees and shrubs 1 m to 3 m high; and a lower stratum of more than two hundred species of grasses and one thousand species of forbs.

8. The Sabie River study area was dominated by *Acacia nigrescens* on undulating terrain and *A. grandicornuta* on the brackish flats. The NRSA was dominated by *Zizyphus mucronata* and *Acacia nigrescens* on undulating terrain and *A. nilotica, A. tortilis,* and *A. grandicornuta* on the brackish flats.

9. Several species of wildlife formerly found in the area were exterminated or reduced by humans in the late 1890s and early 1900s. Others declined as migration routes were blocked by fences, fires were controlled, and diseases swept through the area.

10. Park wildlife management programs include the reintroduction of species, use of fires to control brush encroachment, disease control, and periodic reduction of selected ungulate populations.

11. During each year of the leopard study, more than 360,000 tourists visited the park, but strict regulations reduced their impact on wildlife.

3

Study Methods and Response of Leopards to Humans

INFORMATION on leopards is difficult to obtain by visual observations. Unlike gregarious or open-habitat species, leopards are solitary, prefer dense vegetation, and are often nocturnal. Even where they are abundant, leopards are seldom seen. However, early studies of other solitary and elusive felids (Marshall and Jenkins 1966; Seidensticker et al. 1970; Berrie 1973; Bailey 1974) demonstrated that they could be studied in their natural habitats with radiotelemetry. Because radiotelemetry permits one to collect ecological information without actually observing individuals, I selected it as the principal method to study leopards. I also collected information on ungulates, other leopard prey, and other predators in the study areas in order to better understand the leopard's ecology.

Leopards in the park were also influenced by humans and their activities. My impact on leopards in the study areas was significant because it had the potential to bias information I was collecting. The impact of tourists, roads, and other park development was also important; few, if any, areas exist where humans do not influence the leopard's environment.

Live-Trapping

My study objectives required that I recognize individual leopards residing in the study areas. Because many of the leopards were unlikely to be observed visually, individuals living there had to be captured and identified with radio

3.1. Livetraps for leopards were placed in trees to avoid capturing spotted hyenas and to reduce the captures of lions.

collars or numbers tags. I began trapping leopards in August 1973, briefly curtailed trapping during January and February 1975 (because of poor trapping success), and resumed trapping until I left Africa in July 1975. During the first five months of the study, Lazarus Mangane, a park employee, helped me set traps and weigh leopards. Thereafter, I worked alone, although my wife, Mary, sometimes went along to help me immobilize and weigh leopards.

My daily routine was to check the leopard traps early each morning. I then immobilized unmarked leopards, released previously marked ones, and located radio-collared ones. After checking the traps, I returned to observe feeding leopards, moved and rebaited traps, and collected leopard feces. On selected days of each month I censused prey. If I was able to observe a leopard while on foot, I stayed nearby until it left or night fell. If a leopard was visible from my vehicle, I stayed overnight to observe it. During the study, I drove a Toyota Landcruiser 54,475 km, about 2,270 km/month.

Although leopards may be captured with trained dogs (Turnbull-Kemp 1967), I chose to use livetraps because they were more efficient, could be handled by one person, and were less dangerous to researchers and leopards. Ten cage-type, livetraps with a single sliding door were used to capture leopards (see photo 3.1). I used two sizes of traps made of metal mesh, 9 x 3.8 cm in size and 3.5 mm thick, welded to a framework of square, hollow steel tubing. Six traps were approximately 240 x 80 x 90 cm long, wide, and high, respectively. Later I

3.2. A livetrap set for a leopard. Traps were camouflaged with branches of ever-green trees and a limb positioned to allow leopards to climb into the trap.

had four others made measuring 200 x 60 x 80 cm. They worked just as well and were easier for one person to handle. A sliding door of each trap was held in the "up," or "set," position by a latch connected by a cable to a foot-release in the rear of the trap. The weight of the leopard attempting to reach the bait released a spring attached to the cable, which pulled the latch, permitting the door to slide shut. Small, swinging locks over the doors were later added to prevent escapes. A nylon rope attached to the top of the sliding door was pulled from a safe distance to release leopards as well as nontarget animals from the traps.

Because of their weight and size, the traps were usually set in areas accessible by vehicle. At first I merely set traps at ground level. But spotted hyenas and lions entered the traps so frequently that I eventually had to set traps 1 m to 3 m above the ground in trees. To set traps in trees, a wooden platform built of heavy timber was first wired to an appropriate fork or limb; then the trap was hoisted off the vehicle to the platform using a small block and tackle. After the trap was solidly wired to the platform and tree, it was well camouflaged with brush (see photo 3.2). The traps were painted olive drab to blend into their surroundings.

For bait I used 5- to 10-kg pieces of meat and bone scraps from buffalo, elephant, hippopotamus, and impala that were culled in the park's ungulate reduction program. Pieces of bait used most often were rib cages and vertebral columns because they had the most meat attached; other pieces were almost devoid of meat. The bait was wired securely to the rear of the trap and replenished as often as the culling program permitted. If I had sufficient bait, I would often hang a piece of bait in a conspicuous place near the trap to attract leopards from a distance. When bait was limited, I hung several pieces of dried impala skin nearby for the same purpose.

I attempted to set all traps in prey-abundant areas regularly used by leopards. To determine leopard usage, I examined the firebreak roads, dry stream beds, game paths, and river banks for tracks of leopards each day. After finding an area well traveled by leopards, I selected an appropriate tree and set the trap for a varying period of time depending on trapping success. Initially, trapping success was low, but as I gained experience in selecting trapping sites, trapping success improved. If leopards were not captured at one location, I would move the trap to another site until I discovered a location where trapping success was better. During the first year of the study, I moved the traps quite often. Later, after I had radio-collared a number of leopards and determined their travel routes, I moved the traps into localities that seemed to be particularly attractive to leopards using the area. Because I used only five traps for each study area, I often chose a location for the trap that was situated in an area where the home ranges of two or more leopards overlapped, thus increasing the chances of capturing more than one leopard. After finding a location where trapping success was high, I no longer moved the traps. But this did not occur until the latter half of the study.

I sometimes used more than one trap to capture or recapture a particular leopard. Home ranges of untagged leopards were predicted by noting the spatial arrangement of home ranges of their radio-collared neighbors and the presence of tracks suspected of being those of an untagged leopard. If a vacancy was apparent in a mosaic of known home ranges of radio-collared leopards and leopard tracks were regularly observed there, I invariably captured an untagged leopard in the area. Sometimes a leopard left a distinctive track, such as the huge wide-toed track left by M23, a remarkable leopard that escaped regularly from traps before he was finally captured. Visual observations of untagged leopards also helped to establish their approximate home ranges and to determine locations for traps.

Baited traps were checked at least once daily. At first I baited the traps but did not set them until a leopard had fed on the bait. Although several leopards were captured using this technique, it proved inefficient. Many times the leopard failed to return, or a civet, genet, or hyena found the bait before the leopard returned. I usually checked traps in the morning and again in the evening if I was in the area. Because most leopards were discovered in the traps in the mornings, I assumed they were captured the previous evening or night, or in the early morning. Few leopards were captured in traps in the early evening.

TABLE 3.1 *Capture of Leopards by Season*

Season	Captures		Trap Days/Capture	
	Observed	Expected	Observed	Expected
Dry	89	61.09	15.4	22.41
Wet	23	50.91	45.4	20.49

Captures: Chi-square = 28.05; *df* = 1; *P* < 0.005
Trap Days/Capture: Chi-square = 32.25; *df* = 1; *P* < 0.005

On several occasions I released radio-collared leopards within several hours after they were captured; at other times I purposely avoided capturing nearby leopards that I knew might enter the traps.

TRAPPING EFFORT

Trapping effort for leopards totaled 2,412 trap days during the study. Of these, 1,138 (47%) were in the SRSA, 882 (37%) in the NRSA, and the remaining 392 (16%) adjacent to the SRSA near Skukuza. Trapping effort was greater during dry seasons (1,369 trap days) than wet seasons (1,043 trap days) because bait was readily available and lasted longer. Leopard traps were located on river banks adjacent to the Sabie and Sand Rivers (1,126 trap days), near tributary streams, primarily the Msimuku and Nhlanganeni (in the NRSA) (506 trap days), along the Nwaswitshaka River (494 trap days), and along firebreak roads (286 trap days).

CAPTURE SUCCESS

Seasonal trends.   Thirty leopards were captured 112 times during the study with significantly more leopards captured during dry than wet seasons (see table 3.1). Eighty-nine captures (79%) occurred during dry seasons; only twenty-three (21%) during wet seasons. An average of 15.4 trap days were required to capture a leopard during dry seasons, compared to 45.4 trap days during wet seasons. Although Pienaar (1969) assumed large predators in KNP found prey more dispersed and hunting conditions poorer during wet seasons, my data and that of Smuts (1982) for lions indicated that leopards and lions had a more difficult time capturing prey, and were themselves more easily captured, during dry seasons. Most leopard captures (thirty-six) occurred during the latter half of the dry seasons (July and August). Capture success was lowest (two leopards) during the latter half of the wet seasons (January and February), when newborn prey, especially impala, were abundant. Leopards apparently captured prey more easily during wet seasons because of the dense vegetative cover. The larger home ranges of leopards during wet seasons may also have decreased

TABLE 3.2 *Capture Success of Leopards*

Study Area	Males			Females		
	Leopards *n*	Captures *n*	Captures per Leopard	Leopards *n*	Captures *n*	Captures per Leopard
Sabie River[1]	10	39	3.9	8	23	2.9
Nwaswitshaka River	4	34	8.5	8	16	2.0
Total	14	73		16	39	
Average	-	-	5.2	-	-	2.4

[1] Includes areas south of Sabie River east and west of Skukuza.

the frequency with which they encountered traps. Leopard tracks in the sand and mud confirmed their lack of interest in trap baits during wet seasons. They often passed within 10 m of traps and ignored them.

Individual trends.   Males were captured more often than female leopards. The mean of 5.2 captures/male leopard was more than twice the average 2.4 captures/female (see table 3.2). An average of 33.0 trap days were required to capture a male, compared to 61.9 to capture a female leopard. Males were captured more often in the NRSA (8.5 captures/male) and females more often in the SRSA (2.9 captures/female) (see table 3.2). Hamilton (1976) also reported female leopards were more difficult to capture than males. During 114 trap days he captured eleven males nineteen times, but only one female three times.

I also recorded the sex and age of leopards captured within thirty days of other leopards in the same trap. Subadults were not captured within thirty days of other subadults, The most frequent sequence was an adult female captured after adult males (18% of all occurrences); adult or subadult males were frequently captured (14%) after adult females. Adults were seldom captured within thirty days of another adult of the same sex.

These capture patterns supported a popular belief that leopards, mainly adults, of different sexes are attracted to each other by odor. However, they contradicted another popular belief that smaller leopards, particularly males, are repelled by the odor of larger leopards of the same sex.

Female leopards with dependent young were never captured. Only one known pregnant female leopard (F6) was captured, and even she avoided traps after her cubs were born. Female 11, captured three times before she had cubs in November 1974, and F21, captured only once, avoided traps after they had young. An unmarked female was not captured until her cub (F12) was more than two years old. Leopards less than 1 to 1.5 years old were also difficult to capture. Of 112 captures, only 2 were of leopards estimated to be this age. Because cubs less than a year old were probably left in secluded places while the female was hunting, leopard cubs would have little chance of encountering livetraps. When cubs were mobile and traveled with their mothers, she may have prevented them from entering traps.

Although the average capture/leopard was 3.7, twelve leopards were captured only once, and only three were captured more than nine times each. The most frequently captured leopards were old adult males. Old adult M3 was captured ten times, and adult M14 was captured fifteen times. Subadult M18 was captured three times within five days. Only one female leopard (F4) was captured more than five times.

Emaciated leopards were captured more frequently than healthy leopards. The average interval between captures of emaciated male and female leopards was 13.5 and 9.7 days, respectively, compared to 56.7 and 94.8 days for healthy males and females. Emaciated leopards, especially females, apparently had difficulty capturing prey and became dependent on traps to obtain food.

Trap location and capture success. Leopards were captured more often in traps set adjacent to the seasonal Nwaswitshaka River (14.5 trap days/capture) than in traps along the perennial Sabie and Sand Rivers (22.1 trap days/capture), firebreak roads (26 trap days/capture), or dry tributary streams (31.4 trap days/ capture). Leopards that traveled along the Nwaswitshaka River were exposed to traps on both riverbanks, whereas those that traveled along the Sabie and Sand Rivers were exposed only to traps on one riverbank. Vegetation was also denser along the Sabie and Sand Rivers than the Nwaswitshaka River, thus reducing visibility of baits. Leopards traveling along the Sabie and Sand Rivers could have passed within 10 m of some traps without seeing the baits; those traveling along the Nwaswitshaka River could usually see baits up to 100 m away. Leopards apparently were able to remember the locations of traps. Nine leopards were captured on eleven occasions in previously visited traps, and two male leopards were recaptured six and eight times, respectively, in traps they had previously been captured in.

Bait and capture success. Most leopards were captured with fresh bait and most were captured 2 to 3 days after traps were baited. The average age of bait at time of capture was 3.5 days. In Tsavo National Park most baits accepted by leopards were 4.5 days old, and most baits rejected were 7 days old (Hamilton 1976). Bait deteriorated rapidly during the wet seasons because of the lack of protective crust over the flesh and the feeding by fly larvae. Leopards were captured only three times with maggot-infested bait. Most leopards (45%) were captured in traps baited with meat of buffalo, the most frequently used bait (43% of trap days) (see table 3.3). During wet seasons meat of impala, the most frequently eaten prey of leopards in the park, appeared the least attractive to leopards. An average of 76 trap days, the highest recorded, was required to capture a leopard with impala bait in the wet seasons, compared to 16.4 during

TABLE 3.3 *Trapping Leopards with Different Baits*

Bait	Trap Days		Captures of Leopards	
	n	%	*n*	%
Buffalo	1,043	43	50	45
Elephant	932	39	39	35
Impala	305	13	15	13
Hippopotamus	88	4	8	7
Baboon	44	2	0	0
Total	2,412	—	112	—

dry seasons (see table 3.4). It is also noteworthy that leopards were not captured when some traps were baited with baboon, supposedly a favorite prey of leopards.

The relative ease with which leopards were captured was unexpected. Although monthly capture success varied from 11.5 to 152+ trap days/capture, the 21.5 trap days/capture average was considerably lower than that reported for other carnivores (see table 3.5). Leopards were captured with less trapping effort than bobcats (Bailey 1974), lynx (Berrie 1973), wolves (Peterson et al. 1984), and wolverine (Hornocker and Hash 1981). Furthermore, by prebaiting, Hamilton (1976) increased his trapping success to 5.2 trap days/leopard capture.

If live-trapping is used to study leopard populations, capture effort limited to certain seasons (wet season) may not provide an accurate indication of leopard population composition. Extended trapping periods may be necessary to ensure

TABLE 3.4 *Trapping Effort Required with Different Baits*

Season	Average Number of Trap Days				
	Buffalo	Elephant	Impala	Hippopotamus	Baboon
Wet	13.0	19.4	16.4	11.0	—
Dry	48.7	35.3	76.0	—	44.0+
Total	20.9	23.9	20.3	11.0	44.0+

TABLE 3.5 *Trapping Effort to Capture Various Carnivores*

Species	Average Trap Days per Capture	Type of Trap	Season	Source of Data
Wolverine	1,537	Leghold	Winter	Bailey and Hornocker (1973)
Wolverine	146	Live	Winter	Bailey and Hornocker (1973)
Lynx	238	Leghold	All	Berrie (1973)
Bobcat	149	Leghold	All	Bailey (1974)
Wolf	128	Leghold	Snowfree	van Ballenberge et al. (1975)
Coyote	96	Leghold	Summer	Hawthorne (1971)
Wolf	50	Leghold	Summer	Kolenosky and Johnston (1967)
Leopard	22	Live	All	this study
Leopard	5.2[1]	Live	All	Hamilton (1976)

[1] Includes prebaiting before setting traps.

TABLE 3.6 *Immobilization of Leopards*

Dosage / Sex	Ketamide Hydrochloride No. of Dosages	Average Dosage (mg/kg)	Xylazine Hydrochloride No. of Dosages	Average Dosage (mg/kg)	Reaction Time (in Minutes) To Recumbency No.	Average	To "Head Up" No.	Average
Single dosages:								
Males	13	12.8	13	1.1	10	12.1	5	96.4
Females	18	13.1	18	1.6	16	6.4	4	94.0
Both sexes	31	13.0	31	1.4	26	9.3	9	95.2
Multiple dosages:								
Males	3	12.7	—	—	3	9.7	2	113.0
Females	3	17.3	—	—	1	4.0	3	110.0
Both sexes	6	15.0	—	—	4	6.9	5	111.5

all individuals are identified. For example, although male leopards may be captured with relative ease, females and young leopards appear extremely difficult to capture. Sex ratios, composition, and productivity of leopard populations may then be biased in favor of older males. Locations of traps, as well as age and type of bait, also play an important role in capture success.

Immobilization and Identification

Only untagged leopards or leopards that needed their radio collars removed or replaced were immobilized when captured; others were immediately released. The first three leopards captured were immobilized with phencyclidine hydrochloride (Sernylan; Parke-Davis), averaging 1.4 mg/kg, and promazine hydrochloride (Sparine; Wyeth), averaging 0.7 mg/kg. But thermoregulatory stress, excess salivation, and periodic violent convulsions (with prolonged recovery periods ranging from eight to ten hours) rendered this combination impractical. As an alternative, ketamine hydrochloride (Vetalar; Parke-Davis) and xylazine hydrochloride (Rompun; Bayer) were used with dosages similar to those used on lions (Smuts et al. 1973). Drugs were administered to captured leopards via dart syringes fired by a carbon-dioxide powered Cap-Chur pistol (Palmer Chemical and Equipment). The mean dosage of ketamine for thirty-one single injections was 13 mg/kg of body weight, and the average recumbency and "head up" times were 9.3 and 95.2 minutes, respectively (see table 3.6).

Occasionally leopards required additional injections, if, for example, a dart syringe failed to discharge properly, or if it was deflected by the leopard's movements. The average dosage for six multiple injections was 15 mg/kg; the average recumbency time was 6.9 min. The average dosage of Rompun for thirty-one single injections was 1.4 mg/kg. Although female leopards became recumbent more rapidly than males, no significant difference in reaction times to the drugs was apparent between male and female leopards. Eight leopards, including six females, experienced mild to severe convulsions after immobilization. Female leopards were probably slightly overdosed because I had a tendency to overestimate their weights when estimating desired dosages. Only leopards that were weighed after immobilization are included in table 3.6.

I attempted to immobilize leopards with a minimum of disturbance. First, I approached the captured leopard in the trap to estimate its weight, then returned to my vehicle to prepare the dart syringe. When the dart was prepared, I approached the captured leopard, discharged the dart syringe into the leopard's rump, and retreated quickly out of the leopard's view. After the leopard became quiet and manageable, I returned and either worked on the leopard inside the livetrap or, after removing the dart syringe, lowered the leopard to the ground.

All dart punctures were treated with an antibiotic (Terramycin) and sutured to reduce infection. All immobilized leopards were given routine injections of an antibiotic (Combymycin, Penimycin, or Triplopen) and any previous injuries were treated. External parasites were then collected, and the leopard was measured and fitted with a radio collar. Most leopards were weighed. But if a leopard was too large for me to handle alone or was captured in a trap set high in a tree, I only estimated its weight. Most leopards that were not weighed when initially captured were weighed upon recapture.

Initially, captured leopards were identified by two aluminum, color-coded, numbered ear tags, an ear tattoo, and a neck tag (see photo 3.3). Later, when I realized that leopards could not routinely get rid of their radio collars, I removed one ear tag and the neck tag of recaptured leopards. Leopards whose radio collars were later removed also had their ear tags and neck tags removed. Leopards a year old or less and leopards captured late in the study were identified only with one ear tag.

After immobilized leopards were ear-tagged or fitted with radio collars, they were placed back into the livetraps to recover. This was done to protect them from attacks by other predators while they were under the drug's influence and to reduce the chance of any self-inflicted injuries. Traps were covered with a canvas tarpaulin to prevent sunlight or rain from falling on the recumbent leopard. When the leopard was fully recovered, I opened the door from a distance by pulling the rope attached to it and released the leopard.

Radiotelemetry

The radio-tracking equipment used in this study was similar to that described by Seidensticker et al. (1970). The transmitter collars were 30 MHz with an effective radiated power of about 0.5 mw. The 2.5-cm-wide collar served as the antenna and was adjusted from 33 to 51 cm to fit the neck of different-size leopards. Transmitter components and batteries were embedded in a water-

3.3. An adult male leopard (M3) captured early in the study showing radio collar and ear and neck identification tag. Not all captured leopards were fitted with each marking device.

proof matrix of fiberglass and epoxy resin with an outside opening to a variable tuning capacitor. A magnetic switch embedded within the transmitter collar was used to turn it on and off. A tight-fitting polyethylene hose enclosed the copper band to prevent chafing. Two models of transmitters, a two-cell, mercury-battery model weighing 680 gm and a single-cell, lithium sulfate-battery model weighing 565 gm, were used. The life span of both ranged from 4.9 to 11.7 months.

The portable radio receivers were crystal-controlled, double-conversion, super-heterodyne (A. R. Johnson Electronics). These small (13 x 10 x 8 cm) but rugged receivers used eight 1.5 v batteries or an external 12 v power source. Sensitivity of the receivers was 0.1 mv and audio output was 100 mv. Three types of antennas could be coupled to the receivers. A base-loaded whip antenna mounted on the top of my vehicle was used to scan for signals in the study areas. To determine direction of the signal, I used a tuned loop antenna 30.5 cm in diameter and attached directly to the receiver (see photo 3.4). A similar loop antenna mounted on a Cessna 206E aircraft and was used to locate leopards that could not be found from the ground. All antennas were coupled to preamplifiers that increased gain by 15 db.

Radio-collared leopards were located from the ground and air. On the ground I periodically stopped along my travel route to scan for signals. If a signal was

3.4. A small, hand-held portable radio receiver with attached loop antenna and preamplifier was used to locate radio-collared leopards from the ground.

detected, directional compass bearings were taken from different locations using the loop antenna to determine the position of the leopard. If the topography prevented taking several compass bearings, the distance to the radio-collared leopard, if nearby, was estimated.

When we were radio-tracking from the air, we flew 250 m to 305 m above the ground in a square search pattern, starting over the center of a study area or at the last known location of the sought-after leopard. If a signal was detected, we flew in a semicircular pattern and rotated the loop antenna to determine the approximate location of the leopard. We then dropped to about 150 m and passed directly over the leopard from several directions. Once the location was accurately determined, I plotted the position on a map in reference to a recognizable feature of the terrain.

Radio reception range varied with the transmitter, terrain, and receiving antenna. The average range of reception on the ground was 1.5 km to 2.5 km

TABLE 3.7 *Activities of Leopards*

Leopard Wearing Radio Collar?	Number of Observed Activities					
	Walking	Resting	Eating	Hunting	Hiding	Courting
Yes	23	18	13	2	4	0
No	40	13	6	6	2	3

Chi-square = 12.93; *df* = 5; $P < 0.025$

with the whip antenna.From the top of a hill or koppie, its range increased to 5 km. Reception range using the directional loop antenna was reduced about 30%. The maximum range of reception was about 15 km from an aircraft flying 305 m above ground. The accuracy of the radio-tracking system was checked against visual observations. On the ground the angular error was estimated to be about 3.5°, as the error averaged 5 m to 7 m at 100 m. I was usually within 1 km, often 100 m, when locating radio-collared leopards in the SRSA, but 1 km to 2 km away from those in the NRSA.

The locations of leopards were plotted on 1:50,000 topographic maps and identified by X, Y-coordinates based on a grid size of 0.01 km². The leopard's identification number, habitat, and activity; the date; the time of observation; sunrise and sunset times; and various weather parameters were coded and recorded on computer optical scanning cards. This unit of data was called a "radio-location day." Most leopards were located only once daily. These data were later transferred to magnetic tapes and processed by a digital computer at the University of Idaho.

REACTION OF LEOPARDS TO RADIO COLLARS

To determine if leopards were influenced by their radio collars, I compared activities of leopards with and without radio collars (see table 3.7). Although radio-collared leopards were observed feeding more often than leopards without collars, this was expected; most radio-collared leopards were approached to locate kills. Furthermore, the probability of finding kills of leopards that were not radio-collared was a matter of chance. Otherwise, radio-collared leopards killed, cached, and ate prey, mated, gave birth, and reared young no differently from leopards that were not radio-collared. Only one female removed her collar, probably less than twelve hours after being released, when I fitted it too loosely about her neck.

After six to nine months, the radio collars of one female and four male leopards were bent and twisted.Those of six others (three males, three females) were only slightly scratched or worn. Because the radio collars of seven additional leopards showed little signs of scratching or wear after one to five months, most of the damage was apparently due to wear rather than to attempts to remove the collars. During only one of sixty-five visual observations was a leopard observed scratching its neck and that was probably in response to irritation from mange. Because of these observations, radio collars were assumed not to significantly alter leopard behavior.

Food Habits

Because the diet of leopards in KNP had already been documented (Pienaar 1969), my objectives were to determine the frequency that leopards killed large prey, their utilization of large-prey carcasses, use of small prey, and the impact of leopard predation on prey populations. Two techniques were used to determine kill frequency. The first assumed that a sudden cessation of movements of radio-collared leopards meant they had made a kill. I assessed the validity of this assumption by approaching forty-eight suspected kills on foot and discovered actual kills thirty-six times. The behavior of leopards on four additional occasions suggested a kill was present but hidden, and on four other occasions, the leopards fled before I was able to pinpoint their location. Kills were therefore present at least 75% of the time, and probably from 83% (including four hidden kills) to 91% (excluding 4 occasions when leopards fled far away). I therefore assumed this to be a valid method of estimating kill rates.

The second technique used to determine kill rates was to follow individual radio-collared leopards for extended periods and approach them daily to confirm kills. However, neither technique accounted for kills of smaller prey because such prey could have been eaten in a short period.

To determine carcass utilization, I recorded the time required by leopards to consume their kills and how much of the carcass was eaten. The weights of the consumed portions, and the proportion of the kills lost to other predators and scavengers were estimated. To obtain these data, I periodically returned to kills of radio-collared leopards to determine the amounts eaten during known intervals and the presence of other predators and scavengers. I also recorded the species, diameter, and degree of cover of trees used by leopards to cache carcasses of prey.

The kinds of large prey killed by leopards were also recorded for comparative purposes with other studies. However, the physical condition, age, and sometimes the sex of the prey killed by leopards were impractical to determine because they were often high in trees and could not be examined. I did not want to disturb feeding leopards for fear of causing them to abandon kills. And once leopards finished feeding from carcasses, scavengers quickly consumed or carried off the remains.

The use of small prey by leopards was determined by fecal analysis. Feces of leopards were collected along firebreak roads. To avoid confusion with feces from other predators, I collected only freshly deposited feces near leopard tracks. Feces were pulverized, washed, and preserved in formalin. Later the

TABLE 3.8 *Mean Visibility Limits Used to Estimate Populations of Prey*

Species	Mean Visibility Limit (in meters)
Impala	
Sable River study area:	
Brackish flats:	
Dry and wet seasons	147.0
Undulating terrain:	
Dry seasons	45.4
Wet seasons	40.0
Nwaswitshaka River study area:	
Brackish flats:	
Dry and wet seasons	117.6
Undulating terrain:	
Dry seasons	45.4
Wet seasons	47.3
Giraffe	170.0
Zebra	150.0
Wildebeest	150.0
Warthog	100.0
Greater kudu	75.0
Steenbuck	75.0
Bushbuck	40.0
Gray duiker	30.0
Guinea fowl	30.0
Francolin	25.0
Porcupine	25.0
Hare	25.0

feces were separated into hair, bones, hooves, and feathers, and compared to a reference collection of hair from prey in the study areas. Because my reference collection was limited to the common and abundant species, species identification was not always possible.

Census of Prey

To obtain an index of prey numbers and to estimate seasonal prey densities, I censused prey along firebreak roads within the study areas. A road strip technique (Hirst 1969a) was used to estimate prey abundance. A visibility limit, the mean distance an animal disappears from view, was used to determine the width of the effective strip. Because this changes with each species and vegetation type (Robinette et al. 1974), I stratified my area into two easily discerned vegetation types. One was the sparsely vegetated, relatively flat, brackish depressions dominated by *Acacia grandicornuta, Acacia nilotica,* or *Euchlea divinorum.* The other was denser vegetation dominated by *Grewia* shrubs. Within each type I recorded the distances different species disappeared from view and averaged them for that species mean visibility limit (see table 3.8).

Prey was censused between 6:00 A.M. and 10:00 A.M., the optimum time reported by Hirst (1969a). A 20.5-km route in the SRSA and a 13.3-km route in the NRSA were surveyed. I recorded numbers of game observed on each side of the road as I drove at 5 to 10 kph. I also recorded information on the sexes, ages, and habitat of prey and noted the times and places they drank water, their social and reproductive behavior, and their response to predators.

Vegetation Analysis

Composition and density of vegetation were determined by establishing fifteen transects located at right angles to the firebreak roads crossing the study areas. At representative areas I followed compass bearings for 100 m, recording the number and species of trees and shrubs encountered in a 1-m wide strip.

I also recorded the percentage of bare soil, degree of erosion, slope, average height of grass, number of game trails crossing the transects, and the numbers of pellet groups of impala. I adopted the technique outlined by Van Wyk and Fairall (1969) to classify woody plants. To select transect sites I examined aerial photographs, determined extent of vegetative types, and divided the distance through each type by the number of transects (five to ten) to establish a uniform sample interval along the route. Locations of transects were determined by the vehicle's odometer.

Population Estimate of Leopards in Kruger National Park

The relative abundance of impala and the leopard's preferred habitat, primarily riverine forest and montane habitat, were used as indices to estimate numbers of leopards in the park. During the dry season of 1975, I traveled throughout the park to obtain this information. I recorded numbers and species of prey observed along the major roads in the park, especially those in leopard habitat adjacent to perennial and seasonal rivers. For comparative purposes the counts were conducted in the same manner as those in the study areas. The relationship between leopards and prey in the study areas was then extrapolated to similar areas throughout the park. Estimates of wildlife numbers by park officials were also used as an index in estimating numbers of leopards. Although these methods were crude, no other methods were practical given the time constraints and size of the park.

Response of Leopards to Humans

RESPONSE TO INVESTIGATOR

I observed the behavior of captured and free-roaming leopards when I approached them on foot and by vehicle to determine if there was any unusual behavior that could have biased my information.

Captured leopards. Leopards captured for the first time were usually more disturbed by my presence than previously captured leopards. Several large males and one female leopard that were captured more than five times each, merely retreated to the far end of the trap and remained there motionless and silent until I released them. They apparently became conditioned to the traps, my presence, and the release procedure.

Captured leopards usually attempted to escape by rushing at the closed trap door. This response was later reduced by tying canvas, a visual barrier, over the trap door. When captured leopards were approached, some urinated, fewer defecated, and still fewer bit viciously into pieces of bait in the trap. Captured female leopards were more aggressive and vocal than males, and females captured for the first time were more intent on escaping or attacking than males. The only two young leopards that were captured responded like females when I approached them.

The behavior of released leopards was quite variable. Several male leopards usually fled along the same routes each time they were released. Some females remained in the trap five to ten minutes after the door was opened before they fled; others refused to leave the trap until I left or was out of view. Once an infuriated female leopard appeared to attack upon release. When the trap door was opened, she attacked the upraised door, biting and clawing, but fell 1 m to the ground. She then leaped 2 m up to the trap, scratched and bit the door again, and rushed a nearby vehicle. When she was 3 m away, however, she fled into nearby cover.

Free-roaming leopards. The immobilization procedure did not appear to have a significant influence on movements of leopards, at least within twenty-four hours of their release. Although the average distance between immobilized leopards and trapsites (0.9 km) was less than that of nonimmobilized leopards (1.8 km) one day after their release, the difference was not significant. The physical condition of the leopard appeared to be the most apparent factor influencing movements of immobilized leopards. Healthy leopards moved four to six times the daily distances moved by emaciated leopards.

I purposely attempted to minimize disturbance of radio-collared leopards. Of sixty-five visual observations of radio-collared leopards, thirty-five occurred after I stalked the leopards on foot and thirty occurred from a vehicle. Stalking radio-collared leopards on foot in dense brush was time consuming, frustrating, and frequently unsuccessful. Each attempt to observe radio-collared leopards in this manner required one-half to four hours (average two hours) of careful stalking. Then, the leopards usually saw or heard me from a distance of 50 to 100 m, before I saw them, and fled. In contrast, one large male leopard (M23) often retreated to the nearest patch of dense vegetation and permitted me to circle him from a distance of 15 to 20 m. Male 10 once followed me for 0.5 km as I returned to my vehicle.

Radio-collared leopards were seldom disturbed during the study and did not appear to associate me with a vehicle. Between August 1974 and January 1975, in sixty-nine attempts to observe radio-collared leopards while on foot, I observed them only twelve times (17%). This was probably representative of the entire study period; thus the thirty-five successful observations of radio-collared leopards on foot during the entire study required an estimated two hundred attempts. These two hundred attempts, plus the 112 captures, indicated that radio-collared leopards were potentially disturbed by my presence on foot on 12% of the days I located them by radio, or 6% of the total days they wore functioning radio collars.

Radio-collared leopards usually fled when they detected me nearby. On thirty-five occasions when radio-collared leopards were observed visually, they fled swiftly on eighteen (51%), slunk quietly away on fourteen (40%), and seemed to be undisturbed on only one occasion (3%). Of the eighteen that fled swiftly when they saw me, eleven carefully avoided obstacles in their path, but seven uttered a short, gruff, growl of surprise and ran blindly, crashing through the undergrowth.

Leopards were seldom aggressive toward me. Male 3 once snarled for ten minutes at a distance of 20 m, so I retreated. I was uncertain, however, whether he was threatening me, another leopard, or a nearby hyena. Another male snarled when 10 m from me, but probably at the leopard 15 m behind him. Once, when I stepped from my vehicle, a radio-collared female rushed me from dense cover. But she stopped 15 m away with her ears laid back, front feet widespread and lips curled back, snarling. Then she slowly turned and calmly walked away. The only leopards that did not flee upon my approach were females with cubs. On fifteen to twenty occasions when I approached three different females that had young, they remained motionless and would not leave the area. To prevent aggressive behavior, I did not approach within 10 m of females with cubs.

Radio-collared leopards fled when I was as far as 80 m from them. But on twenty occasions I stalked within 20 m and on three occasions within 10 m of leopards before I observed them. Once I heard a leopard breathing before I could see it, and on another occasion I discovered a leopard sleeping at the base of a tree only 3 m away. All close approaches occurred during the wet season when dense vegetation provided ideal stalking cover. Stalking was more difficult during the dry season, because of increased visibility and noise associated with walking on dry leaves and grass. In Tsavo National Park, where leopards could be approached with greater stealth on foot than with a vehicle, the flight distance of leopards to a man on foot was less than to vehicles, 131 m and 202 m, respectively (Hamilton 1976).

Several observations illustrate the leopards' haste to flee from man. Once I accidentally startled a leopard 20 m from its kill, but instead of fleeing along the densely vegetated riverbank, it ran 60 m across open sand, jumped into the flood-swollen Sand River, and swam 15 m across a rapid current. On another occasion I accidentally disturbed a leopard asleep on a hammerkop nest in a tree. It raced headlong down the tree trunk, leaped 4 m into the middle of a shallow waterhole, landed in a splash, struggled through the shallow water, and then ran away crashing through the undergrowth.

Female leopards were more difficult to stalk and observe than male leopards. Of fifty-one attempts to observe radio-collared females, only 12% were successful compared to 33% for males. Despite numerous attempts to observe female leopards, only sixty-five observations of radio-collared leopards were of females. Hamilton (1976) also reported that female leopards fled more quickly than males and that most of the females were visible for less than one minute. Similarly, Eisenberg and Lockhart (1972) reported seeing male leopards twice as often as females. Because of this, sex ratios of leopards obtained from visual observations are probably biased in favor of males.

The average time per observation of a leopard depended on the technique, time of observation, and the activity of the leopard. Of 2,329.4 minutes (38.8 hours) of observations of leopards, 1,873.4 min (80%) were from a vehicle, and the remainder were on foot. The average period of observation of leopards from a vehicle was 19.7 min compared to 11.4 min while on foot. One leopard sleeping in a large tree was observed three hours from a vehicle before it calmly descended, but two other leopards, also resting in trees, immediately fled when they detected me on foot. Like lions in the park (Stevenson-Hamilton 1947; Pienaar 1968), leopards apparently do not associate vehicles with danger if they are undisturbed. But leopards along seldom-used firebreak roads were much shyer than those living along well-used tourist roads. One male leopard (M23) apparently recognized my vehicle, an uncommon blue pickup truck, and fled whenever I approached him.

My most rewarding observations of leopards were made at night. Thirty encounters of leopards at night from a vehicle averaged 25.4 minutes, compared to 14.9 minutes for 105 daytime observations. At night, leopards allowed the vehicle to approach more closely than during the day, and they quickly became conditioned to a spotlight. Sudden noises caused them to pause, however, and stare at the vehicle in apparent alarm.

The longest observations were of feeding leopards. The average observation of leopards at kills was 32.6 minutes compared to 12.3 minutes for leopards involved in other activities. Because they were less observant of their surroundings, leopards busily eating permitted me to approach the closest on foot during the day. Once I was detected, however, the leopard immediately left its kill, returning after darkness. Of fifty-five kills I approached, only two were moved by leopards after they had been disturbed. Another five leopards moved their kills during the night, away from well-used tourist roads, after they were disturbed by tourists. Changes in leopard behavior as a result of the study were brief and predictable. Captured leopards did not abandon their home ranges, and 60% of the individuals were recaptured up to fourteen times each. Although free-roaming leopards fled when I encountered them on foot, they soon resumed use of the same areas. Leopards did not abandon kills when I accidentally flushed them, but merely returned later to finish eating or to move them to another location. With one exception leopards did not appear to recognize my vehicle. Because of this, it was concluded that the information obtained during the study was not significantly biased by a leopard's occasional encounter with humans.

Response of Leopards to Roads, Tourism, and Development in Kruger National Park

The impact of roads and tourists on wildlife in KNP has previously been addressed by Pienaar (1968). In this section I specifically address the influence of roads and tourists on leopards, realizing that my observations are limited to a small region of the park highly frequented by tourists.

ROADS AND TOURISTS

Three types of roads are present in the park: (1) tarmac roads (2) dirt roads, both of which are open to tourist traffic, and (3) firebreak roads, which are narrow, dirt roads, closed to tourists. I spent 45% of my time in vehicles driving on firebreak roads, 33% on tarmac roads, and 23% on dirt roads.

Contrary to the belief of most tourists, leopards were observed more often from tarmac roads than firebreak or dirt roads (see table 3.9). Each leopard observation on tarmac roads required an average of 313 km of driving compared to 828 km for firebreak roads. Most observations of leopards from a vehicle occurred along two well-traveled roads in the park: the tarmac roads from Skukuza to Lower Sabie and from Skukuza to Tshokwane. To observe a leopard along those roads required driving an average of 188 and 260 km, respectively. Population density of leopards, visibility, and distribution of prey, appeared more important than volume of tourist traffic in observing leopards. Most leopards were seen along riverine areas because of its excellent leopard habitat and because leopards living there were conditioned to passing vehicles. Density of vegetation was extremely important in observing leopards. In my study

TABLE 3.9 *Leopards Observed near Skukuza, Kruger National Park*

Type of Road	Name	Number of Leopards	Average Distance Driven Between Observations (in kilometers)
Tarmac (tourist used)	Tshokwane (Sand River)	16	188
	Lower Sabie	23[1]	260
	Naphe	11	364
	Kruger Gate	3	1,000
	Others	5	500
	Subtotal	56	313
Dirt (tourist used)	Doispan	5	1,000
	Nwaswitshka Water-hole	3	1,667
	Others	2	1,160
	Subtotal	10	1,230
Firebreak (closed to tourists)	Sabie River study area	16	813
	Nwaswitshaka study area	6	1,500
	Others	7	286
	Subtotal	29	828

[1] Includes three observations from dirt loop roads connected to the Lower Sabie tarmac road.

areas, where the density of leopards was high, they were seldom seen because of the dense vegetation. In contrast, leopards along the Lower Sabie Road were often visible in overgrazed areas near the Sabie River. Distribution of prey also influenced the frequency of leopard sightings. Few leopards were seen along the Naphe, Kruger, Diospan, and Nwaswitshaka Waterhole roads, because prey and leopard density were low.

Male leopards were observed more often than females along tourist roads. Only 37% of the leopards observed from tourist roads were females, compared to 68% females observed from roads closed to tourist traffic. This, as well as the difficulty in capturing females, suggested that female leopards, more than males, avoided contact with humans. However, some females were repeatedly seen in the same areas. Three individuals were responsible for 44% of my observations of female leopards along tourist roads, and four accounted for 80% of my observations of female leopards along firebreak roads.

Leopards frequently used roads as travel routes and were merely walking on 66% of the occasions when first observed. Scent marking along roads also suggested leopards used them as travel routes. Some activities such as stalking prey, feeding, and copulation were observed along roads only after tourist hours, although two leopards were once reported copulating along the Lower Sabie Road in view of tourist vehicles.

Most leopards were observed after dark when tourists were prohibited from using the roads. Although less than 25% of the distances I drove on tourist roads were after dark, 36% of all leopard sightings occurred after dark. I drove an average of 530 km between each daytime leopard observation, compared to

310 km for each nighttime observation. Thus, although leopards avoided excessive traffic, they did not avoid roads. Leopards seldom walked more than 100 m along tourist roads during the day, but after dark, they walked up to 750 m. During the day tourists usually forced leopards off roads. If not more than four to five vehicles stopped to watch a leopard, the leopard appeared calm. If twenty to thirty vehicles stopped, they forced the leopard into cover. When only a few vehicles gathered, they usually kept at least 10 m from leopards. When many vehicles arrived, tourists drove nearer. After several minutes, running engines, talking, radios, and movements within the vehicles made leopards flee. On some occasions, leopards bolted through the middle of traffic jams.

The greater the distance from vehicles, the less leopards were disturbed. One leopard rested in a tree three hours while tourists in twenty to thirty vehicles watched from 35 m away. The same leopard fled in twenty minutes when ten to fifteen vehicles stopped within 15 m of its resting tree. If a vehicle's engine was off, leopards would cautiously pass within 3 m, even during daylight.

Leopards seldom fed on kills during tourist hours and often moved them to secluded locations after dark. Of nine kills leopards cached in trees visible to tourists, six were eaten by leopards only after tourists left, five were moved to secluded cover, and only one was fed upon while tourists were watching. I once located a leopard 20 m from an impala cached in a tree. Tourist vehicles were parked near the kill all day but no one observed the hidden leopard. I parked 100 m from the kill and 20 minutes after the last vehicle left, the leopard emerged from the thicket and began feeding. Similar leopard behavior was observed on three other occasions.

OTHER HUMAN DEVELOPMENT

Leopards used concrete causeways to cross the Sabie and Sand Rivers. Three radio-collared leopards periodically crossed the Sabie River and two crossed the Sand River causeways. When crossing, they usually hid in dense cover nearby until tourist traffic ceased. However, M23 once crossed the Sand River causeway in midday near a vehicle. I never received reports of leopards crossing a high bridge over the Sabie River west of Skukuza, nor the railroad trestle at Skukuza, where a troop of baboons slept each night.

Leopards also used road culverts and drain pipes. A female with two cubs periodically stayed in a large drain pipe near Skukuza in late 1972 and early 1973 (G. v. Rooyen, personal communication), and I discovered leopard tracks under a causeway 100 m from the drainpipe. Another female with cubs entered a small road culvert near the SRSA (J. Newman, personal communication), and on one occasion a leopard crawled into a small culvert to avoid tourists. I frequently observed tracks of leopards in culverts under the tourist roads.

A tall fence affected the movements of one radio-collared leopard. This 2.4-m high mesh fence was a barrier for M23, who was never found north of the fence. Once he was unable to climb the airport perimeter fence, but was persuaded to leave via an opened gate (N. de Beer, personal communication). But tracks of leopards inside perimeter fences of the staff village (U. Pienaar, personal communication), suggested that lower fences were easily scaled by leopards, perhaps more agile females or young individuals.

The Skukuza rest camp and staff village influenced the movements of several radio-collared leopards. Three males kept at least 0.5 km and another at least 1 km from the camp when traveling through the area. When a radio-collared female and a younger male passed by the rest camp, they avoided revealing themselves to tourists. Male 23, however, passed in front of the camp during daylight in front of hundreds of tourists on the other side of the river. On four occasions, three radio-collared males were located within 100 m of Skukuza. I once observed a leopard 20 m from our cottage on a moonlit night and saw another walking along the camp fence. Another leopard crossed the road 80 m from the camp entrance gate, and M2 was once located, after dark, beside the natives' living quarters. Leopards occasionally entered the camps at night and stalked and killed impala sleeping inside the fences.

Although leopards living near tourist facilities in KNP were conditioned to the presence of humans, females generally avoided and males ignored tourists. Roads and other structures were often used by leopards, particularly after dark when tourists returned to rest camps. Leopards living near tourist roads were conditioned to overall traffic volume and did not appear disturbed when viewed by tourists in four or five vehicles. However, a large gathering of tourists in vehicles frightened leopards into the nearest cover. Leopards preferred to feed in solitude, away from tourist roads; avoided contact with the Skukuza rest camp; and appeared unable to climb a high, mesh fence. As long as they were not harassed, leopards in the park tolerated tourists and vehicles. Their distribution and abundance, in and adjacent to the study areas, appeared only slightly influenced by the number of roads, tourists, and rest camps in the park. Rest camps and other development had the greatest impact on leopards by reducing the amount of habitat available for their use.

Summary

1. Ten cage-type livetraps baited with 5- to 10-kg pieces of meat and set in trees to avoid hyenas and lions were used to capture leopards.
2. Livetraps were set 2,412 trap days: 1,138 (47%) in the Sabie River study area, 882 (37%) in the Nwaswitshaka River study area, and 392 (16%) near both study areas.
3. Most trapping effort, 1,126 trap days (47%), was along the perennial Sand

and Sabie Rivers. The remaining effort was along the seasonal Nwaswit-shaka River (494 trap days), dry tributary streams (506 trap days), and dirt firebreak roads (286 trap days).

4. Thirty leopards were captured 112 times during the study. Eighty-nine captures (79%) occurred in dry seasons and twenty-three (21%) in wet seasons. Capture success averaged 15.4 trap days/capture during dry seasons and 45.4 trap days/capture during wet seasons. Capture success was highest in July and August and lowest in January and February.

5. Male leopards were captured with less trapping effort (33 trap days/capture) and recaptured more frequently (5.2 captures/individual) than female leopards (61.9 trap days/capture and 2.4 captures/individual).

6. The most frequent individual capture sequence/trap was capturing an adult female after an adult male (18%). Subadults were never captured within thirty days of another subadult and adults were seldom captured within thirty days of other adults of the same sex.

7. Lactating female leopards, females with dependent cubs, and leopards less than one year old avoided livetraps, and only one female known to be pregnant was captured.

8. Although some leopards, especially females, were captured only once, the recapture rate averaged 3.7 captures/leopard. Adult males were captured most frequently, young subadult females the least frequently.

9. Emaciated leopards were recaptured more often (13.5 days/male capture and 9.7 days/female capture) than healthy leopards (56.7 days/male capture and 94.8 days/female capture).

10. Leopard capture success was greatest along seasonal rivers (14.5 trap days/capture), followed by perennial rivers (22.1 trap days/capture), firebreak roads (26 trap days/capture), and dry tributary streams (31.6 trap days/capture). These differences were attributed to leopard travel habits and visibility of traps, baits, and attractants.

11. Individual leopards were frequently recaptured in traps set in one location. On eleven occasions, nine individuals were captured in the traps of their initial capture, and two males were recaptured six and eight times in traps set in the same location. This behavior suggested leopards have excellent memories of locations of traps and baits.

12. The age of the bait averaged 3.5 days at time of capture, and baits were fewer than four days old during 73% of the captures. Although most leopards avoided baits older than seven days, three were captured with decomposed, maggot-infested baits.

13. Leopards were captured with buffalo, elephant, and impala baits in proportion to their frequency of use. However, impala baits were often ignored by leopards during wet seasons.

14. Leopards were captured with relatively little effort (21.5 trap days/capture) compared to other carnivores.

15. Captured leopards were immobilized with ketamine and xylazine hydrochloride. Thirty-one single injections of this combination (average 13 mg ketamine hydrochloride/kg body weight and 1.4 mg xylazine hydrochlo-

ride/kg body weight) resulted in an average recumbency time of 9.3 min and an average "head up" time of 95.2 min. Although there was no handling mortality, some females experienced mild to severe convulsions. Immobilized leopards were left to recover in livetraps before they were released.

16. Captured leopards were identified with radio collars and neck and ear tags. Thirty-five radio collars were placed on twenty-four leopards. Each 30 MHz transmitter collar was 33 to 51 cm in circumference, weighed 565 gm to 680 gm, and functioned 4.9 to 11.7 months. No significant differences were noted among the activities of radio-collared and nonradio-collared leopards.

17. Small, portable radio receivers with three types of interchangeable antennas (vehicle whip, portable loop, and aircraft loop) were used to detect and locate radio-collared leopards. Radio reception on the ground with whip antennas averaged 1.5 to 2.5 km, but was reduced 30% with a portable loop. Flying at 305 m, an aircraft loop had a range of 15 km. Most radio fixes were taken within 1 km to 2 km of radio-collared leopards.

18. Large kills of leopards were located by stalking feeding radio-collared leopards on foot. Small prey occurrence in the leopards' diet was determined by examining leopard feces.

19. A roadside strip census technique was used monthly to estimate numbers of leopard prey in the study areas.

20. The composition and density of trees and shrubs in the study areas were determined by sampling them in fifteen plots (100 m x 1 m) in each area.

21. To estimate the park's leopard population, prey was censused throughout the park and related to prey and leopard densities in the study areas.

22. Captured and free-roaming leopards usually fled from me when I approached them on foot. Only female leopards with young cubs refused to flee.

23. Only 12% of all attempts to visually observe radio-collared leopards after stalking them on foot were successful. Female leopards were more difficult to observe (12% success) than males (33% success).

24. Leopards immediately fled when I approached within 80 m of them on foot, but leopards at night and those conditioned to tourist vehicles could be approached within 10 m to 30 m in a vehicle.

25. The average time leopards were visually observed depended upon mode of observation (vehicle 19.7 minutes, on foot 11.4 minutes); time of day (night 24.5 min, day 14.9 min); and leopard activity (feeding 32.6 min, other activities 12.4 min).

26. More leopards were observed from paved tourist roads than dirt roads used by tourists or firebreak roads closed to tourists. Most leopards were observed along two of the most-used tourist roads in the park. Population density of leopards, vegetative cover, and abundance of prey influenced leopard visibility. Leopards, which avoided tourist roads during the day, used the roads after dark.

27. Although male leopards were visually observed more often than females

along tourist roads, females were observed more often than males along firebreak roads closed to tourists. This suggested male leopards became conditioned to humans and vehicles more readily than females.

28. Leopards conditioned to vehicles did not appear disturbed if tourists in only four or five vehicles watched them. Large crowds, however, eventually forced even the boldest of leopards into cover.

29. Although leopards sometimes killed and cached prey beside tourist roads, they seldom fed there. Instead, they moved the kill, often at night, to secluded locations away from roads to feed.

30. Some leopards used concrete causeways to cross rivers; others used road culverts for travel or to hide in. Although radio-collared leopards generally avoided coming near the Skukuza rest camp, occasionally leopards entered rest camps at night and sometimes even killed impala within the fenced enclosures.

4

The Prey of Leopards

THE kinds, population densities, and behavior of prey influence the quality of a predator's habitat and the health of predator populations. Therefore, before one can understand the ecology of leopards, some knowledge of their prey's ecology is essential. Because leopards live in many diverse habitats, adequately describing prey characteristics is also important. I present information on the kinds, population characteristics, ecological distribution, activity, and social behavior of leopards' prey and of other ecologically important species inhabiting the leopard study area.

Composition of Prey Communities

The most frequently observed species of ungulate prey in the leopard study areas was impala. Impala made up 95% of 12,653 prey observed during systematic prey surveys. Other frequently observed ungulates, in order of decreasing frequency, were kudu (1.9%), giraffe (0.9%), warthog (0.6%), zebra (0.6%), steenbuck (0.5%), gray duiker (0.4%), and blue wildebeest (0.3%) (see table 4.1). Steenbuck, gray duiker, and bushbuck were probably more abundant than my survey methods indicated because of their preference for specific, often dense habitats and their elusive behavior. Although not normally leopard prey, elephant, rhinoceros, and buffalo also inhabited the study areas. One large herd of buffalo occasionally used the SRSA during the dry seasons of 1973 and 1975, and a small herd of bachelor males and several solitary bulls used both

TABLE 4.1 *Ungulates Observed on Monthly Censuses, August 1973 to July 1975*

| | Study Area | | | |
| | Sabie River | | Nwaswitshaka River | |
Species	n	%	n	%
Impala	6,066	96.0	5,894	93.1
Greater kudu	83	1.3	163	2.6
Giraffe	71	1.1	40	0.6
Steenbuck	30	0.5	37	0.6
Warthog	27	0.4	55	0.9
Gray duiker	22	0.3	23	0.4
Bushbuck	8	0.1	1	trace
Elephant	8	0.1	2	trace
Zebra	4	0.1	78	1.2
Wildebeest	4	0.1	30	0.5
White rhinoceros	0	—	7	0.1
Total	6,323	100.0	6,330	100.0

that area and the NRSA. Hippopotamus were abundant along the Sabie River, but were absent along the Nwaswitshaka River.

The relative abundance of many smaller mammals, birds, and reptiles were difficult to determine because of their habitat preferences and behavior. Some species, such as porcupines, hares, and galagoes, were seen only at night. Others, such as civets, genets, and ratels, were rarely observed, although evidence such as tracks, scats, and captures in traps set for leopards suggested civets and genets were common on both study areas. Dwarf and banded mongooses were regularly seen in the study areas, but I could only guess at their numbers. Of the birds, the ground dwelling francolins such as the crested francolin and Natal francolin, perhaps those most vulnerable to leopard predation, appeared equally abundant in both study areas, but Swainson's francolins and coque francolins were uncommon. Of the large reptiles, tree leguans were more common in the NRSA and water leguans more common in the SRSA. Leopard tortoises were regularly seen in both study areas. Although snakes were not frequently observed, most that were seen were poisonous. Of the common snakes, puff adders, pythons, boomslangs, cobras, and mambas were seen in order of decreasing frequency.

Several other differences between prey communities were noted in the leopard study areas. Impala were observed more frequently in the SRSA; kudu, zebra, wildebeest, and warthog were seen more often in the NRSA. Reintroduced red duiker were periodically seen in the SRSA in 1973, and klipspringers were periodically seen only in the NRSA. Cane rats occurred along the Sabie River but were scarce along the Nwaswitshaka River. Crowned guinea fowl were also more abundant along the Sabie than the Nwaswitshaka River. Only the Sabie River supported permanent populations of fish such as catfish, yellowfish, and tiger fish.

Numbers and Composition of Prey Populations

IMPALA

Population size.   Impala were abundant in both leopard study areas. During the two-year study, the average densities of impala in the SRSA and NRSA were 90.7 and 97.4 impala/km^2, respectively (see table 4.2). These estimates were similar to those calculated earlier by Pienaar et al. (1966a) for a 3.2 km strip along the Sabie River (98.4 impala/km^2) and for those I calculated along the areas adjacent to the Sabie River using 1959–60 park census data (97 to 362.3 impala/km^2). My estimates of impala densities in the NRSA were generally higher than those reported earlier by Pienaar for a much larger area that included the NRSA (4.4 impala/km^2). They were, however, similar to densities I calculated using 1959–60 park census data obtained along the Naphe Road bordering the leopard study area (70.4 impala/km^2).

Impala densities in the leopard study areas were higher than those reported for impala elsewhere in Africa. In the nearby Timbavati Game Reserve, Hirst (1975) estimated average monthly densities ranging from 8 to 59 impala/km^2 for fourteen vegetation zones, with the highest monthly density (311 impala/km^2) occurring in the reedbed vegetation zone near water. Elsewhere, estimates

TABLE 4.2 *Estimated Densities (numbers/km^2) of Impala*

Year	Month	Study Area	
		Sabie River	Nwaswitshaka River
1973	August	241.2	—
	September	236.5	31.4
	October	148.1	—
	November	39.6	—
	December	51.2	48.5
1974	January	40.2	35.1
	February	49.9	26.7
	March	64.7	75.5
	April	44.2	70.0
	May	135.4	40.2
	June	177.5	109.2
	July	57.1	89.8
	August	121.9	22.8
	September	57.9	85.4
	October	234.8	290.1
	November	39.2	66.4
	December	42.6	51.2
1975	January	24.0	81.7
	February	21.9	236.9
	March	41.9	302.8
	April	118.1	77.3
	May	38.1	72.7
	June	105.5	87.9
	July	46.3	142.9
	Average	90.7	97.4

TABLE 4.3 *Average Annual Densities (number/ km²) of Impala*

Year	Study Area	
	Sabie River	Nwaswitshaka River
1973	143.1	40.0
1974	88.8	80.2
1975	56.5	163.6

ranged from 1.2 impala/km² in the Umfolozi Game Reserve, South Africa (Mentis 1970), to 16.6 to 32.8 impala/km² in Zimbabwe (Dasmann and Mossman 1962), 1.9 and 19.3 impala/km² in Kenya (Western 1975 and Ables and Ables 1970, respectively) and 2.5 and 30.5 impala/km² in Tanzania (Schaller 1972 and Lamprey 1964, respectively). Low impala densities were usually reported from dry open plains and higher densities from moister brushland and woodland savannas. This relationship is apparently a reflection of a general increase in biomass of ungulates associated with increasing rainfall (Coe et al. 1976).

The population trends of impala between 1973 and 1975 indicated a decline from 143 to 56 impala/km² in the SRSA (see table 4.3). Impala had apparently exceeded the dry season carrying capacity in the SRSA in 1973. After heavy rains fell late in September, I observed 11 dead impala along my census route and estimated that at least 115 had died. In the nearby Sabi-Sand Game Reserve, 487 dead impala were discovered after the same rains (Crabtree 1974). My census data also suggested that impala densities declined 35% in the SRSA after the rains, although some of the difference may have been due to emigration. Official park estimates of impala numbers in the Skukuza Ranger District also supported the data that an impala population decline had occurred. These estimates indicated that the district's impala population declined from seventeen thousand in 1973 to fifteen thousand in 1975 (S. Joubert, personal communication). The population trend of impala in the NRSA was opposite that in the SRSA. There, the population increased from about 40 impala/km² to 164 between 1973 and 1975 (see table 4.3).

Impala populations in both study areas varied considerably between seasons. Impala were most abundant during dry seasons in the SRSA and most abundant during wet seasons in the NRSA. The availability of drinking water undoubtedly influenced seasonal densities, although forage was probably also important. Impala moved into the SRSA as waterholes in adjacent areas, such as the NRSA, dried up. When water and forage became available after the wet season, impala dispersed from the perennial rivers back into seasonal habitats.

In each study area impala population appeared to be composed of a sedentary and migratory segment. Favored habitats in both study areas were occupied year-round by impala, but poorer habitats were occupied only after some needed resource, usually water, became available. Thus impala were seen more often in the dense thickets of the SRSA only after others moved into the area (because of the scarcity of water) from nearby regions. In the NRSA impala moved into upland areas only after water became available during the wet seasons.

Similar seasonal movements of impala in response to forage and water have been documented in other regions. Jarman (1970) reported 52% of the impala she recognized dispersed 1.5 km to 4.1 km when forage became available after seasonal rains. About a month later 75% of those individuals returned to formerly used areas as forage dried up. In the Tarangire Game Reserve, Tanzania, 50% to 70% of the impala left areas to feed on the highly nutritious pods of *Acacia* despite the unavailability of water (Lamprey 1963).

The importance of water to impala is unclear. Although Young (1972) reported that impala were able to survive indefinitely without actually drinking, Jarman (1973) discovered that impala had to drink if the plant's water content dropped below 30%. Lamprey (1963) reported that impala were able to survive without surface water by drinking dew. Movements of impala up to 24 km to obtain water have been documented (Dasmann and Mossman 1962), but most movements appear to be less than 13 km (Young 1972; Western 1975, Dasmann and Mossman 1962). Whyte (1976) reported that movements of impala seldom exceeded 5 km along the Sabie River near the leopard study areas.

Population composition. Adult female impala outnumbered adult males in each leopard study area. Of five hundred accurately sexed and aged impala observed in October 1974, 58% were females for a male:female ratio of 74:100. Male:female ratios of impala ranged from 28:100 to 82:100 in Zimbabwe (Dasmann and Mossman 1962). The sex ratio of impala is apparently even at birth, but social pressure from territorial males force many younger males out of female herds, thus exposing them to differential mortality (Jarman and Jarman 1973b). The male:female sex ratio, therefore, drops from 100:100 at birth, to 60:100 the second year, then to 40:100 the third year following birth.

Most impala observed in the leopard study areas were adults. Before newborn young were present in October 1974, 66% of five hundred sampled impala were adults and 34% were yearlings. Fairall (1969) reported similar findings. Of more than ten thousand impala he sampled in KNP, 33% were less than one year old, 22% were one to two years old, and 45% were more than two years old. Mean life expectancy was 2.6 years, and few were older than 10. One of the oldest impala reported in the literature was twelve years old (Child 1964).

OTHER PREY

Bushbuck. Bushbuck were more abundant in the densely vegetated, riparian habitats of the SRSA than the more sparsely vegetated NRSA (see table 4.4). I often saw bushbuck in the SRSA but regularly saw only two individuals in the

TABLE 4.4 *Ungulates and Leopard Prey Averages Estimated During Dry Seasons (1973–75)*

		Study Area				
		Sabie River			Nwaswitshaka River	
		Densities (numbers/km^2)			Densities (numbers/km^2)	
Species	*n*	Crude	Ecological	*n*	Crude	Ecological
Large Ungulates						
Giraffe	11.8	0.7	0.7	25.4	0.3	0.5
Hippopotamus	16.3	1.0	1.0	0	—	—
Buffalo	8.0	0.5	0.7	4.0	0.05	1.3
White rhinoceros	0	—	—	5.0	0.06	0.06
Elephant	2.0	0.1	0.1	2.0	0.02	0.02
Medium-size ungulates						
Greater kudu	30.0	1.8	1.8	226.7	2.8	4.3
Zebra	2.4	0.1	0.5	28.1	0.4	5.6
Wildebeest	2.1	0.1	0.5	12.5	0.2	2.5
Waterbuck	2.0	0.1	0.1	0	—	—
Small ungulates						
Impala	2,343.5	137.9	137.9	5,229.9	65.4	65.4
Gray duiker	30.4	1.8	2.5	35.0	0.4	1.3
Steenbuck	7.2	0.4	1.5	42.0	0.5	8.4
Warthog	6.2	0.4	1.3	32.0	0.4	6.4
Bushbuck	6.0	0.4	0.5	5.0	0.1	0.2
Klipspringer	0	—	—	4.0	0.05	4.0
Other mammals						
Chacma baboon	106.3	6.3	6.3	97.5	1.2	3.7
Vervet monkey	160.0	9.4	21.3	102.0	1.3	7.7
Hares	187.5	10.9	15.0	1,388.6	17.1	18.3
Porcupine	123.0	7.2	10.0	217.1	2.7	2.9
Cane rat	120.0	7.1	72.7	0	—	—
Small carnivores						
Genet	24.0	1.4	1.4	48.0	0.6	0.9
Civet	12.0	0.7	0.7	12.0	0.1	0.2
Mongoose	85.0	5.0	5.0	200.0	2.5	2.5
Ratel	8.0	0.5	0.3	26.0	0.3	0.3
Birds						
Guinea fowl	178.3	10.5	23.8	150.0	1.9	11.3
Francolins	410.0	24.1	33.3	840.0	10.4	31.6

Crude densities = average number/total area of study area.
Ecological densities = average number/area of preferred habitat.

NRSA. Most reported densities of bushbuck, 0.2 to 0.5/km^2 (Dasmann and Mossman 1962; Foster and Coe 1968; Bourlière 1963a; Petrides and Swank 1965; Field and Laws 1970), were similar to the 0.2 to 0.5 bushbuck/km^2 I estimated for ecological densities in the leopard study areas. The highest reported densities, 66.7 bushbuck/km^2, were in the Sengwa Valley, Zambia (Jacobsen 1974). The male:female sex ratio of the twenty-five bushbuck I observed was 92:100. Others have reported male:female sex ratios varying from 37:100 (Waser 1975) to 120:100 (Jacobsen 1974).

Gray duiker.   Gray duiker, like bushbuck, were most abundant in the SRSA. Although I regularly saw only eight duikers along each census route, I estimated ecological densities of 1.3 and 2.5 duiker/km² in the NRSA and SRSA, respectively. That these duiker were repeatedly seen in the same areas suggested that their home ranges were relatively small. Other reported densities of duiker vary with habitat quality and range from 0.12/km² (Dasmann and Mossman 1962) to 7.0/km² (Bigalke 1974). The male:female sex ratio of the twenty-one duikers I observed was 100:50, but males were probably more conspicuous than females. Mentis (1970) reported a male:female ratio of 92:100 from 3,419 duikers shot in the Umfolozi Game Reserve. In southern Africa duiker populations, most of the individuals (58% to 74%) are less than twenty-two months old (Child and Wilson 1964; Child et al. 1970).

Warthog.   Warthogs were more abundant (6.4/km²) in the NRSA than the SRSA (1.3/km²), presumably because of the NRSA's seasonal abundance of waterholes, aardvark burrows, and forage. In the nearby Timbavati Game Reserve, Hirst (1975) estimated 0 to 14 warthogs/km². Further south in the Umfolozi Game Reserve, Mentis (1970) estimated there were 2.5 to 6.7 warthogs/km². Child et al. (1968) reported that the male:female sex ratio among warthogs was even and that 51% of the warthogs sampled were young-of-the-year.

Steenbuck.   The open, short grass areas of the NRSA supported more steenbuck (8.4/km²) than the more densely vegetated steenbuck habitats of the Sabie River (1.5 steenbuck/km²) (see table 4.4). Although steenbuck may reach densities of 20/km² where there is little competition with other ungulates (Bigalke 1974), densities of 1 to 2 steenbuck/km² are probably more common in KNP (M. Cohen, personal communication). The male:female ratio of the 112 steenbuck I observed was 115:100, and Mentis (1970) estimated 35% of steenbuck populations were young-of-the-year.

Other small ungulates.   Only two klipspringers were regularly seen at a rocky koppie in the NRSA, but they undoubtedly occurred at several other koppies in the interior of that study area. Although no klipspringer habitat were in the SRSA, I once observed a frightened, obviously dispersing individual one day on a firebreak road. The only red duiker I saw in the SRSA was apparently the only individual remaining from a reintroduction prior to my study (G. Smuts, personal communication).

Medium-size ungulates.   Because the young of medium-size ungulates are occasionally prey of leopards, I estimated their population size and composition. Greater kudu were seasonally more abundant in their preferred brushy habitat along the Nwaswitshaka River (2.8 to 4.3 kudu/km²) than along the Sabie

River (1.8 kudu/km^2). Greatest densities occurred in both areas during dry seasons. A sample of 126 kudu I observed had a male:female sex ratio of 117:100, but others (Wilson 1965; Mentis 1970) reported sex ratios from 42:100 to 64:100 in southern Africa.

Waterbuck were rare in the study areas. None were observed in the NRSA and only two were periodically observed in the SRSA. Low waterbuck numbers were also estimated in the Nwaswitshaka River (0.04/km^2) and Sabie River (0.3/km^2) regions by Pienaar et al. (1966a).

Zebra and wildebeest are not abundant anywhere in the Southern District of KNP, including the leopard study area, because of dense brush. I estimated ecological densities of zebra at 0.5 and 5.6/km^2 and of wildebeest at 0.5 and 2.5/km^2 in the SRSA and NRSA, respectively. These estimates are considerably lower than the 8.5 zebra/km^2 and 12.4 wildebeest/km^2 reported for the Tarangire Game Reserve (Lamprey 1964) and Serengeti Plain in Tanzania (Hendrichs 1970).

Large ungulates. No breeding herds of elephant or buffalo resided solely in the leopard study areas. Although two bull elephants, often seen together, used each study area, their ranges obviously comprised much larger areas. Similarly, a large herd of buffalo visited the SRSA only during the dry seasons of 1973 and 1975; another herd once visited the extreme western portion of the NRSA. However, a small group of bachelor bulls and a solitary bull were residents in the SRSA, and a group of bachelor males were sometimes seen near the Nwaswitshaka waterhole.

Two adult males and one female white rhinoceros with a year-old calf resided in the NRSA, but no white rhinoceroses were observed in the SRSA. Using Pienaar et al.'s (1966b) early census data, I estimated that at least sixteen hippopotamus, or 1 hippopotamus/km^2, resided in the SRSA.

Smaller mammals. Places of refuge and food, often in the form of large fig trees, especially in the SRSA, provided ideal habitat for baboons and vervet monkeys. Three troops of chacma baboons averaging 32.5 baboons/troop resided in the NRSA, and four troops averaging 26.6 baboons/troop resided in the SRSA, respectively. Similarly, the number of groups of vervet monkeys in the NRSA was fewer, but larger (17/group) than those in the SRSA (8 vervets/group).

Porcupines were periodically seen in the study areas at night, and their quills were often found lying in the firebreak roads. From observations at night I estimated the ecological densities of porcupines were 2.9 and 10/km^2 in the NRSA and SRSA, respectively. Few reported data on porcupine densities are available. However, Thomson (1974) reported that removal of 5.9 porcupines/km^2 in southwestern Zambia did not significantly reduce porcupine damage to trees.

Hares, also seen mainly at night, were estimated to occur at ecological densities of 15 and 18.3/km^2 in the SRSA and NRSA, respectively. I rarely observed cane rats along the Sabie River, where they were apparently common (Stevenson-Hamilton 1947; Pienaar 1963), but guessed their numbers were at least 120, or 5/linear km of riparian habitat. Cane rats were rare in the NRSA.

Birds.   I assumed that only the larger game birds that fed on the ground were suitable leopard prey. Crowned guinea fowl were especially abundant along the Sabie River, and during dry seasons flocks of up to two hundred individuals were occasionally observed roosting in trees. However, only 10 to 40 birds/flock were usually encountered. I estimated ecological densities in the SRSA were about twice the 11.3 guinea fowl/km^2 estimated in the NRSA. Small game birds such as crested francolin, Natal francolin, and Swainson's francolin also were abundant in the SRSA (see table 4.4).

Biomass of Ungulates and Prey of Leopards

In calculating biomass, I included only those large ungulates that spent most of their time in the study areas, assumed that all leopard prey except impala and kudu were resident, ignored seasonal variations in zebra and wildebeest numbers because they appeared insignificant, assumed that wet season biomass was the same as dry season biomass except for impala and kudu, and used the weights given in tables 4.5 and 4.6.

TOTAL BIOMASS

I estimated the total biomass of large mammals, except major carnivores, and of mammals and birds that were prey of leopards in order to compare the study areas with each other and with other areas of Africa. As noted by Eltringham (1979), biomass, as an ecological measure, is often more meaningful than numbers. For example, six elephants consume more vegetation and have a greater impact on their habitat than do six duikers. Eltringham also cautioned those using biomass as an ecological measure to be aware of the difference between biomass and productivity, the metabolic requirements, and the significance of large, old-age individuals in populations.

The SRSA supported a much higher biomass of large animals (2.3 times) than the NRSA (see tables 4.7 and 4.8). Although impala accounted for most of the biomass of both areas, the principal reason for the difference between the study areas was the presence of hippopotamus, which made up 13% of the 148,129.5 kg of biomass in the 17/km^2 SRSA. In the 81/km^2 NRSA, where hippopotamus were absent, giraffe, another large ungulate, made up 6% of the 310,873 kg of biomass. These data demonstrate (1) the influence of a nearby perennial supply

TABLE 4.5 *Weights Used to Calculate Biomass in Study Areas*

Species	Mean Weight (in kilograms)
Elephant	3,181.8 [1]
White rhinoceros	1,363.6 [1]
Hippopotamus	1,136.4 [1]
Giraffe	681.6 [1]
Cape buffalo	500.0 [1]
Zebra	215.9 [1]
Waterbuck	204.5 [1]
Wildebeest	181.8 [1]
Greater kudu	172.7 [1]
Bushbuck	29.5 [2]
Warthog	29.5 [2]
Porcupine	15.9 [3]
Chacma baboon	15.5 [3]
Gray duiker	14.4 [2]
Civet	13.6 [3]
Steenbuck	11.4 [2]
Klipspringer	10.2 [2]
Ratel	6.8 [3]
Vervet monkey	3.3 [3]
Cane rat	2.7 [3]
Hare	2.2 [3]
Guinea fowl	1.8 [3]
Genet	0.9 [3]
Mongoose	0.9 [3]
Francolin	0.3 [3]

[1] Pienaar et al. 1966b.
[2] Mentis 1970.
[3] Estimated average weight for all sex and age classes.

TABLE 4.6 *Seasonal Weights of Impala Used to Calculate Biomass in Leopard Study Areas*

Sex	Age	Weight in kilograms	
		Wet Season	Dry Season
Male	Adult	60	60
	Juvenile	12 [1]	23 [2]
Female	Adult	44	44
	Juvenile	12 [1]	23 [2]

SOURCE: Hirst 1975
[1] Average weight at about six to seven months old.
[2] Average weight at about three months old (extrapolated from growth curves).

of water in maintaining high biomass of ungulates and other prey and (2) that a significant portion of the area's large animal biomass was incorporated into large, relatively predation-free ungulates.

The biomass of large animals in the study areas, especially the SRSA, was

TABLE 4.7 *Estimated Biomass of Ungulates and Leopard Prey in the Sabie River Study Areas*

Species	Biomass (in kilograms) Dry — All Species	Dry — Only Leopard Prey	Wet — Only Leopard Prey	Average — Only Leopard Prey	% of Total
Hippopotamus	18,522.7				
Giraffe	8,045.5				
Elephant	6,363.0				
Buffalo	4,000.0				
Greater kudu	5,181.9	375.0	89.4	232.2	0.30
Zebra	518.2	38.4	38.4	38.4	0.05
Wildebeest	381.8	23.1	23.1	23.1	0.03
Waterbuck	409.1	28.1	28.1	28.1	0.04
Impala	98,261.5	98,261.5	43,222.8	70,742.2	91.30
Gray duiker	345.3	345.3	345.3	345.3	0.45
Warthog	183.2	183.2	183.2	183.2	0.24
Bushbuck	177.3	177.3	177.3	177.3	0.23
Steenbuck	81.8	81.8	81.8	81.8	0.11
Porcupine	1,956.9	1,956.9	1,956.9	1,956.9	2.52
Baboon	1,645.5	1,645.5	1,645.5	1,645.5	2.12
Vervet monkey	528.0	528.0	528.0	528.0	0.63
Hare	418.8	418.8	418.8	418.8	0.54
Cane rat	327.3	327.3	327.3	327.3	0.42
Civet	163.3	163.3	163.3	163.3	0.21
Mongoose	77.4	77.4	77.4	77.4	0.10
Ratel	54.6	54.6	54.6	54.6	0.07
Genet	21.8	21.8	21.8	21.8	0.03
Guinea fowl	324.5	324.5	324.5	324.5	0.42
Francolins	139.4	139.4	139.4	139.4	0.18
Total	148,129.5	105,171.2	49,847.0	77,509.1	100.00
kg/km²	8,713.5	6,186.5	2,932.2	4,559.6	—

greater than that reported for many other areas of Africa. Eltringham (1979) recalculated data presented earlier by Leuthold and Leuthold (1976). Eltringham's data revealed that the biomass in the Serengeti (6,340kg/km²), Mara-Loita (2,470kg/km²), Nairobi National Park (4,150kg/km²), Isiolo (2,000kg/km²), Mkomazi Game Reserve (1,200kg/km²), and Tsavo National Park North (4,116kg/km²) and South (4,454kg/km²) (all in East Africa) was less than that in the SRSA (8,713kg/km²). In only four instances did they exceed the lower 3,838kg/km² estimated for the NRSA. Similar conclusions were drawn by Hirst (1975) after he calculated biomass values ranging from 2,290 to 8,710kg/km² in the nearby Timbavati Game Reserve. He noted that his area supported five times the biomass measured for East African *Acacia* and *Brachystegia* woodlands and other woodlands and forests in Zambia.

Most African regions where reported biomass estimates exceeded those in the leopard study areas were in regions such as Rwenzori National Park, Lake

TABLE 4.8 *Estimated Biomass of Ungulates and Leopard Prey in the Nwaswitshaka River Study Area*

	Biomass (in kilograms)				
	Season			Average	
	Dry		Wet	Only Leopard Prey	% of Total
Species	All Species	Only Leopard Prey	Only Leopard Prey		
Giraffe	17,318.2				
White rhino	6,818.8				
Elephant	6,363.0				
Buffalo	2,000.0				
Greater kudu	39,157.3	2,834.0	1,146.01	1,990.0	0.64
Zebra	6,067.1	450.0	450.0	450.0	0.14
Wildebeest	2,272.1	137.5	137.5	137.5	0.04
Impala	219,286.2	219,286.2	376,352.8	297,819.5	95.50
Gray duiker	397.7	397.7	397.7	397.7	0.13
Warthog	945.4	945.4	945.4	945.4	0.30
Bushbuck	147.7	147.7	147.7	147.7	0.05
Steenbuck	477.3	477.3	477.3	477.3	0.15
Klipspringer	40.9	40.9	40.9	40.9	0.01
Porcupine	3,454.1	3,454.1	3,454.1	3,454.1	1.11
Baboon	1,509.0	1,509.0	1,509.0	1,509.0	0.48
Vervet monkey	336.1	336.1	336.1	336.1	0.11
Hare	3,155.9	3,155.9	3,155.9	3,155.9	1.01
Genet	43.7	43.7	43.7	43.7	1.01
Civet	163.3	163.3	163.3	163.3	0.05
Mongoose	182.0	182.0	182.0	182.0	0.06
Ratel	177.3	177.3	177.3	177.3	0.06
Guinea fowl	272.7	272.7	272.7	272.7	0.09
Francolins	285.6	285.6	285.6	285.6	0.09
Total	310,873.0	234,296.8	388,675.1	311,985.8	100.00
kg/km ^{2}	3,837.9	2,829.5	4,798.5	3,851.7	—

Manyara National Park, and Tarangire Game Reserve, where most of the biomass was buffalo (Field and Laws 1970; Watson and Turner 1962; Lamprey 1964) or hippopotamus (Bere cited in Bourlière 1963a). However, if larger areas surrounding my relatively small leopard study areas were included in biomass estimates, such as Pienaar et al.'s (1966a) estimate for the region including the NRSA, many more buffalo—making up 35% of the biomass—would have been included. In comparing the biomass of all ungulates in KNP to other areas in Africa, Pienaar et al. (1966a) believed the factor limiting population growth in the park was the lack of water.

BIOMASS OF LEOPARD PREY

The biomass of leopard prey was assumed to include (1) only the young of medium-size ungulates, (2) all small ungulates, (3) all smaller mammals, and (4) game birds (see tables 4.7 and 4.8). Large ungulates were excluded, even though a leopard was once observed feeding on a young giraffe (Pienaar 1969).

Impala accounted for the bulk of leopard prey biomass in both study areas, contributing 91.3% and 95.5% of the average 4,560 and 3,852 kg/km^2 in the SRSA and NRSA, respectively. Other important ungulate prey of leopards, in decreasing order of significance, were gray duiker, warthog, and bushbuck in the SRSA and steenbuck, warthog, and gray duiker in the NRSA. Excluding impala, the greater proportion of prey biomass in duiker and bushbuck in the SRSA made that area superior to the NRSA, in terms of prey, because steenbuck and warthog prefer open habitats where they are less vulnerable to leopard predation. The second most important group of prey to leopards in both study areas was small mammals, particularly species such as baboon, porcupine, and vervet monkey in the SRSA.

The seasonal fluctuations in biomass of prey suggested that the SRSA was more favorable to leopards in dry seasons and the NRSA was more favorable in wet seasons. During dry seasons the SRSA supported 6,186 kg/km^2 of leopard prey; only 2,829 kg/km^2 occurred in the NRSA. Conversely, during wet seasons prey biomass in the NRSA increased to 4,798 kg/km^2, and that in the SRSA declined to 2,932 kg/km^2.

A comparison of the biomass of leopard prey in the SRSA—the higher-quality leopard habitat area—to other African areas suggested that this area was perhaps one of the better for leopards in Africa. Savannas and savanna woodlands of most African regions apparently provide only 5% to 25% of the prey biomass available to leopards in the SRSA, even excluding small mammals and game birds (see table 4.9). Regions supporting 30% to 32% of the prey biomass found in the SRSA were usually the more densely vegetated *Mopane* or *Acacia* woodlands. In open savannas most prey, such as zebra or wildebeest, are unavailable to leopards because they are too large, are organized in social herds, and inhabit open cover. In dense rain forests most potential prey of leopards appear to be arboreal or highly social, thus reducing their vulnerability to leopard predation. This is particularly true of the primates whose biomass may exceed more than ten times that of the small forest ungulates (Bourlière 1963a). Thus actual prey biomass available to leopards may be very low in dense rain forest regions.

Ecological Distribution of Prey

IMPALA

Seasonal changes in population density, forage, and water influenced impala distribution. Impala in the SRSA preferred the open, brackish flats during wet seasons and increased their use of the more densely vegetated undulating terrain only during dry seasons (see fig. 4.1). In the NRSA progressively fewer

TABLE 4.9 *Biomass of Leopard Prey (kg/km²) in Various Regions of Africa*

Species	Woodland-Savanna					Woodland		Forest	Rain forest
	(1)	(2)	(3)	(4)	(5)	(6)	(7)	(8)	(9)
Zebra		94.1	67.9		91.5	93.4		8.8	
Wildebeest		176.8	24.2		99.5	14.1			
Hartebeest		7.1	90.5						
Topi		8.5							
Thompson gazelle		84.6	35.9		780.0				
Grant's gazelle		8.0	139.4						
Kob	147.0			581.0					
Waterbuck	20.1	1.7	11.0	17.7	0.4	3.9		2.3	
Greater kudu						16.0			
Impala		81.6	225.1		100.0	1,008.0	107.6	314.6	
Common reedbuck			5.7				22.6	24.0	
Nyala						191.3			
Bushbuck			5.1			5.0	8.2	8.1	
Warthog	63.0		39.5	70.0		98.0	102.2	8.1	
Gray duiker						6.0			
Steinbuck						24.0			
Klipspringer						1.0			
Others		23.5							75.6
Total	230.1	497.7	644.3	688.7	1,071.0	1,269.4	431.9	365.9	75.6
Percentage compared to the Sabie River leopard study area	5.0	12.0	15.0	16.0	25.0	30.0	10.0	9.0	2.0

SOURCES OF DATA: (1) = Queen Elizabeth N.P. (1,670 km²) (Bere 1969 cited by Bourlière 1963a); (2) = Serengeti Ecological Unit (25,500 km²) (Schaller 1972); (3) = Nairobi N.P. (116 km²) (Ellis 1961 cited by Bourlière 1963a); (4) = Albert N.P. (600 km²) (Bourlière 1963a); (5) = Serengeti N.P. (10,000 km²) (Grzimek 1958 cited by Bourlière 1963a); (6) = Henderson Ranch, Rhodesia (=Zimbabwe) (125 km²) (Dasmann and Mossman 1960); (7) = Umfolozi G.R. (1,445 km²) (P. Brooks, pers. comm.); (8) = Lake Manyara N.P. (91 km²) (Schaller 1972); (9) = Tano Nimri Forest Reserve (250 km²) (Collins 1959 cited by Bourlière 1963a).

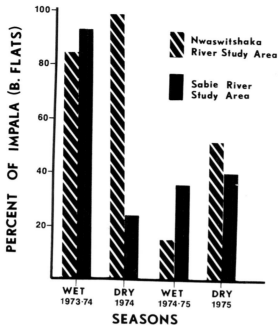

FIGURE 4.1 Proportions of impala observations in the open, sparsely vegetated brackish flats habitat.

impala were seen in the brackish flats each dry season of the study. Seasonal and long-term changes in population density influenced impala distribution. The annual influx of impala during the dry season into the SRSA appeared to force later-arriving impala into the less preferred habitat of the densely vegetated areas. Similarly, as the overall impala population increased during the study in the NRSA, additional impala were forced into the less preferred brushy habitat. When the population peaked in 1974–75, the majority of impala were no longer seen in the brackish flats but in *Acacia* thickets. After the population declined, a greater proportion of impala were again seen in the brackish flats (fig. 4.1).

Impala apparently preferred the brackish flats because of the available forage, presence of waterholes, and sparseness of stalking cover for predators. Impala preference for short grass areas, especially during wet seasons, was demonstrated by pellet counts, which showed fewer pellet groups as grass height increased (see fig. 4.2), and seasonal observations of males (see table 4.10). Similar habitat selection was noted for impala by Jarman (1970) and Hirst (1975). Brackish flats preferred by impala in the leopard study areas were characterized by the evergreen shrub *Euclea divinorum*. According to Hirst (1975) this vegetation zone occurs on deep, acidic clay soils, which have superior water-retaining capacity. Thus, during the most critical time of year—the dry season—forage

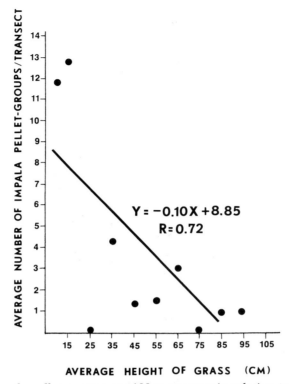

FIGURE 4.2 Impala pellet-groups per 100 m transect in relation to grass height in the leopard study areas.

productivity was probably best in the brackish flat areas. Brackish flats also provided security from stalking predators; impala repeatedly returned to these areas to spend the night.

Because impala are "generalists" (Ferrar and Walker 1974), they are able to switch from grazing to browsing, although they are grazers by choice. Because grasses usually make up at least 90% of impalas' diet (Lamprey 1963; Hirst 1975), areas heavily used by impala often appear only lightly browsed (Dasmann and Mossman 1962). This information helps explain several other observations regarding ecological distribution of impala. One was that impala strongly

TABLE 4.10 *Habitats Used Seasonally by Male Impala*

| Season | Number of Males | | Total |
	Brackish Flats	Undulating Terrain	
Dry	55	157	212
Wet	225	195	420
Total	280	352	632

Chi-square = 43.6, *df* = 1, *P* < 0.01

preferred recently burned areas where grasses and forbs were sprouting. Twenty-six days after a fire in the NRSA, I counted 99.7 impala/linear km of burned area, compared to 10.7 impala/linear km of unburned habitat. Another observation was that even after the rains ceased impala remained in the brackish areas to feed on the nutritious pods of *Acacia grandicornuta* and *Acacia nilotica*. Browsing by impala in the *Grewia* thickets increased late in the dry season, only when more nutritious forage was unavailable elsewhere and after the denser cover was safer from stalking predators.

Sex of impala also influenced ecological distribution. Female herds and territorial males used the brackish flats more often than lone males and bachelor herds. Nonterritorial males are apparently forced by territorial males into inferior habitat during the rut and become conditioned to remain there (Anderson 1972). Thus, unlike females who occupy habitats of their own choosing, males are forced socially as well as nutritionally to seek marginal habitats in order to survive (Jarman and Sinclair 1979).

OTHER PREY

Bushbuck preferred the densest habitats in the leopard study areas, especially the riparian vegetation frequently used by leopards. Because bushbuck are "specialist" feeders on a mixture of grasses, forbs, and woody plants, they prefer woody, shrubby habitats (Ferrar and Walker 1974), avoid open habitats (Pienaar 1974), and spend most of their lives in small areas. Bushbuck water requirements appear flexible. Waser (1975) seldom saw bushbuck drink, but Jacobsen (1974) saw bushbuck drink frequently and even dig in moist sand to obtain water.

Although favoring dense habitats similar to bushbuck, gray duiker appeared more habitat adaptable. Gray duiker appear to favor woody habitats regardless of the availability of water (Dorst and Dandelot 1970). It was not unusual to see gray duiker in the southwestern corner of the NRSA more than 4 km from the nearest water.

Steenbuck preferred the open brackish flats in the leopard study areas. Ferrar and Walker (1974) characterized steenbuck habitat as homogeneous grasslands, and Eloff (1959) reported they were not dependent on drinking water. Warthog, like steenbuck, also preferred the study areas' open brackish flats. Warthogs, however, are dependent on drinking water (Lamprey 1964), favor short grass savannas (Hirst 1975), and depend for protection on burrows dug by other animals (Frädrich 1974). Klipspringers are habitat specialists and always remain near rocky hills or outcrops where they browse and graze on vegetation among the rocks (Lamprey 1963).

Kudu occurred in all habitats in the leopard study areas, were mainly browsers, and appeared independent of drinking water. Female kudu, however, moved into the dense riparian vegetation during dry seasons. These highly adaptable ungulates are one of the most successful in South Africa (Pienaar 1974). Although I seldom saw waterbuck, their preferred habitat is apparently riverine galley forest with a dense tree canopy, tall grass, sandy soil, and nearby water (Hirst 1975). Zebra and wildebeest prefer open, short grass savannas, a habitat that was limited in both of the leopard study areas.

Baboon and vervet monkey preferred areas near riparian forest in the study areas where they could escape to and sleep in large trees for protection from predators. Denser vegetation in undulating terrain was used by these primates only after deciduous leaves had fallen, thus increasing visibility and offering protection from surprise attack by predators. Little was known about the ecological distribution of hares, porcupines, genets, civets, ratels, and mongoose, but guinea fowl and francolins were usually seen near water.

Prey Activity

Prey activity patterns can influence their vulnerability to predation. Most antelope drink when predator activity is low (Leuthold 1977), and those relying on vision for predator detection are often active only during the day. Those relying on cryptic coloration are either crepuscular or nocturnal.

IMPALA AND OTHER PREY

My observations of impala activity appeared to be similar to those reported by Jarman and Jarman (1973a). Impala are most active during the day although they may sometimes feed between 11:00 P.M. and 4:00 A.M. Usually, impala become active at dawn, feed until 10:00 or 11:00 A.M., ruminate while standing for several hours, feed again until about 4:00 P.M., ruminate once more and feed again before sunset. During the nights most impala ruminate while lying down, but few actually sleep with their eyes closed. Peak movements occur before dusk with a second peak after dawn. Movements were greater during dry than wet seasons and distances moved at night were only 33% to 50% of daytime movements. High winds caused nervousness among impala, and during heavy rains they stopped feeding until the rain stopped. During the hottest part of the day (10:00 A.M to 4:00 P.M.) impala were inactive and often ruminated in the shade. Most drinking occurred between 9:00 and 11:00 A.M.

Steenbuck activity was similar to that of impala. They were active during the day, primarily from 8:00 A.M. to 12:00 noon (64% of 110 observations). Their activity declined during the hottest part of the day, increased again after 4:00 or 5:00 P.M., and then abruptly declined after 7:00 P.M. Warthogs, baboons, vervet monkeys, guinea fowl, and francolins also avoided activity after sunset.

The activity patterns of bushbuck and gray duiker were unlike those of impala and steenbuck. Most bushbuck activity occurred between 4:00 and 7:00

P.M., when they began leaving cover to feed—the same period during which impala moved into open areas to spend the night. Few bushbuck were seen between 12:00 noon and 4:00 P.M., although Jacobsen (1974) reported bushbuck were active from 5:00 to 11:00 A.M. and from 1:00 to 6:00 P.M. Waser (1975) reported bushbuck spent 81% of their time from 9:00 A.M. to 5:00 P.M. in dense thickets. Like bushbuck, gray duiker were most active in the mornings and evenings. Peak daytime activity occurred between 4:00 and 5:00 P.M., and few duiker were seen during midday (1:00 to 4:00 P.M.). Other prey active at night were porcupines, hares, genets, and civets. Cane rats, mongoose, and ratel were seen during the days, as well as at night.

Social Organization of Prey

The social organization of prey is not only an interesting aspect of their ecology, it can also determine their vulnerability to predators (Kruuk 1972). Ungulates in open habitats may be easy to locate but difficult for leopards to capture because of their social cohesiveness. Likewise, solitary prey in dense cover may be more easily captured by stalking predators, but they may be difficult to locate. From one viewpoint, the social organization of prey in the leopard study area was but another measure of their vulnerability to leopard predation and a reflection of their antipredator strategies.

Impala

Three types of socially distinct units occur in impala populations: (1) females and their offspring, (2) males in bachelor herds, and (3) lone territorial males during the rutting season (Jarman and Jarman 1973b). Most impala in the leopard study areas congregated in female herds. Of 4,178 impala observed during the 1975 rutting season, 86% were in female herds, 10% in bachelor male herds, and 4% occurred as lone territorial males. The average number of impala in 206 female herds was 17.5; in 118 bachelor male herds, it was 3.5. Most female herds were composed of fewer than ten individuals, and most bachelor male herds of fewer than five. The largest herds were formed during the late wet season (January through March) when up to 216 individuals/female herd and 49 individuals/bachelor herd were observed.

Water, forage, cover, and antipredator behavior influenced the size of impala herds. Female herds increased in size soon after the rains arrived, water became available, and forage conditions improved. In 1974 female herd sizes in the NRSA declined from 36.7/herd in the wet season to 24.1/herd in the dry season. SRSA numbers for the same year were 51.7/herd in the wet season and 32.7/herd in the dry. Increased herd size appeared to be an antipredator strategy because herd use declined in dense cover and increased in open habitats during the wet season. The average distance between female herds also increased from 3.2 km to 5.1 km between dry and wet seasons. This distance and the movements of female herds suggested that female herds were highly aggregated during wet seasons. Female herd size also declined as territorial males drove away younger males from the herds. After the rutting season, however, younger males previously driven from female herds are allowed to return (Jarman and Jarman 1973b).

In summary, during dry seasons, most impala occur in small, evenly dispersed herds that use most of the available habitat within walking distance of drinking water. During the rutting season territorial males drive away younger males from female herds. These younger males then form bachelor herds. Later, after the rut and during the wet seasons when water and forage become available, female herds increase in size as some younger males rejoin them. As stalking cover for predators increases, impala herds congregate into even larger herds and avoid dense cover as an antipredator strategy.

OTHER PREY

Most of the smaller ungulates in the study areas were solitary or occurred in family groups. Gray duiker, steenbuck, and klipspringer maintained pair territories, with the adult male and female excluding conspecifics from a small area (Bigalke 1974). I rarely saw more than one duiker/observation; 84% of the steenbuck were alone when observed; and klipspringers were usually seen in pairs. Whether bushbuck maintain territories is unknown (Leuthold 1977; Bigalke 1974). I seldom saw more than one bushbuck/observation, usually a male. But loose social groups, up to twelve individuals, have been reported (Jacobson 1974). Warthogs usually occur in family groups, although old males may become solitary (Frädrich 1974). Most warthogs in the study area were in family groups of up to nine members. The larger ungulates in the study area were organized into more complex social units. Greater kudu have one of the most adaptable social organizations of the Tragelaphine antelopes (Leuthold 1977), with group size changing with the seasonal food supply (Simpson 1972). I usually saw kudu in groups of two to four, occasionally up to thirteen, which consisted primarily of females with young. Waterbuck, like impala, are also organized into female and offspring herds, bachelor male herds, and territorial males (Spinage 1974).

Baboons in the study areas lived in social groups averaging 26.6 baboons/troop in the SRSA and 32.5 baboons/troop in the NRSA. Group sizes were larger in the NRSA apparently because of that area's more open habitat. Baboons' social groups are complex and dominated by one or more adult males that warn and protect the group from danger. Although baboon home ranges overlap, intensively used core areas surrounding sleeping places are seldom visited by neighboring groups (Altmann and Altmann 1970).

Vervet monkeys were seen in social groups ranging from three to twenty individuals. They were dominated by older males and seldom ranged far from the escape cover offered by trees. With the exception of the social banded and dwarf mongoose, the remainder of potential mammalian prey of leopards were seen alone (ratels, civets, genets, cane rats, and porcupines) or in pairs (hares). Guinea fowl were usually seen in family groups during wet seasons, but during dry seasons flocks of up to two hundred individuals roosted nightly in trees. Crested, Natal, and Swainson's francolins were usually seen in family groups of up to 13 birds/group.

Antipredator Responses of Prey

Each species of prey reacts differently to different predators. Thompson's (or Thomson's) gazelles flee at 300 m or more when they see wild dogs trotting or bounding and may run up to 1.5 km even though not pursued (Walther 1969). When they see lions, however, they may permit the lions to approach within 40 m to 60 m. From their reactions, they obviously feel themselves immune to attack (Schaller 1972). The gazelles apparently are aware that unlike wild dogs the lions will not run after them. In both instances the gazelles see the predator. They may at times even follow lions to keep them in view. That prey will follow a predator stresses the importance of vision—a visible predator, particularly a stalking predator like a leopard, is not as dangerous as a hidden one.

Several observations suggest that prey will permit visible leopards to approach within 25 m without feeling endangered, but will flee if they are closer than 10 m. Thompson's gazelles did not flee when an approaching leopard came within 25 m, but did flee when another was only 6 m away (Schaller 1972). Axis deer followed a leopard that was 100 m away when first detected and allowed another to approach within 45 m before falling back to a safer distance (Eisenberg and Lockhart 1972).

Solitary antelope use a concealment strategy to avoid detection. Jacobson (1974) reported that a leopard walked by a thicket containing two adult bushbuck, and Waser (1975) watched a lion pass within 10 m of a concealed bushbuck without seeing it. Solitary antelope often associate with other prey that may detect leopards. Feeding bushbuck sometimes associate with feeding baboons (Elder and Elder 1970; Jacobson 1974) or vervet monkeys, which provide a warning in case of danger. Another strategy of solitary antelope is to remain motionless, or to flee into dense cover and then remain motionless, if approached by danger. I often saw steenbuck use such a strategy when I approached them on foot.

Impala use several behavior strategies to detect predators. These include gregariousness, association with other herbivores, constant vigilance, sudden phases of immobility of head to detect movements, visually searching for hidden predators, sudden surveillance of surroundings after lowering their head to feed, alarm barking, and jumping high over bushes to confuse pursuing predators (Schenkel 1966). Impala may remain motionless if detected by a predator in dense cover, but move away if they are in the open. When attacked, impala in a group explode in all directions, apparently to confuse the predator, and later reunite (Jarman and Jarman 1973b; Jarman 1974; Dasmann and Mossman 1962). Impala in the leopard study areas seldom fled from visible leopards. Rather than fleeing, they alarm barked and attempted to keep the leopard in view without approaching too closely. Impala did not watch passing leopards very long, merely keeping the leopard in view until obviously it was no longer considered a threat. They did not follow leopards into dense vegetation. Certain impala alarm barked longer than others when they observed leopards. Lone territorial male impala were especially reluctant to leave their areas even if a leopard was visible. I once watched such a male impala alarm bark for twenty-five minutes and approach within 30 m of a hidden leopard before finally leaving the area.

Avoidance of dense cover was a frequently used antipredator strategy of impala, especially during wet seasons. Later, after the leaves fell and grasses dried, impala ventured more frequently into *Acacia* and *Grewia* thickets to browse. Dense beds of reeds along the Sabie, Sand, and Nwaswitshaka Rivers were especially avoided by impala, and impala frequently used tourist and firebreak roads as travel routes to pass through dense vegetation.

Impala were especially cautious when selecting drinking sites, and waterholes surrounded by dense vegetation were seldom used. Impala would not use one waterhole until most of the adjacent grass cover was trampled down by a herd of buffalo, and they reluctantly drank at the Sabie River only after the surrounding waterholes became dry. Rather than walk through dense vegetation to reach fresh water flowing in the Nwaswitshaka River, impala continued to drink muddy, stagnant water at waterholes in open, short grass habitats.

Impala were especially cautious while drinking. Often several hours elapsed before the first impala of a group approaching a waterhole actually drank. The slightest disturbance, such as the loud call of a bird, often sent them fleeing. Drinking periods were frequent but brief. The longest I recorded an impala drinking continuously with its head down was seventy-six seconds, Most drank only a few seconds and then raised their heads as if expecting danger. One impala drank eight times for an average period of 10 sec/drink and another drank seven times, averaging 6 sec/drink. Most drinking periods were less than twenty-five seconds.

Arboreal prey such as baboons and vervet monkeys immediately fled to trees when they encountered leopards. By their alarm calls I could often monitor the movements of radio-collared leopards through dense vegetation. Vervet monkeys usually fled to the outermost branches when a leopard was nearby and, if possible, they followed above the leopard, jumping from tree to tree alarm calling. Female gray langurs also escape in trees from leopards, but subadult male langurs sometimes fall in their excitement to escape and are captured (Eisenberg and Lockhart 1972).

Alarm calls by one species of prey often alert other prey in the vicinity. During 135 visual observations of leopards, I recorded impala alarm calling at leopards on fourteen occasions (10%), vervet monkeys on twelve (9%), and baboons on two (1%). I also heard waterbuck, black-backed jackals, guinea fowl, Natal francolins, and red-billed hoopoes alarm call in response to leopards. On four occasions impala and vervet monkeys simultaneously alarm called at leopards, and on one occasion vervet monkeys and baboons simultaneously alarm called. Vervet monkeys were usually the first to see and alarm call at leopards moving through dense cover along the rivers.

Another antipredator strategy used by prey under attack is retaliation. Warthogs are formidable prey and have been known to maim leopards with their sharp tusks (Pienaar 1969) or run them up trees (Stevenson-Hamilton 1947). Baboons are also dangerous because large males can kill (Marais 1939) or disembowel leopards (Stevenson-Hamilton 1947). Smaller prey can also be dangerous. I once watched a ratel wound a leopard. And snakes, particularly cobras, occasionally kill leopards (Turnbull-Kemp 1967). Adults of large or gregarious prey will also sometimes attack or mob leopards. Eisenberg and Lockhart (1972) described how a herd of water buffalo chased a leopard into a tree, and Brander (1923) once discovered a sambar deer holding a leopard at bay in a tree.

Summary

1. Ungulates were censused monthly in the study areas. Of 12,653 ungulates observed, impala were most abundant (94.6%), followed by greater kudu (1.9%), and giraffe (0.9%). Other ungulates observed were warthog, zebra, steenbuck, wildebeest, and gray duiker.

2. Impala population densities averaged 90.5 and 97.4/km^2 in the Sabie River and Nwaswitshaka River Study Areas, respectively. Monthly impala densities ranged between 22 and 303 impala/km^2. Impala densities declined in the SRSA but increased in the Nwaswitshaka River area. Impala were most abundant during dry seasons in the Sabie River Study Area and during wet seasons in the Nwaswitshaka River Study Area.

3. The sexual composition of the impala population was 58% females and 42% males. Seventy-four percent of the males and 61% of the females were adults.

4. Other important ungulate prey of leopards were bushbuck (0.2 to 0.5/km^2), warthog (1.3 to 6.4/km^2), gray duiker (1.3 to 2.5/km^2), and steenbuck (1.5 to 8.4/km^2).

5. Leopard prey biomass averaged 4,559.6 and 3,851.7 kg/km^2 in the Sabie River and Nwaswitshaka River Study Areas, respectively. Ninety-one per-

cent of the prey biomass in the SRSA and 96% in the NRSA was impala.

6. The biomass of leopard prey in other regions of Africa suggest that savanna habitats support only 5% to 25% of the prey biomass estimated for the *Acacia* habitats of the KNP study areas.

7. Impala in the Sabie River Study Area preferred open, brackish flats habitats during wet seasons. During dry seasons more impala were seen in densely vegetated habitats in undulating terrain. Impala were attracted to burned areas during wet seasons, and male impala occupied inferior habitats, relative to females, during dry seasons.

8. The highest population densities of leopard prey occurred in dense riparian habitats adjacent to the Sand, Sabie, and Nwaswitshaka Rivers.

9. Most (86%) impala associated in female herds; 10% were in bachelor male herds; and 4% were territorial, or lone, males. The average sizes of female and bachelor male herds were 17.5 and 3.5, respectively. Other ungulate prey of leopards occurred in small groups, pairs, or as solitary individuals.

10. Impala were least active during midday and night. Bushbuck and gray duiker were most active at dawn and dusk.

11. The response of impala to visible leopards included alarm calls; keeping leopards in view; and approaching leopards, or allowing them to approach, within 10 m before flight. Responses of other prey to visible leopards varied.

PART TWO

The Leopard and Other Predators

5

Leopard Population Characteristics

POPULATION dynamics are an important aspect of leopard ecology, especially in regard to their conservation and management. Birth and mortality rates, population composition and densities are not only indicators of habitat and population quality and stability, they also help to explain observed social interactions, land tenure systems, and interspecific competition among leopards and other predators in an area.

Several points should be remembered while reading this chapter. First, because leopard populations in KNP were not exploited by humans, the impact of additional, human-caused mortality on leopard populations was not addressed. Second, the data were obtained during a period when rainfall was abundant and the leopard's principal prey, impala, were numerous. Third, leopards were studied in only two habitats, both of which were high quality relative to other leopard habitats in the park and elsewhere in Africa.

Age Classes

Five relative age classes of leopards could be distinguished during the study: old adults, prime adults, subadults, large cubs, and small cubs. I could not determine the precise ages of leopards that were captured or observed in the field. Stained sections of roots of canine teeth from two dead leopards, one obviously young and the other old, failed to show distinct cementum annuli, which are characteristic in many temperate-latitude carnivores. I relied, therefore, on differences in size and appearance of leopards to assign them to age classes.

OLD ADULTS

Old adult leopards were large, scarred, faded in coloration, and often had badly worn or missing teeth. Old males usually had tattered ears, with torn or split edges, and scars on their face, neck, or shoulders. Old females had fewer facial scars than males, but usually had scars on their necks. The background coloration of old adult leopards, especially on the head, neck, and shoulders, was a richer, and darker yellow than that of young individuals. The black spots on old adults were often faded. The teeth of old adults were dark yellow, and the tips of the canines or incisors were usually well worn. Canine and incisor teeth were sometimes missing. The actual age of these leopards was unknown. Although leopards in captivity may live for twenty-one years (Crandall 1964), in the wild, leopards that live ten to eleven years are probably old (Turnbull-Kemp 1967).

PRIME ADULTS

Prime adult leopards were smaller and weighed less, and their coats were brighter than old adults. Their teeth were usually white or only slightly yellow, and the tips of their canines and incisors were only slightly worn. Their background coloration had a whiter and brighter appearance than that of old adults, and the deep yellow of the head, neck, and shoulders was absent. Prime adults had few if any facial scars. The darkly pigmented nipples of old and prime adult females were usually greater than 9 mm compared to the shorter, pinkish-colored nipples of subadult females. Adult male leopards generally had the largest scrotums, but those of nearly mature subadults were often of similar size.

SUBADULTS

Subadult leopards were lighter and smaller than adults. Their general body build was slender and delicate in appearance compared to adults; their teeth were white and sharply pointed; and facial and body scars were rare or absent. Their dark black spots contrasted vividly with their light background coloration. Sometimes male leopards were difficult to classify as adult or subadults on the basis of appearance only. Therefore I took into account the individual's past history, if known, and his apparent social status. A male initially captured as a subadult retained that classification until his behavior or movements indicated a change of status. This behavior included scent marking, vocalizing, and courtship. Subadult leopards were probably 1.5 to 3.5 years old.

LARGE CUBS

Only two large cubs, both females, were captured during the study. Each weighed less than 20 kg and lacked permanent teeth. The incisors and canines were deciduous, although the tips of the permanent canines were protruding beyond the gum line behind the deciduous canines of one female. The upper and lower molars of one cub were permanent, but only the tips of the molars were evident in the other cub. Although one was obviously older than the other, I estimated both were approximately one year old. Both of these cubs appeared to be traveling alone when captured because no sign or tracks of adults were seen nearby or on earlier occasions near the traps.

In the Serengeti National Park a male leopard did not become fully independent until twenty months of age and continued its association with its mother for an additional two months. In the same area a female leopard began traveling alone when thirteen months old but was not independent until twenty-two months (Schaller 1972). A subadult male in Wilpattu National Park, Sri Lanka, estimated to be 2.5 years old was also seen with an adult female, presumably its mother (Muckenhirn and Eisenberg 1973).

SMALL CUBS

Small leopard cubs, those less than six months of age, were not captured during the study. Females apparently hid small cubs the first several months of their lives, and later, when the cubs were mobile, they remained behind while the female hunted. I did not see any one of F6's young cubs, despite many attempts, until they were about six months old, and I never saw the small cubs of two other radio-collared females.

Size of Leopards

Adult male leopards are larger and more muscular than females. The average weights and body measurements of thirty captured leopards (see table 5.1) showed that old adult males were 70% heavier and 10% longer than old adult females. Leopards in KNP were also larger than those measured from more northern African countries. Although Meinertzhagen (1938) reported an average weight of 63 kg for six males from Kenya, his largest male weighed only 65 kg, compared to 70 kg for two ungorged males I weighed. Similarly, the heaviest male weighed in Zambia by Robinette (1963) was 56 kg and by Wilson (1968) 59.5 kg. A male from Tanzania weighed by Sachs (Schaller 1972) from Tanzania was 35.8 kg, and a large male killed in India weighed 44.1 kg (Schaller 1967).

Stevenson-Hamilton (1947) claimed he could distinguish two types of leopards in the KNP region: a small, long, lanky and rufous-colored leopard found

TABLE 5.1 *Average Sizes of Captured Leopards*

| | | | Weight (in kilograms) | Body Measurements (in Centimeters) | | | | | | |
| | | | | Lengths | | | | | | |
Age	Sex	Number		Ear	Hind Foot	Body	Total	Height at Shoulder	Circumference of Neck
Old adult	Male	5	63.1	8.5	27.2	141.3	219.4	77.6	49.4
	Female	6	37.2	7.7	24.7	121.9	200.1	69.1	38.7
Prime adult	Male	3	58.2	8.5	27.3	131.7	211.5	77.3	48.5
	Female	5	37.5	7.7	23.8	122.0	198.4	66.3	37.9
Subadult	Male	6	50.9	8.2	26.8	130.8	208.2	72.8	43.0
	Female	3	30.1	7.8	23.3	119.3	192.7	60.0	34.7
Large cub	Female	2	18.6	7.4	21.5	106.0	155.5	54.8	28.3

in the hot, lowlands, and a larger, lighter-colored leopard living in the hilly, high country above an elevation of 610 m. Pienaar (1969) reported the leopards seldom exceeded 59.1 kg in the park, but a 85 kg male was once reported killed in the neighboring Sabi-Sand Game Reserve (Turnbull-Kemp 1967). The size of the skin from this particular leopard, however, cast some doubt on its reported size.

Stomach contents can significantly add to a leopard's weight. Wilson (1968) reported that the stomach contents of a 43 kg female leopard weighed 6.6 kg or 18% of its body weight. Of seven males that reportedly weighed more than 91 kg, only one apparently had an empty stomach (Turnbull-Kemp 1967). Male leopards more than 70 kg in weight and 230 cm in length are exceptionally large individuals.

The largest female leopard I captured weighed 43.2 kg, compared to 58.2 kg for a female in Kenya reported by Meinertzhagen (1938). The average weight of six female leopards from Zambia was 33.6 kg with the largest at 36.4 kg (Wilson 1968). Some variation in the size of females occurred in the study areas. Adult female leopards from the NRSA 5 km upstream from Skukuza were considerably smaller then those captured elsewhere. These females, definitely adults, weighed 26% less and were 14% shorter than other adult females. Two prime adult females from this area averaged 30.9 kg and 180.5 cm long compared to 41.8 kg and 210.3 cm for prime females captured elsewhere. An old adult female captured near the Nwaswitshaka Waterhole weighed only 33.6 kg and was 192.5 cm long compared to the average 38.3 kg and 201.6 cm for all other old adult females.

The type and abundance of prey was different in the areas inhabited by these smaller females. More duiker, steenbuck, and smaller mammals (although fewer impala) appeared to be in this region. Also, despite numerous attempts, I never located these smaller females with an impala carcass. This suggested the possibility that females in this area took smaller prey. The apparent fidelity of females to their natal areas relative to the widely dispersing males may also explain the localization of such physical characteristics. Males living in the same area used larger areas and were occasionally located with carcasses of impala they had killed. No significant differences in size were noted among male leopards living in the same region and males living elsewhere.

Female leopards appeared to mature more rapidly than males. The increase in average weight of females between the subadult and prime adult, and the prime adult and old adult age classes averaged 25% and 0%, respectively, compared to 14% and 8% for males. Thus rapid growth of females appeared to occur earlier than among males. A male cub reared in captivity increased from 0.6 kg at birth to 1.2 kg in 4 weeks, 2.7 kg in twelve weeks, 32 kg in fifty-two weeks, and 45 kg at 96 weeks, when he was considered mature (Crandall 1964). In Zambia a tame twenty-month-old male and twenty-seven-month-old female weighed 41.8 kg and 32.7 kg, respectively (Wilson and Child 1966). A male approximately two years old, reared in captivity near KNP, weighed 44 kg (I. Crabtree, personal communication).

Population Composition

Female leopards were only slightly more abundant than males in the leopard study areas. Of thirty captured leopards, sixteen were females for a sex ratio of 1.1 females/males, and the sex ratio of forty-four accurately sexed, visually observed untagged leopards was also 1.1 females/males. Based on sixty-five observations, however, the sex ratio of visually observed tagged leopards was only 0.6 females/males, and the combined sex ratio for all visually observed tagged and untagged leopards was only 0.8 females/males. In the Serengeti National Park, Schaller (1972) reported that females outnumbered male leopards two to one, but in Wilpattu National Park, the sex ratio of observed leopards was only 0.1 females/males (Muckenhirn and Eisenberg 1973).

These apparent differences in reported sex ratios in leopard populations are undoubtedly related to observational bias and differences in behavioral responses of male and female leopards to humans. When observers are on foot, their data are probably biased in favor of males, because females are more elusive. Two years of trapping in my study areas revealed that although females were present, they were seldom visually observed. Even trapping itself may give a biased sex ratio because females are more difficult to capture. The cumulative sex ratios of captured leopards were 0.6, 1.1, and 2.5 females/males during the first, second, and third calendar year of trapping. Hence, most males were captured early in the study but many of the females were not captured until later.

The number of female leopards captured in the SRSA equaled the number of males but outnumbered males two to one in the NRSA. The average sex ratio of captured leopards of all ages was 1.4 females/males in both study areas. If only adults were considered, females equaled males in the SRSA and outnumbered them three to one in the NRSA. The average sex ratio for adults in both areas was 1.8 females/males.

If we assume an equal sex ratio at birth, male leopards appear to have higher a mortality rate than females. This appears to be characteristic of large wild felids. For example, female tigers outnumbered males 4 to 1 in India's Kanha National Park (Schaller 1967). And female lions outnumbered males 1.7 to 1 in Kafue National Park and 2 to 1 in Lake Manyara National Park (Mitchell et al. 1965; Makacha and Schaller 1969). In Serengeti National Park the number of males fluctuated seasonally as young nomadic males wandered about searching for prey (Schaller 1972).

Most leopards captured in the study areas were adults. Of thirty individuals captured, 63% were adults, 30% subadults, and 7% large cubs (see table 5.2). Because large cubs were rarely captured and small cubs never captured, these

TABLE 5.2 *Captured Leopards*

		Residence			
		Leopard Study Areas		Outside Main Study Areas	
Age	Sex	Sabie River	Nwaswitshaka River		Total
Old Adult	Male	2	1	2	5
	Female	2	2	2	6
Prime Adult	Male	1	1	1	3
	Female	1	4	0	5
Subadult	Male	3	2	1	6
	Female	2	1	0	3
Large Cub	Female	1	1	0	2
Total		12	12	6	30

age ratios were characteristic of leopards more than two years old. Excluding two large cubs, 39%, 29%, and 32% of the captured leopards were old, prime, and subadults, respectively.

Older leopards were more common in the SRSA (33%) than the NRSA (25%), and most older leopards were females. Old adult females outnumbered old adult males two to one in the NRSA, but the ratio was equal in the SRSA. Similarly, prime adult females outnumbered prime adult males 4 to 1 in the NRSA, but equaled the number of prime males in the SRSA. But subadult male leopards outnumbered subadult females three to two in the SRSA and two to one in the NRSA.

The observed sex ratios of leopards indicated an influx of surplus males into the study areas' populations. Other data supported this view. First, most radio-collared subadult males wandered widely, up to four times the distances of the subadult females. Second, radio-collared subadult males either remained in the SRSA or moved there from the NRSA. None moved directly away from the Sabie River. Third, two males, one subadult, and one prime adult suddenly appeared in the SRSA in 1975, soon after the deaths of a resident subadult and adult male. These data indicate a much greater influx of male than female leopards into the study areas.

The sex ratios of leopards also suggested that adult males were replaced more rapidly than were females. The equal sex ratio of subadult to adult males (1:1) was apparently maintained by the influx of younger males into the study areas' populations. No influx of young females was documented. Young females appeared to have been born and reared in the study areas, whereas many younger males seemed to have immigrated into the areas.

Population Density

Estimates of the population size of leopards in the study areas were based on several assumptions. First, I assumed that I accounted for all leopards more

TABLE 5.3 *Estimate of Leopard Populations*

Area	Year	Adults		Subadults and Older Cubs		Total
		Marked	Unmarked	Marked	Unmarked	
Sabie River study area	1973	2	3	2	2	9
	1974	3	2	4	1	10
	1975	4	0	4	0	8
Nwaswitshaka River study area	1973	1	8	1	2	12
	1974	7	2	3	0	12
	1975	8	1	2	0	11
South side of Sabie River, west of Skukuza	1973	1	3	1	?	5+
	1974	0	3	1	?	4+
	1975	0	3	0	?	3+
South side of Sabie River, east of Skukuza	1973	0	4	?	?	4+
	1974	2	2	?	?	4+
	1975	1	2	?	?	3+
North side of Sand River	1973	1	1	?	?	2+
	1974	1	1	?	?	2+
	1975	1	1	?	?	2+

than one year old in the study areas, whether they were tagged or not (see table 5.3). I based this assumption on the fact that leopards of the same sex and age had distinct and fairly exclusive home ranges, that the entire study areas were used by leopards and any vacancies in home range mosaics were actually occupied by unmarked leopards, and that all untagged leopards observed early in the study were eventually captured or accounted for by visual observations. Second, I assumed that I was able to account for most of the reproduction, mortality, and dispersal occurring among leopards in the study areas' populations. Third, I assumed that I properly accounted for individuals that only used the peripheries of the study areas and accordingly adjusted the densities.

The first assumption was supported by evidence from capture success, visual observations, and radio-tracking. As the study progressed fewer untagged leopards were observed in the study areas, and the capture of each new untagged leopard decreased the probability that another leopard, especially of the same sex and age, would be captured in the same area. I was also able to photograph untagged leopards and compare their features to leopards I later captured. I was therefore able to eventually account for all leopards observed within the study areas. Radio-tracking leopards also allowed me to identify regions within the study areas that were not used by tagged leopards. These were areas where I invariably observed untagged leopards.

The second assumption was supported by observations of tracks of female leopards and cubs in the dry sandy beds of streams or firebreak roads even if I was unable to capture them initially. Reproduction was also accounted for by the condition of captured females who were obviously pregnant, by changes in the behavior and movement patterns of radio-collared females after they had

TABLE 5.4 *Leopard Densities*

		Leopard Study Area					
		Sabie River			Nwaswitshaka River		
		Number		Square Kilometers per Leopard	Number		Square Kilometers per Leopard
Age	Sex	Mean	Adjusted		Mean	Adjusted	
Adults only	Male	2.0	0.7	25.3	3.0	1.7	47.7
	Female	2.7	2.2	8.1	6.0	6.0	13.5
	Both sexes	4.7	2.9	6.1	9.0	7.7	10.5
All over one year old	Both sexes	9.0	5.3	3.3	11.7	9.7	8.3

(Adjusted numbers are corrected for measured or estimated portions of home ranges of leopards outside study area boundaries.)

cubs, and by the visual observation of one cub. Mortality was determined by finding dead radio-collared leopards or their radio collars, or by failing to recapture previously tagged leopards after periods of intensive trapping. Dispersal was primarily accounted for by radio-tracking individuals off the study areas or by the sudden appearance of new individuals in the study areas.

The assumption that leopards living only on the peripheries of the study areas were accounted for was supported by visual observations of untagged leopards that were never captured within the study areas. These leopards were invariably seen along the study areas' boundaries but not within the study areas. Examples include leopards seen along the Naphe and Lower Sabie roads. The monitoring of two such individuals captured along the boundaries of the study areas revealed that they confined all of their subsequent movement to areas outside the study areas.

The population density of leopards in the SRSA was 2.5 times that estimated for leopards in the NRSA (see table 5.4). An average of 4.7 adult leopards/year estimated in the SRSA gave a density of 6.1 km²/adult. In the NRSA an average of 9 adults/year resulted in a density of 10.5 km²/adult. If all leopards more than one year old were considered, the average density was 9 leopards/year (3.3 km²/leopard) in the SRSA and 11.7 leopards/year (8.3 km²/leopard) in the NRSA.

Although only six leopards were tagged outside the main study areas, I also estimated leopard densities in several areas I knew well. For a 28 km² region south of the Sabie River west of Skukuza, there were 3.3 adults (8.4 km²/adult); for a 34 km² area east of Skukuza, 3.6 adults (9.4 km²/adult); and for a 12 km² area north of the Sand River, 2 adults (6.0 km²/adult).

I also estimated leopard densities using the Schnabel (1938) and Lord (1957) capture-recapture methods (see table 5.5). Each calendar year was considered a separate trapping period, with known mortality excluded in two of the three methods. When mortality was excluded the estimates obtained by these methods were high, but estimates were similar to those obtained in table 5.3 when mortality was included. The greatest difference between population estimates

TABLE 5.5 *Leopard Population Estimates*

		Method of Estimate					
		Unadjusted Mortality				Adjusted Mortality	
		Schnabel (1938)		Lord (1957)		Lord (1957)	
Leopard Study Area	Year	*n*	SD	*n*	SD	*n*	SD
Sabie River	1974	8	3	8	3	8	3
	1975	13	14	24	14	12	7
Nwaswitshaka River	1975	19	3	13	2	11	2

Standard deviations are based on numbers of recaptures of marked leopards.

using these methods was for the SRSA in 1975. The leopard population was overestimated using the capture-recapture methods that excluded mortality because during that period four untagged leopards were captured and two tagged leopards died.

Leopard densities in the KNP study areas were higher than those reported elsewhere. In Serengeti National Park, Schaller (1972) estimated 1 resident leopard/28.6 km^2 of woodland habitat and 1 leopard/22 to 26.5 km^2 of the entire park. From the biomass of ungulates in Wilpattu National Park, Eisenberg and Lockhart (1972) estimated 29 km^2 could support one adult leopard. In Tsavo National Park (East), Hamilton (1976) estimated a density of 1 adult leopard/13 km^2. The density of leopards in the Kalahari Gemsbok National Park, Republic of South Africa, was estimated at 1 leopard/160 km^2 (Bothma and Le Riche 1984). In the mountainous Cedarberg Wilderness Area, Cape Province, Republic of South Africa, Norton and Henley (1987) estimated about 6 to 9 adult leopards/100 km^2. Although different methods were used to estimate leopard numbers in these studies, capturing and marking individuals is undoubtedly the most accurate census method for such an elusive carnivore. The accurate census method, as well as the high prey biomass available to leopards, probably accounted for the high leopard population densities recorded in the KNP study areas.

Reproduction

Because females concealed their young in dense cover, I could not observe and count the number of leopards born in the study areas. I was, however, able to determine the number of females annually rearing young. Early in the study, tracks of small cubs and captured large cubs were used to ascertain productivity. Later, after most adult females were radio-collared, changes in their behavior and movement patterns, particularly three months after suspected breeding, indicated young had been born. Most attempts to observe small cubs were unsuccessful. While females were off hunting, I diligently searched dense patches of cover and rocky koppies suspected of harboring their cubs, but could not find them. The return of the females indicated the cubs were still present, but well hidden.

AGE AT SEXUAL MATURITY

Limited information suggested that most female leopards in the study areas became sexually mature at about three years of age. Of five captured females, none had young before they were about three years old. Even when radio-collared subadult females shared the same areas with an adult male, they were never located together courting. Of eleven captured adult females, eight apparently had young before the study began, and two had young, perhaps for the first time, during the study.

All tagged and untagged male leopards known or suspected of breeding were old or prime adults. Thus, although adult F4 was located with old M3 on eight occasions, and with subadult M1 on five occasions, she was never located with subadult M2, whose range partially overlapped hers. Her associations with the old male differed from those with the subadult male. She spent an average of 2.1 days/association with the old male but only 1 day/association with the subadult. Similarly, adult F11 and M14 also spent an average of 2.3 days together during four associations. Subadult M1 may have been F4's offspring: he was initially captured within her range, his home range was within hers, and he later made exploratory movements outside that range. He was an estimated 3.5 to 4 years old when he left the study area. Similarly, subadults M2, M7, M18, M19, and M26, all estimated to be less than three years old, did not appear to be sexually mature.

ESTROUS PERIOD

Female leopards are polyestrous. The average interval between thirty-five estrous periods of three captive leopards was 45.8 days with a range of 20 to 50 days, and the average duration of estrus was 6.7 days (Sadlier 1966). The intervals between periods during which I located pairs of adults together ranged from 4 to 96 days, with an average of 32.9 days. Pairs of adults were located together on thirteen occasions and remained together for an average of 2.1 days/association. Because I could not determine when females were in estrus, nor how long estrous lasted, these associations probably represented the period when males were most attracted to females. The longest period males and females were located together was four days; the shortest was one day.

MATING SUCCESS

Most courtship associations between leopards in the study areas appeared to be reproductively unsuccessful. Of thirteen suspected courtship associations between radio-collared adult males and females, only two (15%) were known to result in the birth of cubs. Of eight suspected courtship associations between M3 and F4, only the last appeared successful. Similarly, of four suspected courtship associations between F11 and M14, only one resulted in the birth of cubs. The only known association between F13 and M14 did not result in the birth of cubs. Among lions, only 20% of the known sexual contacts resulted in conception (Schaller 1972).

Captive leopard data indicate that if a female fails to conceive or loses a litter, she is capable of remating within fifty days. Only once during the study did I suspect that a female remated after losing cubs. Female 11 was located with M14 on August 11, 1974, had cubs in early November, apparently lost her cubs in late November or early December, and was again located with M14 on December 28–30. Assuming she lost her litter, remating would have occurred within eleven to twenty-seven days.

Number of Females Producing Young

Of eleven adult females captured during the study, ten (91%) apparently had young prior to or gave birth during the study (see table 5.6). During some years no females gave birth to cubs in the study areas; during others up to one-half of the females produced young. The average proportion of adult females producing young each year was 27.7%. In Wilpattu National Park five females produced seven litters in two years (Muckenhirn and Eisenberg 1973). In Serengeti National Park two of four females had young one year, and two years later, both females had young again (Schaller 1972). Forty-one percent of all captured adult female cougars had young at the time of capture (Robinette et al. 1961), and according to Hornocker's (1969) information, about 39.2% of the adult female cougars in his study area annually produced young. Forty-five percent of all lioness observed by Schaller (1972) were accompanied by cubs.

Table 5.6 *Breeding Female Leopards and Cubs*

	Leopard Study Areas					
	Sabie River			Nwaswitshaka River		
	Females Producing Cubs		Number of Cubs	Females Producing Cubs		Number of Cubs
Year	*n*	% of Breeding-Age Females		*n*	% of Breeding-Age Females	
1973	1[1]	33	2	3[2]	50	6
1974	0	—	0	2[3]	33	4
1975	1[4]	50	2	0	—	0

[1] Leopards F16.
[2] Leopards F6, F17, and F25.
[3] Leopards F11 and F21.
[4] Leopards F4.

Birth Intervals and Dependence of Cubs on Female

Births of leopards in the wild appear to be separated by at least 16.5 months, with three litters averaging 17.1 months apart (Le Roux and Skinner 1989). The intervals between two litters of leopards observed by Schaller (1972) were twenty-four and twenty-five months. Five females in my study areas could not have successfully reared young within two years of their previous litters; at least two females had litters separated by at least thirty months; and another older female did not raise a litter in thirty-six months. If these five females bore young at intervals of 24, 24, 30, 30, and 36 months, the mean interval between births would have been 28.8 months.

Leopard cub development and independence have been documented by Le Roux and Skinner (1989) who observed a habituated, free-ranging female rear successive litters of cubs over a three-year period in an area near the KNP. Unweaned leopard cubs were frequently moved between secluded spots. They were presented meat by the female at sixty-five days, but were not observed to eat meat before seventy-two days. Suckling ceased after 101 days. Cubs were then led to kills until they were 9.5 months of age, after which they accompanied the female on hunts. The cubs killed their first impala at eleven months of age. Cubs from two litters were totally dependent on this same female for 348 and 414 days ($n = 2$ litters).

Birth Rates

Leopards in captivity are born after a gestation period ranging from 98 to 105 days (Crandall 1964); two litters in the wild averaged 106 days (Le Roux and Skinner 1989). The number of young/litter for twenty-seven litters born in the Zoological Gardens in London ranged from one to three cubs (Zuckerman 1953) and for ten litters born in the San Francisco Zoo from two to three (Reuther and Doherty 1968). In the wild, one female averaged 2.2 cubs/litter for five litters (Le Roux and Skinner 1989). Schaller (1972) reported seeing two females with two small cubs each, but three females had only one larger cub each.

I saw a six-month-old cub with F6 and the tracks of only one small cub with those of another female. Unconfirmed reports of small cubs in the park ranged from 1 to 2/litter. J. van Deventer (personal communication) once saw a female with three small cubs crossing the Naphe Road in February 1974, and a female destroyed in the Malelane District was carrying two near-term fetuses on June 27, 1974 (T. Whitfield, personal communication). Stevenson-Hamilton (1947) reported two near-term fetuses in a female killed in August in KNP and claimed two cubs was the average litter size. Turnbull-Kemp (1967) reported litters of up to six cubs, but observed that one or two cubs were most common.

I estimated the number of young leopards produced in the study areas by assuming an average litter size of two at birth (see table 5.7). In the SRSA, F16

TABLE 5.7 *Annual Recruitment*

	Leopard Study Area						Average Annual Increase (%)
	Sabie River			Nwaswitshaka River			
	Number			Number			
Year	Cubs	Adults	% Increase	Cubs	Adults	% Increase	
1973	2	3	40	6	9	67	57
1974	0	5	0	4	9	44	29
1975	2	4	50	0	9	0	15
Mean	1.3	4.7	28	3.3	9.0	37	34

probably had young in 1973, and F4 in 1975. No cubs were produced in the study areas in 1974. Large cub F20 was probably F16's offspring from 1973; the small cubs of F4 were not observed. Visual observations and radio-tracking data suggested that subadult F12 was adult F29's offspring from 1972 and that subadult M1 was F4's offspring from 1971. This information suggested the average number of young produced per year in the SRSA was 1.3, or an average potential increase of 30% of the adult population.

The estimated number of cubs born per year in the NRSA was 3.3. Female 6 had cubs late in 1973, and F25 apparently had cubs earlier that year. A female cub (F15) captured within adult F17's home range indicated she had cubs in 1973. In 1974 radio-collared F11 and F21 had cubs. Cubs were not born in the study area in 1975. The average annual increase was estimated to be 37% of the adult population.

If we assume an average litter size of two, no cub mortality, and a birth interval of twenty-four months, each adult female in the study areas should have been able to rear 1 cub/year. If the birth interval was 28.8 months, each female would have reared an average of 0.8 cubs/year. Actual productivity measured by the number of larger cubs and subadults captured in the study areas, indicated adult females reared about 0.5 cubs/female/year in the SRSA and 0.5 cubs/female/year in the NRSA. These estimates agree with observations indicating that although two cubs are usually born per female, only one cub is successfully reared.

BIRTH SEASON

Although leopards appear capable of giving birth during any time of the year, cub survival is probably related to period of birth. In East Africa, where there are two rainy periods, most hyenas (Kruuk 1972), cheetahs (Eaton 1974), wild dogs, and perhaps lions (Schaller 1972) are born in the early long-rain period. For hyenas, Kruuk (1972) suggested that births peaked approximately one gestation period after a peak in prey abundance. Frequency of sexual activity in lions also appeared related to the availability of prey (Schaller 1972). In Seren-

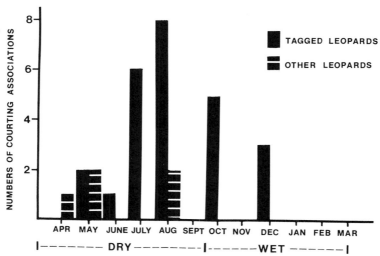

FIGURE 5.1 Cumulative monthly distribution of observations of courting leopards.

geti National Park two litters of leopards were born at the beginning and one near the end of the dry season (Schaller 1972).

I estimated periods of birth of leopards in the study areas by observing peaks in breeding activity. I located radio-collared males and females together on twenty-eight days and documented reports of leopards courting on five additional occasions (see fig. 5.1). Most courtship among leopards occurred during the late dry season, from July through September (49%), and the least occurred during the wet season, from January through March (9%). Thus more cubs would have been born during the early wet season than at any other period.

The peak in breeding activity of leopards in the study areas coincided with maximum concentrations of impala in the late dry seasons. Mortality and condition data suggested leopards had a difficult time capturing prey during most of the dry season. Presumably, poorer hunting success in the dry seasons was caused by the lack of stalking cover.

The peak period of leopard births occurred during the peak period of impala births. Newborn impala were first seen in November and births peaked in December and January. Leopards actively preyed on young impala. I watched a female leopard pursue a young impala on December 10, and M14 killed a several-week-old impala on November 25. In late November I also found the remains of an adult female and a young impala cached in the same tree by a leopard.

More cover would have been available for female leopards to conceal their cubs from predators during the wet season. Small cubs are unable to take evasive action to escape from predators. Turner (cited in Schaller 1972) described how a pride of lions treed a female leopard and then killed her two small cubs because they were too small to climb the tree. In the Kalahari Desert leopard cubs are often born in aardvark burrows, stay in the burrows or under bushes during the female's absence, and use burrows as daytime rest sites (Bothma and Le Riche 1984). If trees are present, older leopard cubs can escape from predators. When I encountered F6 and her six-month-old cub on the ground at a kill, she quickly sent the cub up a slender tree for refuge, but she remained on the ground. The cub would have been safe from predators, even lions, in the tree.

Mortality

Leopard mortality was assessed by two methods: confirmation and probability. Confirmed observations were when a carcass, remains, or radio collar were recovered. Without radio transmitters, I would have found only one dead leopard in two years. Even using radiotelemetry, confirming mortality was not easy because leopards that died were quickly torn apart and consumed, or their remains were scattered by hyenas or other scavengers.

Probable instances of mortality included several categories. If a severely emaciated radio-collared leopard suddenly stopped moving and the transmitter failed thereafter, I assumed the leopard had died and scavengers had destroyed the transmitter. Crocodiles were sometimes responsible for such occurrences. One radio collar and a carcass of a leopard wearing a radio collar were recovered from the Sabie River. Hyenas were known to bite and twist the radio collars of dead leopards, and they could also have broken collars.

If a non-radio-collared but tagged, severely emaciated leopard suddenly could not be recaptured in the study area, I also assumed it was dead. Two leopards (F12 and F20) were in this category. When I failed to recapture healthy, previously tagged adult male leopards during the dry seasons, I assumed they had died because they had previously been recaptured. Such mortality could not be assumed for previously tagged healthy female leopards because females were always difficult to recapture.

CAUSES OF MORTALITY

Most older leopards died from starvation, natural violence, or were killed by poachers. Ten confirmed and four probable deaths revealed that nine leopards (64%) died of starvation and the other five (36%) from other causes (see table 5.8). Of the latter, at least one died from natural violence, one was killed by poachers, and three died from unknown causes.[1]

1. Although not suspected during this study, infanticide is now a known cause of mortality among leopards (see appendix C).

TABLE 5.8 *Mortality Among Radio-collared Leopards*

Cause of Death	Number		Total	% of Total
	Confirmed	Suspected		
Starvation	6[1]	3[2]	9	64
Natural violence	1[3]	0	1	7
Poaching	1[4]	0	1	7
Unknown	2[5]	1[6]	3	21
Total	10	4	14	100

[1] Leopards M5, F8, F13, F16, F24, and F25.
[2] Leopards M2, F12, and F20.
[3] Leopard M23.
[4] Leopard M9.
[5] Leopards M7 and M19.
[6] Leopard M3.

Although leopards of different sexes and ages died of different causes (see table 5.9), an equal number of males and females died during the study. The age at death was similar for male and female leopards: three males and three females were old adults (43%), one male was a prime adult (7%), three males and three females were subadults (43%), and one female was a large cub (7%).

Male and female leopards died from different causes. The deaths of all radio-collared female leopards were attributed to starvation, but only 25% of the adult males and 33% of the subadult males died in this manner. Two males died of starvation (29%), one from natural violence (14%), one from poachers (14%), and three from unknown causes (43%). The significantly different mortality patterns among tagged male and female leopards (Chi-square = 11.14, *df* = 1, $P < 0.001$) will be discussed later.

Starvation.   Starvation was the leading cause of leopard mortality. D. Gradwell, provincial veterinarian in Skukuza, conducted postmortem examinations on three leopards, and V. de Vos, veterinary ecologist, conducted a postmortem on another leopard. Both veterinarians reached similar conclusions regarding the deaths of the four leopards: death occurred from apparent starvation, no infectious disease was evident, all the leopards were heavily infested with external and internal parasites, and no debilitating injuries were apparent.

The appearance of these and other leopards that died from starvation was similar. The leopards were underweight, their heads appeared abnormally large

TABLE 5.9 *Sex and Age of Leopards at Time of Death*

Age	Males	Females	Total
Old Adult	3	3	6
Prime Adult	1	0	1
Subadult	3	3	6
Large Cub	0	1	1
Total	7	7	14

5.1. A leopard (F16) that died from starvation. Most leopards dying from starvation were very young, very old, or had been injured.

because of the sparse fur on the neck, their limb bones and ribs prominently projected through an obviously thin muscular frame, their coats were dull, and the density of their fur was sparse. Their fur was often falling out, sometimes in small patches, and the skin on the toes and pads was peeling. The skin was often darkly pigmented, giving the leopards a dark appearance, and their eyes were sunken with yellowish discharges in the corners. The leopards required considerable effort to walk, and their hindquarters swayed from side to side.

Leopards that died from starvation were found in different locations. Female 8 died beside a well-traveled tourist road, and F16 died at the base of a large tree 30 m from a tourist road (see photo 5.1). I did not know where F24, F25, and M5 died because their bodies were dragged by scavengers, probably hyenas, into relatively open areas. Female 13 died in a trap where she was being held overnight after being experimentally treated for mange. The carcasses of two leopards that were recovered within twenty-four hours of death had not been discovered by scavengers; two recovered within forty-eight to seventy-two hours were partially eaten by scavengers; and one carcass recovered at least five days after death was torn apart and the remains carried off by hyenas.

Cause and effect were difficult to distinguish when interpreting cause of death. Did leopards become emaciated because of parasitic infestations, or were those infestations a result of poor condition? Did parasites first weaken the leopard, or was the leopard already weakened before acquiring the parasites? Did injuries prevent leopards from capturing prey, or were injuries obtained in reckless attempts by emaciated leopards to capture desperately needed prey? Did lack of hunting experience contribute to a leopard's poor condition?

Several observations suggested that emaciated leopards were unable to capture prey before their deaths. One leopard attempted to obtain bait from a trap in the NRSA for five days between November 12 and December 18, 1973. At first I assumed F6 was responsible, and not wishing to capture her because she was pregnant, I wired the trap door open. However, after F6 left the vicinity, a leopard continued to take the trap bait. Several days later an extremely emaciated subadult female (F8) was captured, fitted with a radio collar, and released. Two days later she took bait from another trap 1.8 km away, but I had wired the trap door open to prevent her capture. She was still alive 3.2 km from the initial capture site on the evening of December 25, but was dead the next morning. Heavy rains (143 mm), which fell on seven of nine days during this period may, through exposure, have hastened her death.

Healthy leopards did not visit traps as frequently nor were they recaptured as often as emaciated leopards. From June 17 to 20, 1974, emaciated F16 took bait from two different traps on three occasions, and was captured once. She died on July 22. Her death may also have been hastened through exposure; a violent hailstorm struck the area on July 1, depositing 82.8 mm of rainfall in less than one hour and covering the ground with hailstones up to 17 mm in diameter. Emaciated F24 was captured four times and took trap bait two additional times in twenty-two days; F12 was captured three times in eight days; and F25 was captured twice in nine days. Emaciated M2 was captured five times and attempted to get the trap bait six more times in sixty days, and emaciated M5 was captured two times and took trap baits five additional times in thirty-six days. Slightly emaciated M26 was captured six times in twenty-one days.

The hunting behavior and feeding habits of emaciated leopards also suggested they were unable to capture large prey. The only kill subadult M2 made after he became emaciated was a civet. This was cached in a tree and fed upon for five days, an extremely long period for small prey. Never known to have killed an impala, even before he became emaciated, he was once found scavenging on an impala carcass. An attempt by another emaciated leopard to kill a ratel and its later killing of a civet also suggested that emaciated leopards were unable to capture large prey.

Other evidence supported this belief. The feces of emaciated female cub F20 contained only vegetation and the hairs of a small mammal, probably a mongoose. Several feces collected in the home range of emaciated F13 contained only the remains of small mammals, perhaps genets. At death, her stomach contained 1.5 kg of impala hooves, bones, and hide—parts usually not eaten by healthy leopards. This suggested that some emaciated leopards were scavenging from other predators. Slightly emaciated subadult M26 waited near a tree

TABLE 5.10 *Emaciated Leopards Before Death*

Leopard No.	Weight (in kilograms)		% Loss	Days Between Last Capture and Death	Weight Loss per Day (in kilograms)	% Weight Loss per Day
	Last Capture	At Death				
F8	32.3	20.0	38	7	1.8	5.4
F13	38.6	32.3	16	31	0.2	0.5
F24	29.3	20.9	29	39	0.2	0.7
F25	33.6	24.6	27	32	0.3	0.8
Mean	33.5	24.5	27.5	27.3	0.6	1.9

while an adult male fed on an impala. Within minutes after the adult abandoned the kill, M26 rushed up the tree to claim the scanty remains.

Defensive attacks from small but dangerous prey or from other predators may have been the source of wounds frequently observed on emaciated leopards. The leopard that attacked the ratel received severe bites on the forelegs. Emaciated F16 had an 8.5-cm laceration on her right shoulder, a 7.5-cm diameter bloody abscess around another puncture wound, and four deep puncture wounds on her hindquarters. She appeared to have been bitten several times by a large predator.

Emaciated leopards lost weight rapidly before their deaths. Four female leopards that lost an average of 27.5% of their body weight in 27.3 days, lost 0.6 kg/day (see table 5.10). One female (F8) lost 1.8 kg/day, but three other females lost an average of only about 0.2 kg/day. If one assumes an average weight of 37.2 kg for adult females and 30.1 kg for subadult females (see table 5.1), a loss of 0.2 kg/day would reduce their average weight 50% in ninety-three and seventy-five days, respectively. Although loss of weight varied, most emaciated leopards had apparently been losing weight thirty to sixty days before capture. At time of death leopards had not lost more than 40% of their last recorded weight.

Information from five individuals indicated that healthy leopards could become severely emaciated in one to three months. Female 13, who was in good condition when her radio collar was permanently removed on June 29, was recaptured thirty-one days later—6.3 kg lighter and in poor condition. Male 26 was just beginning to exhibit signs of poor condition thirty-nine days after he was first observed, and M5 became emaciated thirty-seven days after he was wounded. Female 12, who was in excellent condition when observed on a kill nine months after being fitted with a radio collar, became emaciated sometime within the following three-month period. Male 2, who was not interested in trap baits before May 1974, was in poor condition when captured at the end of the month. Four out of five leopards (80%) became emaciated within thirty-nine days, and one within ninety days. Although little information is available on the weight loss of wild carnivores in tropical or subtropical environments, lion

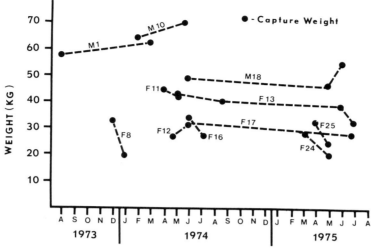

FIGURE 5.2 Changes in capture weights of individual leopards during the study period.

cubs in Serengeti National Park rapidly lost weight in 1 month and died 2 to 2.5 months later (Schaller 1972).

Most leopards lost weight in the late wet and early dry seasons (see fig. 5.2). Seventy-nine percent of the monthly periods of weight loss were between January and June, and only 21% between July and December. For leopards captured in the SRSA, 89% of these monthly periods were between January and June, and for those captured in the NRSA, 67%. If poor condition and weight loss were a result of inadequate food, these seasonal trends indicate that leopards had difficultly capturing prey at the end of the rainy season.

Seasonal changes in the abundance and availability of impala, the leopards' principal prey, probably influence the condition of leopards. In the SRSA weight loss of leopards coincided with annual changes in impala abundance. Between late 1973 and early 1974 the average density of impala declined 45%. Most (78%) leopards became emaciated during early 1974. Similarly, between late 1974 and early 1975, when the average density of impala declined another 29%, three additional leopards became emaciated. Seasonal densities of impala were highest in the early wet season (116 impala/km^2) and lowest in the late wet season (70.1 impala/km^2). Leopards rapidly lost weight from January through March and April through June, when impala abundance was lowest (see fig. 5.2).

Factors other than prey abundance may also have influenced leopard condition. Despite an increase in impala abundance in the NRSA during the study, some leopards still lost weight during the late wet and early dry seasons. These two periods were preceded by a decrease of only 6% in the average density of impala, compared to a 35% decrease for the same period in the SRSA. Thus the magnitude of the change in impala density appeared less important than the

TABLE 5.11 *Infected Leopards*

Condition of Leopard When Last Captured	Number Captured	Leopards Infected with *Notoedres cati*	
		n	%
Severely emaciated	10	10	100
Slightly emaciated	3	3	100
Healthy	17	3	18
Total	30	16	53

period in which the changes occurred. Whatever the actual cause, leopards lost weight and became emaciated in both study areas during the same period.

Annual changes in vegetative cover may have reduced leopard hunting success in both study areas. By April most deciduous trees and shrubs cast their leaves (Van Wyk 1971) and by May most grasses and forbs cease to grow (Hirst 1975). Increased visibility may have made it more difficult for leopards to stalk impala. During this period impala were still dispersed because drinking water was available in many small pools. It was during these months (April and May) that most leopards lost weight (see fig. 5.2). Although no direct correlation was established, circumstantial evidence suggested that during the early dry season leopards had a difficult time capturing the widely scattered impala. Increased trapping success and attraction of leopards to baits also suggested that leopards were having a difficult time obtaining food during the early dry season. Later in the dry season, when water became scarce, impala were probably forced to travel longer distances to obtain water (Young 1972). Leopards may then have regained a hunting advantage as impala passed through the densely vegetated riparian zones to drink in rivers.

Another factor may have aggravated weight loss among leopards. Thermoregulatory energy demands may have increased the leopards' metabolic requirements as night temperatures dropped during the dry season. Because leopards are primarily nocturnal, energy expenditure may have exceeded intake during periods of low prey vulnerability. This seasonal negative energy balance may also have contributed to the seasonal weight loss exhibited by leopards.

Parasites and diseases.   Emaciated as well as dead leopards were heavily parasitized by mange. Of thirty captured leopards, sixteen (53%) had mange before or developed mange during the study (see table 5.11). All emaciated leopards, as well as three apparently healthy individuals, had mange. Eight leopards (27%) had mange before they were initially captured, six (20%) developed mange after capture, and two (7%) had mange but apparently recovered before the study began. Fourteen leopards (47%) were unaffected by mange.

The mite responsible for mange was a feline ear mite, *Notoedres cati* (D. Argo, personal communication). According to Young et al. (1972a), who described the same mange among cheetahs released in KNP, this parasite causes severe and

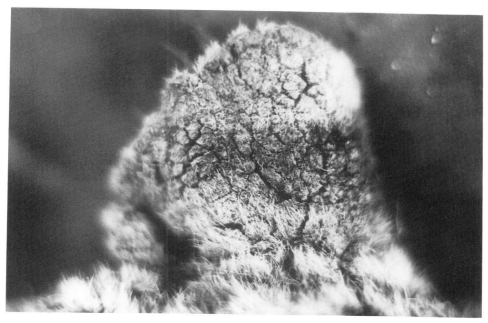

5.2. The crusted skin of the ears of a dead leopard that was heavily infested with the ear mite *Notoedres cati*.

irritating dermatitis crustosa on the head and neck, as well as other parts of the body. Infected cheetahs became emaciated and dehydrated; their thickened skin became dry and wrinkled; and the skin was hairless and darkly pigmented. Leopards infected with mange exhibited the same symptoms. In addition the ears of leopards were crusted (see photo 5.2) and their skin became crusted, wrinkled, and cracked (see photo 5.3). Fly larvae were present between the crusted plates of skin on one leopard (F13). The dark appearance of emaciated, mange-infested leopards was apparently caused by the increased pigmentation of skin, which resulted from hair loss.

Two leopards with mange were experimentally treated by park veterinarians. Severely emaciated, mange-infested M19 was treated with a solution of 5% Malathion (Young et al. 1972) on August 8, 1974, and released. When captured thirty days later, his condition was the same, if not worse. He was assumed dead after his radio collar was recovered from the Sabie River on October 2. Another severely emaciated leopard (F13) was treated similarly, but died while being held in a trap overnight for further examination. The exceptionally cold night (-1° C) may have enhanced her death through exposure (A. de Vos, personal communication).

Mites of the genus *Notoedres* are usually transmitted by direct contact with an infected individual, or under certain circumstances, by contact with larvae, nymphs, or adults that may still be alive in the host's environment. Gordon

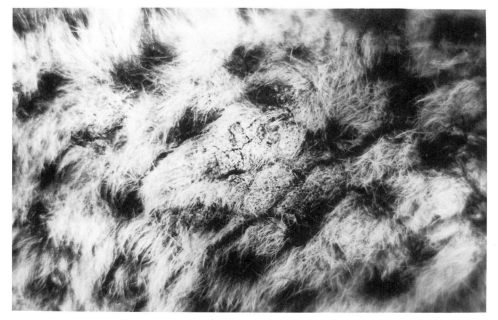

5.3. Crusted body skin of a dead leopard heavily infested with the ear mite *Notoedres cati.*

(cited in Lapage 1968) believed larval *Notoedres muris* infected new individuals because the female parasites seldom leave the safety of the burrows in the host's skin. If larvae are unable to find a new host while off the former host's body, they die within twenty-four hours. Gordon believed that other stages (eggs, nymphs, adults) were seldom responsible for infections, but that under certain climatic conditions these stages might survive without a host for one to two weeks.

I could not determine how mange was transmitted among leopards. Of fourteen leopards that became infected during the study, eight (57%) were already infected before capture; six (43%) became infected after capture. Infection with mange before and after capture was not significantly different, assuming leopards had an equal chance of becoming infected whether or not they were captured (Chi-square = 0.29, *df* = 1, *P* < 0.500). Thus the capture and handling process did not appear to be the causative factor. If *Notoedric* mange was spread via larvae, an uninfected leopard would have to be captured within twenty-four hours in the same trap as an infected individual. This did not occur during the study. If mange was spread by the less likely mode of adults, nymphs, or eggs, uninfected leopards would have to be captured in the same trap within two weeks of an infected individual. This occurred only once during the study, when M10 was captured in the same trap five days after F25. Although skin samples were not taken from M10, he did not appear to have mange at the time. When M10 was recaptured forty-two days later, a skin scraping revealed he had mange. Whether he contracted mange from F25, via the trap, or from some other source was unknown. The only other leopard (M26) that could have contracted mange in this manner was captured seven and a half weeks after an infected leopard (F24). These data suggested that traps were not a significant source of infection for the six leopards that contacted mange after capture.

Mange could have been transmitted from female to offspring or between pairs of courting adults. In the western section of the SRSA all five captured leopards had or developed mange during the study. One was an adult male (M23), one an adult female (F16), and three were either subadults (M2 and F24) or large cubs (F20). The old adult leopards probably spent time together, when courting, although I did not locate them together during the study. It was also possible that subadult M2, and perhaps subadult F24, were adult F16's offspring from 1971. Large cub F20 was probably her offspring from 1973. All were captured in the same locality and had overlapping home ranges. Of four leopards with mange in the NRSA, two were old adults (M10 and F25) with overlapping home ranges and two were subadults (F8 and M19) captured in the same area.

The transfer of mange mites from female to offspring was likely among leopards because cubs have frequent and prolonged body contact with their mothers. In Kenya, Murray (1967) reported that a four- to-six-month-old leopard cub had sarcoptic mange. Because a cub this size probably only came into contact with its mother or litter mates, interfamily transmission of mange appeared likely. Although contraction of mange from prey may also be possible, especially with sarcoptic mange, transmission among socially related carnivores may be common. Among lions, members of prides can become infected with sarcoptic mange, which may lead to mortality, especially among small cubs (Young et al. 1972b). Mange has also been reported among members of the same pack of wolves (Cowan 1947). Leopards may also spread mange during courtship when rubbing of the head and neck, places usually heavily infested with mites, is common.

Because mange is most common in individuals already in poor condition (Lapage 1968), mites may have been present on most leopards but only become virulent during periods of physiological stress. During periods of food shortage, leopards unable to maintain body weight, may be especially vulnerable to heavier infestations of the mites. Evidence supporting this view was observed among cheetahs that developed *Notoedric* mange up to six weeks after being captured, even under strict isolation (Young et al. 1972). These cheetahs apparently had mites prior to capture, but the mange became virulent only after the cheetahs were subjected to the stress of captivity.

Two leopards (F6 and F27) appeared to have recovered from mange before they were captured initially. The skin on their necks was thick, the fur on the

5.4. A large male leopard (M23) with past evidence of mange *Notoedres cati* lies in shade in the cool sand of the Sand River.

neck sparse, and the edges of their ears were bare and thickened. However, their body weights were normal, and they appeared in good health. These observations suggested that even if mange becomes virulent, some leopards recover. Other leopards may have had mange for extended periods. Male 23, although underweight, did not appear to be handicapped by mange, at least for the six-month period after his capture. Previous observations of this distinctive and well-known leopard by park personnel suggested that he may have had mange up to a year before he was captured. When I first saw him three months before he was captured, he already appeared to have mange (see photo 5.4).

Although the relationship between leopards' physical condition and parasitic infestations was unclear, poor nutrition may have predisposed leopards to mange. Nutritional complications are frequently implicated in mite infestations (Sweatman 1971). Animals often become susceptible to mange if deficient in vitamin A (Cross 1964) or a variety of minerals (Kutzer and Onderscheka 1966; Onderscheka et al. 1968). The occurrence of mange in populations of chamois was reduced 75% after providing them with mineral-enriched salt licks (Onderscheka et al. 1968). Felids have a high vitamin A requirement and must obtain all their vitamin A from prey (Scott 1968). But vitamin A is restricted to specific tissues, mainly the liver, and also the lungs, adrenals and kidneys. Body fat and muscle tissue are virtually devoid of this vitamin.

If certain leopards survived on a diet of small prey, or scavenged remains from large prey for prolonged periods, they could have become vitamin A deficient. The internal organs of small prey probably do not provide the same amount of vitamin A as the liver or other organs of impala-size prey. Scavenged skin and bones would provide little, if any, vitamin A.

Mange was most evident among leopards captured in the early dry season. The physiology of mites could explain the apparent increase of mange during this period. Water balance in mites is critical because they must maintain an equilibrium humidity that is dependent on the relative humidity and temperature of their external environment (Wharton 1960). The feeding rate of mites is also related to their equilibrium humidity. Mites tend to feed more in desiccating environments because they must obtain fluids from their host's tissues to replenish the water diffusing through their cuticle. Damage to a host is usually associated with feeding (Cheng 1964). When the humidity is high, mites feed less and become less harmful. In KNP high humidities during wet seasons may reduce the impact of mite infestations on leopards. As temperatures and relative humidity decline during the dry season, mites could begin to feed more rapidly, thus causing increased tissue damage to their hosts. If leopards also experienced a temporary food shortage and lost weight during the same period, they may have become particularly susceptible at a time when the mites were most actively feeding. Conversely, the mites' increased feeding on the host's tissues could have preceded, rather than followed loss of weight.

Seasonal variations in mite numbers are often marked and significant in mite pathology (Sweatman 1971). Many mites, like *Notoedres*, that burrow into the stratum corneum of nonhibernating hosts, become numerous at certain times of the year. In temperate latitudes, mites frequently become numerous during the autumn and winter (Green 1951; Kutzer 1966; Honess and Winters 1956), and in tropical or subtropical latitudes during the dry seasons (Timoney 1924; Schaller 1972). Another wild felid, the bobcat, sometimes becomes seriously infected with ear mites in the winter and develops crusty lesions similar to the leopard (Penner and Parke 1954; Pence et al. 1982).

All captured leopards, especially emaciated ones, were also infested with ticks. Most leopards had at least ten ticks, usually on their heads, ears, necks, forelegs, paws, and base of tail. Occasionally, 2 cm x 5 cm clusters of ticks were found under the forelegs. Emaciated leopards (F8, F16, F24, and F25) were heavily infested with ticks. More than one hundred were removed from the head alone of F8, and I estimated at least five hundred were on her entire body.

A sample of 223 ticks collected from five leopards from each study area revealed *Rhipicephalus simus simus* was the most common species (36%), with *R. appendiculatus* second most abundant (see table 5.12). Other collected ticks included *R. simus senegalensis, Haemaphysalis leachii leachii, Amblyomma hebraeum* and *Hyalomma truncatum.* Significantly more *R. appendiculatus* were collected from leopards in the NRSA than the SRSA (Chi-square = 32.66, $df = 1$, $P < 0.001$) perhaps because of that area's abundant grass. *Rhipicephalus appendicula-*

TABLE 5.12 *Ticks Collected from Leopards*

| | Leopard Study Areas | | | | Total | |
| | Sabie River | | Nwaswitshaka River | | | |
Species of Tick	*n*	%	*n*	%	*n*	%
Rhipicephalus simus	17	29	63	38	80	36
Rhipicephalus appendiculatus	6	10	48	29	54	24
Rhipicephalus senegalensis	17	29	23	14	40	18
Haemaphysalis leachii	16	27	21	13	37	17
Amblyomma hebraeum	3	5	7	4	10	4
Hyalomma truncatum	0	0	2	1	2	1
Total	59	100	164	100	223	100

tus is most commonly found on ungulates, including impala, bushbuck, steenbuck, duiker, and klipspringer (Hoogstraal 1956). In Zambia adult *R. appendiculatus* are rarely found on ungulates during dry seasons, but they quickly reach a peak of abundance after the first rains fall in November. Impala were most abundant in the NRSA during the wet seasons. Perhaps leopards also fed more often on impala then and acquired ticks from this prey.

Most of the 84 female (42%) and 139 male (32%) ticks were *R. simus simus*. More male than female *A. hebraeum* (10 males, 0 females) and *R. appendiculatus* (43 males, 11 females) were collected from leopards, despite the fact that female ticks were more conspicuous and more likely to be collected than males. The most frequently occurring tick on leopards, *R. simus simus*, is a common tick on carnivores in Africa. Adults are most commonly seen during wet seasons on carnivores such as lions, leopards, cheetahs, hyenas, foxes, jackals, civets, genets, and ratel (Hoogstraal 1956). *Rhipicephalus simus senegalensis*, like *R. simus simus*, is also found on carnivores and, less often, on antelopes. Although smaller than other ticks, *H. leachii leachii* is a common tick on carnivores, particularly canids. This tick is responsible for transmitting canine and feline babesiosis. *Amblyomma hebraeum* is the vector of tick typhus in man, but the adults rarely feed on carnivores. *Hyalomma truncatum* is adapted to drier climates than most African ticks, is commonly found on ungulates, and is seldom found on wild carnivores. Leopards probably acquired these ticks from vegetation as well as from their prey.

Turnbull-Kemp (1967) listed other species of ticks collected from African leopards: *Ixodes cavipalpus, I. cumulatimpunctatus, I. moreli, I. muniensis, I pilosus, I. oldi, I. rasus, I. vandicus, Amblyomma nuttali, A. tholloni, A. variegatum, Haemaphysalis aciculifer, H. leachii muhsami, H. paramata, Rhipicephalus armatus, R. capensis, R. compositus, R. pravus, R. sanguinus, R. sulcatus, R. tendeiroi, R. tricuspis* and *R. ziemanni*. Ticks collected from Asian leopards included *Amblyomma spp., Haemaphysalis dentipalpus, H. hystricis, H. kongisbergeri, H. papuana, Rhipicenter nuttali, R. bicornis, Rhipicephalus haemaphysaloides haemaphysaloides, Haemaphysalis leachii leachii, H. bispinosa* and *H. concinna*. Other external parasites include the fowl flea, *Echidnophaga gallinacea,* the pig flea, *E. larina,* and the chigger, *Gahrliepia (walchia) rustica.*

Although leopards probably support many endoparasites, I collected only a few from dead leopards. The small intestines of three dead leopards contained tapeworms. Leopard F16 had five tapeworms more than 10 cm long and many small ones; F24 had ten large and many small tapeworms; and F25 had at least fifty tapeworms about 10 cm long. Feces from two of four other captured leopards contained *Taenia spp.* ova. One was emaciated F12; the other was a healthy adult male (M14). Thus, of seven leopards specifically examined for tapeworms, five (71%) were positive. The fish tapeworms *Taenia pisiformes, T. ingewi,* and *Dibothriocephalus latus,* have been reported in African leopards. *Spirometra decipens,* and *S. felis* are two tapeworms found in Asian leopards (Turnbull-Kemp 1967).

The feces of four leopards contained eggs of a trematode suspected of being the lung fluke *Paragonimus sp.* Unidentified worms, or flukes, were removed from the bronchii of F13. *Paragonimus westermanii* have been found in the lungs of Asian leopards (Turnbull-Kemp 1967). Because the intermediate host of these flukes are fresh water crustacea, usually a crab or crayfish, its occurrence suggests that leopards were feeding on crustacea.

Schistosoma flukes were recovered from the blood vessels of the mesenteries of F13, and unidentified filarial worms were collected from the connective tissue under the skin. Filarial worms reported in leopards include *Dirofilaria granulosa* and *Dracunculus medinensis* (Turnbull-Kemp 1967). Larval *Dirofilaria* are transmitted by the bites of infected arthropods.

Parasitic nematodes reported from leopards include *Toxascaris leonina, T. mystax, Ancylostoma braziliense* and *A. caninum* in the small intestine; *Broncostrongylus subcrenatus* in the lungs; *Galoncus perniciosus* and *G. tridentatus,* or hookworms, in the small intestine; *Cyathospirura chevreuxi* and *Gnathostoma spinigerum* in the stomach, and *Physaloptera praeputiale* in the penis (Turnbull-Kemp 1967). A pentastomid, *Neolinguatula nuttali,* has been found in the leopard's nasal passage (Haffner et al. 1969).

Two of six (33%) fecal samples I collected from different leopards contained sporocysts of an Isosporan-type coccidial parasite. *Isospora felis* is a common coccidia found in felidae, but coccidia do not usually significantly effect wild carnivores (Pellerdy 1965). Little is known about coccidia in wild felids, although coccidia are usually host specific.

In tsetse fly regions of Africa, wildlife often harbors trypanosomes, or protozoan parasites of the blood. These parasites, transmitted by the mouthparts of biting flies, seldom affect wild indigenous mammals (McDiarmid 1962). The tsetse fly no longer occurs in KNP, but in Serengeti National Park, where 85% of the examined lions and all the examined hyenas were infected, the only leopard checked was negative (Baker 1969). Carnivores may be bitten by tsetse flies; they may also contract the disease by eating infected prey.

A more potentially dangerous parasite that may occur in the blood of leopards is *Babesia,* which is transmitted by the bites of ticks. This parasite breaks down the red blood cells, causing anemia. Apparently two types of *Babesia* may occur in leopards (Barnett and Brocklesby 1969), a large and smaller form. *Babesia felis,* which is commonly found in lions (Schaller 1972), is probably also found in leopards. Adult leopards are probably similar to adult lions, which are immune to *Babesia* unless resistance is already lowered by some other means. Lion cubs may suffer, however, before they acquire immunity.

Still another blood parasite found in leopards is *Hepatozoon spp.* (Keymer 1964 cited in Krampitz et al. 1968). This protozoan occurs in ticks that, when ingested by carnivores, migrate from the digestive tract to the liver where they develop and destroy liver cells (Schaller 1972).

Leopards are susceptible to a number of diseases. During the 1960–61 anthrax epizootics in the northern part of KNP, four leopards died from the disease after feeding on anthrax-infected carcasses (Pienaar 1969). Several tests for brucellosis in leopards were negative (Sachs et al. 1968), but a leopard in Nairobi National Park exhibited severe pulmonary hemorrhaging and necrosis that were similar to those of "Nairobi bleeding disease" in dogs (Murray 1967). A form of distemper in leopards was described by Turnbull-Kemp (1967), and a small leopard cub appeared to have subacute enteritis (Murray 1967). Turnbull-Kemp (1967) also reported several apparent cases of rabies among leopards. According to Q. Hendey of the South African Museum in Capetown, the limb bones of subadult F8 exhibited a pathological condition involving an osteitis of undetermined origin.

Natural violent deaths.   Leopards are sometimes killed by other predators, less often by prey, and rarely by other leopards. Leopard cubs left behind while mothers are hunting are probably most vulnerable to predation, but adult leopards are also no match for a lion, several hyenas, or a pack of wild dogs. During my study at least one adult leopard was killed by another predator.

Adult M23 was killed by a crocodile in the Sabie River. During the night of June 1, 1975, his distinctive tracks indicated he moved 5.4 km from his previous location. After visiting a nearby trap without finding any bait, he reversed direction 1 km and followed a dusty game trail to the Sabie River. He left the game trail, climbed down a muddy bank and walked to the edge of a pool toward some tall reeds. His tracks abruptly ended in a muddy disturbed area at the end of the pool where a struggle had apparently taken place (see fig. 5.3). After detecting an unusually rapid signal from his transmitter, I searched the area on foot but could not locate him. After dragging the water with a grappling hook without success, I finally located his body as I probed under the bank. It had been cached by a crocodile about 0.6 m underwater, 0.5 m from the edge of the bank (see fig. 5.4). Several tooth holes were in one shoulder, and one hind leg had been torn off. Whether the leopard attempted to drink at the pool or

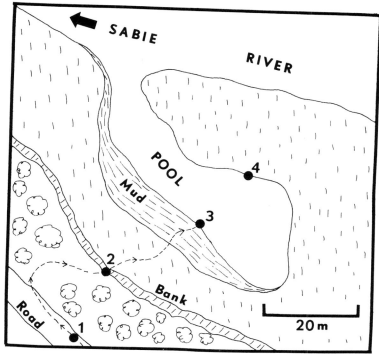

FIGURE 5.3 Locations (numbered black dots) and movements (dashed line) of male leopard M23 before he was killed by a crocodile along the Sabie River. Leopard M23 traveled along firebreak road (1), left it and moved down an embankment (2), and walked over a mud flat (3) to the edge of the water, where his tracks disappeared. He was attacked by a crocodile either at the water's edge or while swimming across the pool. His carcass was found cached underwater under a bank on the opposite side of the pool (4).

was swimming or wading across the pool when attacked by the crocodile was unknown.

Another leopard (M19) was also eaten and may have been killed by a crocodile. Twelve days after he left his home range, M19's radio transmitter was recovered from the bottom of the Sabie River in an area he had never previously visited. Whether he was killed as he attempted to cross the river or whether he died beside the river and was scavenged by a crocodile was unknown.

Crocodiles are undoubtedly an important predator in KNP; all perennial rivers and pools harbor considerable numbers (Pienaar 1969). Individuals up to 4.3 m long are common, and they have been known to kill full-grown buffalo, giraffes, hyenas, and lions in the park. Crocodiles up to 4 m long inhabited the Sabie River adjacent to the leopard study areas, and one was often seen sunning on a rock where M19's radio collar was recovered. Turnbull-Kemp (1967) believed that where leopards were forced to drink in crocodile infested waters, crocodiles were a significant mortality factor.

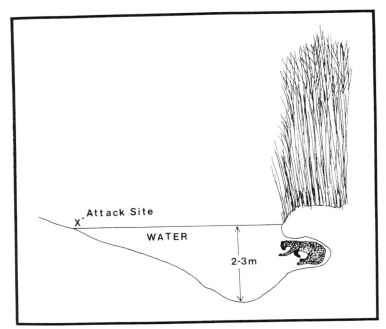

FIGURE 5.4 Cross section of the pool where the carcass of male leopard M23 was found, showing where the crocodile cached the leopard under the bank.

Leopard cubs are especially prone to predation. In KNP the remains of young leopards have been found in the stomachs of hyenas (Stevenson-Hamilton 1947), and a hyena was once observed carrying a young leopard in its mouth (Pienaar 1969). Turner described how several lions treed a female leopard and then devoured her two small cubs in Serengeti National Park, (Schaller 1972). Eaton (1974) believed the greatest mortality factor among cheetah cubs was predation. During my study, F11's cubs may have been eaten by a python that inhabited the same reed bed occupied by the young cubs. The snake, at least 3.5 m long. was first seen crawling from a hole only 4 m from their place of concealment. Thirteen days later, I encountered the same snake in the same place (in the female's absence) and nine days later, the female abruptly stopped using the reed bed. Because her movement pattern changed after this move and she began associating with resident M14 in the area, I assumed she had lost her cubs. Reticulated pythons in Asia have been known to eat leopard cubs, and an African python was once reported to have disgorged a mature leopard (Turnbull-Kemp 1967).

Mammalian predators, particularly lions, also kill leopards. Tourists in Serengeti National Park once found a pride of lions around the carcass of a female leopard that was bitten through the lower back and throat (Schaller 1972), and Turnbull-Kemp (1967) reported that lions occasionally killed leopards. In KNP Stevenson-Hamilton (1947) once saw a leopard treed by three lions, and on

June 6, 1974, I watched a pride of lions pursue a female leopard that fled up a tree. She remained in the tree for thirty minutes after the lions left before cautiously climbing down. Another time a lioness with blood on her fur walked out of cover at the base of a tree where a leopard was feeding on a bushbuck. Hyenas have also been observed chasing and treeing leopards (Turnbull-Kemp 1967; Smith 1962; Kruuk 1972). J. Mayer (personal communication) once watched three hyenas tree one of my radio-collared leopards, probably adult M14. The leopard jumped to the ground as Mayer's vehicle approached, but the hyenas pursued the leopard until it climbed another tree. The hyenas were still under the treed leopard when Mayer left the area. Packs of wild dogs also have been observed chasing leopards in KNP but have never been observed actually killing a leopard (T. Whitfield, personal communication; Stevenson-Hamilton 1947; Pienaar 1969). In Asia, wild red dogs will also pursue and tree leopards (Turnbull-Kemp 1967).

Leopards are sometimes killed by their prey. Marais (1939) described how two large male baboons killed a leopard they had previously injured, but one of the baboons was also killed. A seriously wounded leopard and another one that had been disemboweled were once found beside the carcasses of baboons in KNP (Stevenson-Hamilton 1947) Another leopard was seriously mauled by a baboon along the Lower Sabie Road (Pienaar 1969). Pienaar also reported that leopards that try to take baboons from troops during broad daylight could be severely mauled or torn apart by large male baboons. Once when I approached a trap containing a leopard, a troop of highly excited baboons ran away from the trap. The condition of the nearby vegetation and the highly agitated state of the leopard indicated the baboons had been harassing the captured leopard. On other occasions the alarm calls of nearby baboons informed me that a leopard had been captured in a livetrap.

Warthogs can also injure leopards and are probably capable of killing leopards. One leopard that attacked a large male warthog in the park was seriously injured (Pienaar 1969), and Stevenson-Hamilton (1947) saw a female warthog with young attack and tree a large male leopard. Wild bush pigs in India will also attack and tree leopards (Turnbull-Kemp 1967) as will other ungulates. A sambar deer once chased a leopard up a tree (Brander 1923), and a giraffe attacked two leopards that attempted to approach its calf (Selous 1920). Along the Lower Sabie Road, J. van Deventer (personal communication) once watched a large male impala rush head down toward an approaching leopard, which promptly stopped and then walked away.

Porcupines, whose quills are sometimes found in leopards (Stevenson-Hamilton 1947), may also be responsible for the deaths of leopards, and I observed how a ratel severely wounded a leopard. Poisonous snakes have not been reported to kill leopards but may do so. I once discovered a large cobra in a leopard trap and saw a puff adder near hidden F11's cubs.

Instances of leopards killing other leopards are rare, but leopards do occasionally fight. I observed only one fight, when a large male feeding on a kill in a tree was suddenly attacked by another smaller leopard. After a brief struggle, the attacking leopard climbed down. On another occasion an adult male leopard snarled threateningly at a subadult male, but no fight occurred. Both encounters I observed were associated with kills.

Leopards sometimes fight over land tenure rights. Corbett (1947) described an extended fight between the famous man-eater of Rudraprayag and (supposedly) another male. Corbett believed such fights were unusual and that most disputes were solved without fighting. Of eight encounters documented among radio-collared male leopards, Hamilton (1976) believed six resulted in possible conflicts, and on at least two of these occasions, one of the leopards was severely bitten.

Although many leopards in my study areas had scars, the cause of their injuries was unknown. All captured adults had scars, and four out of fourteen males (29%) and four out of sixteen females (25%) had been wounded within several weeks before their capture. In four of the eight injuries (50%), the wounds were infected. Adult M23 had twelve puncture wounds on his neck and a long shallow laceration on a hind leg. Male 10 and M18 had minor lacerations on their hind legs but no apparent bite marks. Male 5 had three deep puncture wounds on a rear leg and wounds on his front legs. The distal 15 cm of his tail was missing and bleeding when I captured him. Although captured leopards have been known to bite their own tails when enraged (Hamilton 1976), this particular male was calm when previously captured and appeared to have been in a fight with another predator.

The four injured female leopards I examined had puncture wounds or lacerations on their shoulders and backs. One had bite wounds on her neck, another had similar wounds at the base of her tail, and three appeared to have bite wounds on their forelegs. One emaciated female (F16) later died, but her previously treated wounds were already healed. The three other wounded females were in good condition despite their wounds. During the postmortem examination of leopard F16, a broken but healed rib was discovered, and M23 had a broken but healed zygomatic arch on the left side of his face. Although leopards I examined had no eye damage, Turnbull-Kemp (1967) reported that blindness in one eye among leopards was not uncommon. Among ninety-two resident lions Schaller (1972) examined, six were blind in one eye, four had the terminal parts of their tails missing, and one had a missing ear—all probably from fighting.

Poaching.   Leopards in KNP and adjacent game reserves appeared fairly well protected from poaching during my study. However, poaching appeared to be significant on adjacent farmlands and native trust lands. During 1973 only two leopards were known to have been poached in the park (Kruger National Park

1974). One was captured in a foot trap near the western boundary at Pretorius-kop and was killed as it attacked a ranger. Although there is little information on the number of carnivores destroyed outside the park boundaries, of 347 lions ear-tagged in the park between 1973 and 1974, seven were shot by farmers or hunters outside the park within one year (Smuts 1982).

During my study, only one radio-collared leopard was killed by poachers. Male 9, captured east of Skukuza, never left his home range during 3.5 months of monitoring. He then suddenly disappeared and several weeks later his radio transmitter's signal was detected from an aircraft over a citrus estate adjacent to the park. A lengthy search of the estate revealed the collar, well hidden by man, under a dense mat of vegetation. Many wire snares had previously been confiscated nearby. Apparently poachers captured the leopard, skinned it, and hid the transmitter. The actual place where the leopard was killed (that is, whether inside or outside the park) was unknown.

Undetermined causes of death.   The causes of three confirmed deaths of radio-collared leopards were undetermined. One leopard (M19) was probably killed by a crocodile. Hyena-scattered remains and the damaged radio collar of M7 found in the adjacent Sabi-Sand Game Reserve in August 1974 suggested a violent death. Another leopard (M3), in excellent condition when last captured, probably also met a violent death.

Mortality Rate

Actual mortality data were available only for tagged leopards that were more than one year old. Cub mortality was estimated from known and suspected births and from known cub survival. Confirmed and probable instances of mortality were combined to obtain an estimate of total mortality over the two-year study period.

ADULT LEOPARDS

The average annual mortality rate estimated for all tagged adult leopards was 18.5% (see table 5.13). Adult male leopards exhibited a higher annual mortality rate (25%) than adult females (13.5%), and old adults had a higher mortality rate (23%) than prime adults (13%). Thirty percent of the old adult males died each year, compared to only 17% of the old females. The average annual mortality rate for prime males was 17%, compared to 10% for prime adult females. Thus, each year, twice as many males died as adult females, and twice as many old adults died as prime adults.

Mortality rates differed among adults in each study area. In the SRSA three

TABLE 5.13 *Annual Mortality Rate of Radio-collared Leopards*

Age	Sex	Dead Leopards			Total Leopards Captured	Annual Mortality Rate [1]
		Confirmed	Suspected	Total		
Adult	Male	3	1	4	8	0.25
	Female	3	0	3	11	0.135
	Both sexes	6	1	7	19	0.185
Subadult	Male	2	1	3	6	0.25
	Female	2	2	4	5	0.40
	Both sexes	4	3	7	11	0.32

[1] One-half of the mortality rate for the two-year study.

of six adults died in two years, but in the NRSA only two of eight adults died in two years. Two adults that died in the SRSA were male; no males died during the two years in the NRSA.

Although sample sizes were small, the mortality rates for adult leopards in the study areas appeared similar to those for other solitary predators. During four winters the annual mortality rate of adult cougars varied from 9% to 22% with a mean of 14.8%. Four of the five adults were killed by hunters (Hornocker 1970). The annual mortality rate for spotted hyenas more than one year old was 16.7% (Kruuk 1972), and Jonkel and Cowan (1971) estimated 14% of the adult black bears in their study area died each year. Annual adult mortality in bobcats, a smaller solitary felid, was 5.7% including man-caused mortality, and 3% for natural mortality only (Bailey 1972). Among social carnivores, Schaller (1972) estimated an annual mortality rate of 5.5% to 12.5% for adult lions, and Mech (1970) calculated that about 22% of gray wolves more than two years old died annually. Schaller (1972) described how an injured lioness was saved from starvation by eating food killed by other members of the pride during a nine-month period. A similarly injured leopard would probably have died from starvation.

SUBADULT LEOPARDS

An average annual mortality rate of 32% for tagged subadult leopards was nearly twice as great as for adults. About 40% of the subadult females died each year, compared to 25% for subadult males. In two years four of five subadults (40%/year) tagged in the SRSA died, as well as two of four (25%/year) tagged in the NRSA. The high mortality rate among subadult leopards may have been related to their poorer hunting success. Of seven deaths among tagged subadult leopards, five (71%) were attributed to starvation and two to undetermined causes. Four of the five subadults that died from starvation were females, and two that died from undetermined causes were males. Thus the mortality pattern of subadult leopards was similar to that of adults; most females died of starvation and most males died from other causes, particularly natural violence.

Subadult males were probably exposed to more dangers than females because they wandered widely in unfamiliar surroundings. Four of six radio-collared males wandered sometimes up to 30 km outside their normal home ranges, whereas subadult females were sedentary. With smaller home ranges, however, subadult females would be more vulnerable to food shortages than males. Males could take advantage of seasonal changes in prey distribution, but females had to survive on whatever prey remained in, or were resident in, their home ranges. Supporting this view was the fact that most exploratory movements of subadult males occurred during wet seasons when impala were widely distributed.

CUBS

I assumed that females had 2 cubs/litter at birth and estimated the cubs' survival by the number of large cubs later captured or known to accompany females. The minimum average annual mortality rate for cubs, based on these assumptions, was 50% in both study areas (see table 5.14). More cubs may have been born than I estimated, but it is unlikely that more survived.

Leopards seldom appear to rear more than 1 cub/litter to maturity despite litter sizes of two to three at birth. Stevenson-Hamilton (1947) also noted that leopards seldom were accompanied by more than one well-grown cub and that it was exceptional for more than 1 cub/litter to reach maturity. Turnbull-Kemp (1967) believed mortality among leopard cubs was high except under extremely favorable circumstances.

Mortality rates among cubs are generally high for most carnivores. The estimated annual mortality rate for lion cubs in Serengeti National Park was at least 67% (Schaller 1972) and at least 50% for lions in KNP (Pienaar 1969). Cheetah cub mortality was estimated from 43% (McLaughlin 1970) to 50%

TABLE 5.14 *Estimated Mortality Among Leopard Cubs Less than One Year Old*

	Study Area						
	Sabie River			Nwaswitshaka River			Average Annual Mortality
Year of Birth	Cubs		Mortality Rate	Cubs		Mortality Rate	
	Born	Surviving		Born	Surviving		
1973	2[1]	1[2]	0.50	6[3]	3[4]	0.50	0.50
1974	—	—	—	4[5]	2[6]	0.50	0.50
Average	2	1	0.50	10	5	0.50	0.50

Assumes an average litter size of two at birth.

[1] Cubs of female F16's

[2] Large cub F20 captured

[3] Cubs of females F6, F17, and F25

[4] One cub observed with F6; large cub F15 captured (assumed F17's); tracks of one small cub observed with F25

[5] Cubs of females F11 and F21

[6] Radio-tracking suggested F11 lost her entire litter; F21 still had her cubs when study was completed.

(Graham 1966; Pienaar 1969), and among wolves, pup mortality may be from 28% to 88% depending on food sources and rate of exploitation (Mech 1970).

Leopard Population Regulation

The following characteristics suggested that the breeding population of leopards in the study areas was self-regulated (Hornocker and Bailey 1986): (1) food resources were more than adequate to maintain a stable population of breeders, (2) the number of breeders remained stable during the period of observation, and (3) the density of breeding female leopards was relatively low, but stable compared to the total population. Stability of the breeding population was achieved despite a high rate of turnover in the population because the turnover was confined primarily to cubs and subadults. Furthermore, stability was achieved with little apparent fighting among breeding resident leopards.

ADEQUACY OF FOOD RESOURCES

Abundant prey was available to breeding resident leopards during the study period. In the SRSA an average of 2.9 breeding resident leopards had 2,430 medium- to small-size ungulates available per year, or 838 ungulates per leopard. If only impala were considered, each breeding resident had 808 impala available per year. Based on an annual kill rate of 52 large prey/leopard/year, there were sixteen times more ungulates available to breeding residents than they actually killed. If prey biomass and daily kill requirements of adult leopards were considered, about sixteen times more prey biomass were available than breeding resident leopards actually killed.

Prey in excess of the breeding resident leopard requirements was also available in the NRSA. An average of 7.7 breeding residents had 5,615 medium- to small-size ungulates per year, 5,230 of which were impala. Thus the 729 ungulates or 679 impala available per leopard per year was fourteen times the minimum number of ungulates they required. Prey biomass was also twenty-six times greater than that killed by resident leopards each year. If all leopards more than one year old were considered, the number of impala in the SRSA and NRSA still exceeded that required by a factor of 8 and 10, respectively. Similarly, when prey biomass was considered, it exceeded that killed by all leopards by a factor of 9 and 21 in the SRSA and NRSA, respectively.

Leopard-impala ratios were lower but probably still adequate when impala populations declined in the SRSA (1975 = 960 impala) and NRSA (1973 = 3,240 impala). During low-impala-years the resident leopard:impala ratios were 1:331 and 1:421. For all leopards more than one year old, the leopard:prey ratios were 1:181 in the SRSA and 1:334 in the NRSA. Even if resident lions and spotted hyenas were taken into consideration, sufficient prey appeared to be available for resident leopards in both study areas. As previously discussed, however, not all leopards appeared capable of utilizing their prey resources because some died from starvation. This emphasized that only a proportion of the available prey was "catchable." Although experienced resident leopards in good condition caught sufficient prey, inexperienced young leopards and dehabilitated old ones did not. Food, therefore, regulated the total number of leopards the areas could support, but this factor acted primarily upon young or very old leopards. Thus the relationships between social behavior and the food supply were closely interwoven with adequate food resources supporting a stable population of breeding residents.

STABILITY OF THE RESIDENT LEOPARD POPULATION

The number of breeding resident leopards using the study areas remained relatively constant. Using known population parameters, I estimated the average annual rate of change was between -1% and $+3\%$/year for a hypothetical population of one hundred adults (see table 5.15). Although at least ten confirmed, perhaps fourteen, leopards died and fourteen cubs were born during the study, most of the confirmed or suspected leopards that died were young (six) or very old (six). Only one prime adult leopard died during the study. The number of resident leopards using the SRSA varied by only one individual each year of the study (1973 = 5, 1974 = 5, 1975 = 4). In 1973 no resident leopards died after the study began, and I assumed no deaths occurred that year before the study began. In 1974 one resident female (F16) died but was not replaced, and in 1975 another adult male (M23) died. Vacancies left by the deaths of

TABLE 5.15 *Population Trend in a Hypothetical Population of One Hundred Adult Leopards*

	Birth Interval	
	24 months	28.8 months
	n	*n*
Adults	100	100
Females [1]	64	64
Cubs [2]	64	54
Surviving subadults [3]	32	27
Surviving adults [4]	22	18
Adults dying [5]	19	19
Annual rate of change	$+3.0\%$	-1.0%

Based on population parameters documented in the Kruger National Park Leopard study areas.
[1] Adult sex ratio = 1.8 females/male
[2] Average litter size = 2 cubs/female
[3] Mortality rate for cubs = 0.50/year
[4] Mortality rate for subadults = 0.32/year
[5] Mortality rate for adults = 0.185/year

adults F16 and M23 could easily have been filled by other leopards present in the study area when the study terminated. Subadult M26 could have replaced old M23, and F24 could have replaced F16 had she herself not died. Adult M3, who disappeared from the study area, was apparently replaced by M30.

In the NRSA the estimated number of breeding resident leopards (nine) remained constant each year. This again suggested that space and resources were available for only a limited number of resident leopards and that social interactions determined which leopards in the population gained and retained breeding status.

The higher mortality rate of resident leopards in the SRSA could have been merely a consequence of that population's age structure. Sixty-seven percent of that area's adults were old, compared to 38% of those in the NRSA. During the study the older leopards in the SRSA were being replaced by younger individuals immigrating into the study areas from adjacent regions. The number of available younger leopards appeared adequate to compensate for the mortality of residents in the study areas. However, in situations where young leopards reaching maturity exceed the number of residents that die, they could either disperse to other areas or remain to compete with the adults for resources.

NUMBERS OF BREEDING RESIDENT FEMALE LEOPARDS

Resident female leopards, those sexually mature individuals producing young, made up only a small proportion of the adult population of leopards inhabiting the study areas. In the SRSA only three of ten leopards (30%) and in the NRSA six of twelve leopards (50%) were resident females. This limited the total productivity of the leopard populations in both study areas. However, as discussed in chapter 9, the land tenure system was such that sufficient numbers of "floaters," or nonbreeding individuals, were available to occupy vacancies left by the deaths of resident leopards.

DISPERSAL AND MORTALITY AMONG LEOPARDS

Although the exact social mechanisms regulating the density of resident leopards in the study areas were unclear, a substantial proportion of both study areas' populations were evidently composed of nonproductive individuals. These individuals did not contribute genetically to the area's population and were apparently prevented from doing so by the leopard's social system. Most nonproductive leopards died or dispersed before they had an opportunity to enter the area's social system. The proportion of nonproductive individuals, or floaters, in the population was probably influenced by the habitat adjacent to and the food supply within the study areas.

Because the leopard population throughout KNP was not exploited by humans, all available leopard habitats within the park were presumably occupied by resident leopards. Thus leopards dispersing from the study areas were unlikely to find major new unoccupied habitat within the park. Outside the park, excluding the private Sabi-Sand Game Reserve and similar reserves northwest of the leopard study areas, there was little opportunity for leopards to discover suitable habitat. Most of the lands outside the park were in private farms or native trust lands, where natural prey was scarce or absent. Leopards venturing outside the park did so at considerable risk from poachers and from land owners protecting their livestock. The chances of long-term survival for these leopards was probably low, as exemplified by the death of leopard M9, assuming he ventured outside the park.

The movements of dispersing leopards, usually males, within the park, suggested that they found adjacent park habitats already occupied by resident leopards. Their periodic return visits to their familiar home ranges strongly suggested that favorable unoccupied habitat was difficult to discover. This behavior, in conjunction with the strong attraction of most mammals to familiar areas (Ewer 1968), may have been responsible for the high number of floaters among the study areas' leopard populations. Also, if the adjacent habitats were already occupied by leopards, perhaps a disperser's chances of survival were higher in familiar, but already occupied, areas. If adjacent habitats had been available and unoccupied, dispersing males might not have returned repeatedly to their natal areas.

The study areas' abundant food resources may also have influenced subadult leopards to remain in or periodically return to their natal areas. But because of their inexperience, subadults' hunting success was probably much lower than that of adults. Young bobcats (Bailey 1972) and lions (Schaller 1972) are not as successful in capturing prey as are adults. Furthermore, mortality rates among dispersing carnivores are generally much higher than among residents (Storm 1965; Peterson et al. 1984). Therefore, in situations where dispersal is low, subadult mortality rates would be higher than those of the residents occupying the same area. This occurred in the study areas, where the mortality rate of subadult leopards was twice as great as that of adult resident leopards. In poorer quality habitat the mortality rate among subadults would probably be even greater.

Another aspect of dispersal was that the few vacancies left by the deaths of adult males in the SRSA appeared to be filled by newly arriving males rather than by males reared in the study area. Adult M3 was apparently replaced by M30, a newcomer, instead of M1 who was reared in the SRSA. A similar pattern was observed in the western portion of the SRSA. There, adult M23 was probably replaced by M26, another newcomer, instead of M2 who was reared in the area.

The rate of influx of new female leopards into the study areas appeared to be less than that for male leopards. Subadult female leopards were highly sedentary, compared to subadult males. Of six radio-collared male leopards, three explored distant areas while being monitored. None of three radio-collared females explored distant areas. Dispersing subadult males were also more strongly attracted to the SRSA than the NRSA. None of the nine radio-collared subadults dispersed out of the Sabie River region. Six subadults initially captured along the Sabie River remained in that region even while exploring, and they never explored deep within the NRSA. In contrast a subadult male captured well within the NRSA explored and later died in the SRSA.

The resident status of leopards was of utmost significance with regard to the relationship between food and social behavior in regulating leopard densities. Although the food supply established an upper limit to the total number of leopards inhabiting the study areas and acted primarily through mortality of subadults, social behavior among leopards determined how many subadults were to become residents. Resident status and a land tenure system determined the number and probably the characteristics of leopards that survived and contributed offspring to future generations. The importance of dispersing individuals, or floaters, within a leopard population should not be overlooked, however. Dispersers constantly replace residents that die in poor and high-quality habitats, and they colonize new habitats.

Estimate of Leopards in Kruger National Park

Kruger National Park probably has more leopards now than at any other time since it was proclaimed a reserve. The increase in numbers of leopards is undoubtedly related to the dramatic increase in numbers of impala throughout the park (Pienaar 1969). In 1915 only 6,800 impala were in the park. By 1968 the number had risen to 97,400. At the same time, brush encroachment into open habitat provided better concealment and stalking cover for leopards. A predator control policy, in effect until 1960, probably had little impact on leopard numbers because the few leopards that were killed were probably destroyed near the few scattered visitor camps and ranger posts. Between 1954 and 1960 only forty-two leopards were killed by park rangers, or 7 leopards/year (Pienaar 1969).

Intensively censusing leopards throughout the park was not feasible because of their secretiveness and preference for dense cover; therefore I assumed there was a relationship between numbers of leopards and their principal prey, the impala. From data obtained in the SRSA, I assumed each leopard more than one year old required 3.3 km^2, which contained an average of 298.7 impala. Prey biomass estimates for the SRSA indicated each leopard required an area supporting 15,047 kg of prey, 91% of which were impala.

Based on these assumptions, the park's leopard population was not likely to be underestimated, because impala were the most abundant prey of leopards in the park. Other prey, such as bushbuck, gray duiker, steenbuck, and warthog occurred at much lower densities than impala (Pienaar 1969). Smaller mammals, such as baboon, cane rats, and hares, even if numerically more abundant elsewhere than in the SRSA, probably did not exceed the biomass of impala, especially in the drier regions of the park. The effect of competition on leopards from lions, hyenas, wild dogs, and cheetahs was assumed to be similar throughout the park. Finally, I believed an estimate of leopard numbers based merely on land area would be inferior to the prey estimate method because such a population estimate would not account for differences in prey abundance.

During the dry season of 1975, I censused impala in each district of the park using the same census techniques I used in the leopard study areas. After estimating impala populations in each district, I divided the estimates by 298.7 impala/leopard to estimate the number of leopards more than one year old. To estimate the number of leopards less than one year old, I assumed 68% of the leopards more than one year old were adults, 64% of the adults were females, and 27.7% of the adult females had two cubs annually.

I estimated the park's leopard population to be 669 during the dry season of 1975 (see table 5.16). Forty-nine percent of the park's leopard population was estimated to occur in the central district and 28% in the southern district, and 23% in the northern district. My estimate of 160,987 impala for the entire park was slightly more than the official park estimate of 153,000 (S. Joubert, personal communication) for 1975, but the park's impala estimate only reduced the leopard population estimate by thirty-one individuals. Both estimates, however, were similar to the 650 leopards estimated for the park in 1963 (Pienaar 1963).

According to my estimates 50% of the park's leopard population inhabited perennial river riparian zones, even though riparian zones made up only 16% of the park's total area. Leopard densities were probably also higher around the park's seasonal rivers, dams, and waterholes because prey was more abundant and resident there throughout the year. In order of decreasing rank, the riparian zones of the Sabie, Crocodile, Olifants, Letaba, and Levubu Rivers appeared to support the highest impala populations and therefore the highest leopard populations (see table 5.17). Assuming the park's size was 19,166 km², the average density was 1 leopard/28.7 km², or 3.5 leopards/100 km². Leopard densities varied widely throughout the park. In the short grass plains of the central district and the low prey density areas in the northern district, each leopard may have required 23 to 167 km². Along the high-prey-density corridor along the Sabie River, densities averaged 1 leopard/3.3 km².

Relatively few estimates of leopard populations are based on documented biological data. Schaller (1972) estimated 1 leopard per 22 to 26.5 km² in Serengeti National Park, and in Wilpattu National Park, Eisenberg and Lockhart (1972) estimated 1 leopard/29 km². In Tsavo National Park, Hamilton (1976) estimated 1 adult/13 km². An estimated fifteen to twenty-five adult leopards probably inhabited the 800 km² Cedarberg Wilderness Area in the mountainous

TABLE 5.16 *Impala and Leopards in Kruger National Park in 1975*

District	Habitat	Impala				Estimated Leopards[2]			Area per Leopard (km²)
		Area (km²)	Number Observed	Length of Route (in kilometers)	Total Number[1]	Older than One Year[2]	Cubs[3]	Total	
Northern	Riparian	1,631	529	116.0	24,808	83	20	103	15.8
	Other	8,495	357	411.9	12,271	41	10	51	166.6
	Subtotal	10,127	886	527.9	37,079	124	30	154	65.7
Central	Riparian	824	1,154	105.2	30,129	101	24	125	6.6
	Other	4,693	966	157.2	48,064	161	39	200	23.5
	Subtotal	5,517	2,120	262.4	78,193	262	63	325	17.0
Southern	Riparian	669	551	48.4	25,179	84	20	104	6.4
	Other	2,853	485	112.3	20,536	69	17	86	33.2
	Subtotal	3,522	1,036	161.1	45,715	153	37	190	18.5
	Total	19,166	4,042	951.1	160,987	539	130	669	28.7

[1] $N = nA/2DR$ (N = estimated numbers, n = observed numbers, A = area, D = distance of census route, R = mean visibility limit (= 0.150 km in riparian and 0.300 km in other habitats.)
[2] Assume 298.7 impala/leopard older than one year; 68% of leopards are adults; 64% of adults are females; and 27.7% of the female leopards bear two cubs/year.

TABLE 5.17 *Impala and Leopards in Major Perennial River, Riparian Zones in Kruger National Park*

Major River	Area of Riparian Zone (km²)	Number of Impala	Older than One Year	Cubs	Total	Area per Leopard (km²)
Levubu	238	3,800	13	3	16	15.5
Letaba	673	8,000	27	7	34	19.8
Olifants	1,088	20,000	67	16	83	13.1
Sabie	560	50,000	167	40	207	2.7
Crocodile	389	35,000	117	28	145	2.7
Total	2,958	116,800	391	94	485	6.1

(Estimate is based on the relationship between numbers of adult leopard and numbers of impala in the leopard study areas (298.7 impala/adult leopard). Areas of riparian zones and impala numbers from Pienaar 1966b.)

region of the Cape Province in the Republic of South Africa (Norton and Henley 1987). No more than one hundred leopards were estimated to occur in the 9,500 km² Kalahari Gemsbok National Park in the northwestern desert region of the Republic of South Africa (Mills 1984). Thus the densities of leopards in the SRSA and NRSA in KNP were higher than those reported elsewhere. But the leopard densities I estimated for the entire KNP were similar to those reported for the Serengeti and Wilpattu National Parks.

Summary

1. Leopards captured in the study areas were classified, in order of decreasing age, as old adults, prime adults, subadults, large cubs, and cubs less than one year old.
2. Weights and total lengths of captured leopards indicated old adult males averaged 63.1 kg and 219.4 cm and females averaged 37.2 kg and 200.1 cm. Old males were 70% heavier but only 10% longer than old females. The largest captured male and female leopards weighed 70 kg and 43.2 kg, respectively.
3. Average weights of leopards indicated that females attained physical maturity more rapidly than males. Old males were heavier (8%) than prime males, but prime females were only slightly heavier (1%) than old females.
4. The average sex ratio of all captured leopards was 1.1 females/male but 1.8 females/male for adults only. Sex ratios based on visual observations were biased in favor of males apparently because of their greater boldness and the elusive behavior of females.
5. The population composition of captured leopards in the study areas was 63% adults, 30% subadults, and 7% large cubs. Cubs less than one year old were not captured. The female:male ratio ranged from 2:1 to 1:1 among old adults, 4:1 to 1:1 among prime adults, and 3:2 to 2:1 among subadults in the Nwaswitshaka River and Sabie River study areas, respectively.

6. The population density of leopards was 2.5 times greater in the Sabie River study area than in the Nwaswitshaka River study area. The density of adults was 1 leopard/6.1 km² in the SRSA and 1 leopard/10.5 km² in the NRSA. Of all leopards more than one year old the population was 1 leopard/3.3 km² in the SRSA and 1 leopard/8.3 km² in the NRSA.

7. Leopards in the study areas did not appear to successfully reproduce until they were approximately three years old.

8. Courting male and female leopards stayed with each other an average of 2.1 days/courtship period before separating.

9. Breeding success among leopards appeared to be low. Only two of thirteen (15%) suspected courtship associations among radio-collared males and females resulted in the birth of cubs.

10. Only 27.7% of the adult females in the study areas successfully produced young annually, and the interval between litters averaged 28.8 months (range = 24 to 36 months).

11. The annual potential increases in leopard populations was estimated to be 30% of the adult population in the Sabie River study area, and 37% in the Nwaswitshaka River study area.

12. Most cubs (five of six litters) produced by leopards were born during the early wet seasons; of these, three were born in December. The seasonal peak in leopard births coincided with the peak in impala births. Abundance of impala fawns as prey and abundant cover for concealment of cubs appeared important in the timing of births of leopard cubs.

13. At death, 43% of the leopards were old adults, 43% were subadults, 7% were prime adults and 7% were large cubs. Starvation was the only cause of death documented among radio-collared female leopards, and starvation was responsible for the deaths of 25% and 33% of the radio-collared adult and subadult males, respectively. Most radio-collared male leopards died from violent causes.

14. Lack of hunting experience among young leopards, and poor condition and physical handicaps among old leopards appeared to be principal reasons leopards in these age classes died of starvation.

15. Before their deaths, four emaciated female leopards lost an average of 0.6 kg/day (range = 0.2 to 1.8 kg/day).

16. The greatest loss of weight among emaciated leopards occurred in the late wet and early dry seasons, when impala were widely dispersed, low in numbers, and difficult to stalk because of poor stalking cover.

17. Mange (*Notoedres cati*) was a significant external parasite in leopard populations of both study areas. Sixteen of thirty (53%) captured leopards had mange before or contracted mange during the study. Mite infestations were believed related to nutritional deficiencies among leopards and seasonal changes in humidity.

18. Many ticks were found on captured leopards. Six species were identified from 223 ticks sampled, the most common being *Rhipicephalus simus simus* (36%) and *R. appendiculatus* (24%). Sex and numbers of ticks on leopards varied with species of tick and study area. Endoparasites of leopards

included *Taenia spp., Paragonimus spp., Schistosoma spp.,* and an Isosporan-type coccidial parasite.

19. Crocodiles killed one, perhaps two, radio-collared male leopards in the Sabie River. Although most adult leopards exhibited scars and wounds, these injuries did not appear to seriously handicap leopards. Poaching did not appear to be a serious mortality factor during the study period, at least in the vicinity of the study areas.

20. The annual mortality rate for radio-collared adult leopards averaged 18.5%. Twice as many adult males died as adult females, and twice as many old adults died as prime adults.

21. The annual mortality rate for radio-collared subadult leopards averaged 32%. More subadult females died annually than subadult males.

22. The annual mortality rate of small cubs in the study areas was estimated to average at least 50%.

23. The leopard population in both study areas appeared to be self-regulated because food resources were adequate and the numbers of breeding adults stable. The loss of adult females appeared to be compensated for by females produced within the study areas. The loss of males was compensated for by an influx of younger males from adjacent areas.

24. The leopard population in Kruger National Park during the 1975 dry season was estimated to be 669 leopards. Fifty percent of the population was estimated to inhabit only 16% of the park's total area, adjacent to perennial rivers. The overall density of leopards in the park averaged 1 leopard/28.7 km^2. Throughout the park's diversity of habitats and prey communities, leopard densities were estimated to range from 1 leopard/6.6 to 166.6 km^2.

6

Activity and Habitat Use
Patterns of Leopards

THE feasibility that an animal will follow a routine activity and habitat use pattern depends to a large extent on local environmental conditions such as day-to-day variations in weather, interactions with other species, social factors, and the availability of resources (Leuthold 1977). Flexibility should be viewed as adaptive, allowing an animal to take advantage of its changing environment. Although leopards often appear nocturnal, in Wilpattu National Park, Sri Lanka, where leopards are the dominant carnivore, they are often active during the day in open habitats (Eisenberg and Lockhart 1972). This suggests that an important factor influencing leopard activity and habitat use may be the presence of other large predators. In this chapter, I relate the activity and habitat use patterns of leopards to major environmental factors in the study areas.

Nocturnal Versus Diurnal Activity Patterns

Variations in the signals (Lord et al. 1962) from radio-collared leopards, when monitored for the first three to ten minutes of each daily observation, were used to classify activity patterns. This method indicated that leopards were more active during the night (78%, $n = 284$) than day (58%, $n = 1,397$). Differences between locations of leopards each morning, as opposed to morning and evening observations, also suggested that most long-distance movements occurred during the night. Although monitoring leopards only once a day sometimes indicated no net movements, more frequent monitoring re-

TABLE 6.1 *Leopard Activity Based on Visual Observations*

Form of Activity	Day		Night	
	Observations		Observations	
	n	%	n	%
Walking	39	62	24	38
Resting	30	97	1	3
Eating	12	63	7	37
Hunting	5	63	3	37
Hiding	6	100	0	0
Courting	1	33	2	67
Total	93	—	37	—

vealed that leopards were moving about during the night. This was especially evident for females with cubs, who stayed near their cubs during the day but moved at night, a finding also reported by Seidensticker (1977). In addition most of the impala killed by leopards were killed between sunset and sunrise. Leopards in Tsavo National Park were less active during the day (20%) (Hamilton 1976) than KNP leopards, but leopards in the Cape Province were most active during the day, with peaks in the late morning and late afternoon (Norton and Henley 1987). In the southern Kalahari Desert most leopards rested during the heat of the day (Bothma and Le Riche 1984). Most leopards (97%) observed during the day in Wilpattu National Park were moving (Eisenberg and Lockhart 1972).

Although leopards were most active during the night in KNP, 58% of the daytime radio observations of leopards indicated some form of activity. I visually observed leopards stalk prey, feed, scent mark, and vocalize during daylight hours. Thus, although leopards exhibited similar forms of activity during the day and night, the frequency of those activities increased significantly at night. This behavior may have been a reflection of the leopard's opportunistic method of hunting. If unsuspecting prey appeared near a resting leopard during the day, it often seized the opportunity and attempted to ambush or stalk the prey.

Forms of Activity

Although the radiotelemetry technique did not allow me to determine what leopards were doing, visual observations of leopards revealed significant differences in forms of activity between day and night (see table 6.1). During the day leopards were observed, in decreasing frequencies, walking, resting, feeding, hiding, hunting, and courting. During the night they were observed walking, feeding, hunting, courting, and resting, in order of decreasing frequency. Sixty-five percent of the night observations, but only 42% of the daytime observations were of traveling leopards; thus, traveling was primarily a nocturnal activity and resting a daytime activity. Feeding, hunting, and courtship also occurred most often during the night, although leopards sometimes engaged in these activities during the day. Other forms of activity, such as female family rearing behavior and play among cubs, were not observed. Because resting and traveling were the most frequently observed forms of activity, I describe them in greater detail.

RESTING

Resting was the dominant activity of leopards during the day. Leopards rested on the ground, in or adjacent to dense cover, or in trees. On the ground, leopards often rested in deep shade, especially during midday in wet seasons. Leopards resting on the ground usually lay on their sides with their front and hind legs stretched outward. Those resting in trees selected large limbs to lie on lengthwise, with their legs dangling over for stability (see photo 6.1). If the tree limbs were small, leopards often lodged themselves in a forked limb for support.

Resting leopards were alert and usually detected me despite my efforts to stalk them silently under cover. Several leopards in deep sleep, however, allowed me to approach them closely before they awoke and fled. In contrast to the lions I encountered when I was on foot, leopards were more alert and ready to flee quickly. Lions often watched me until I retreated, or they slowly got up

6.1. A female leopard lies stretched out on a limb of a tree on the bank of the Sabie River to escape the hot afternoon sun.

and walked away. When leopards detected me, they fled immediately, often running noisily through the brush to escape.

Resting leopards were not stationary for extended periods. The longest I observed a resting leopard remain stationary was about twenty minutes. Periodically they changed their body position or merely looked around. They immediately became alert if they heard an unusual noise such as the alarm call of a bird or mammal. During the day, leopards often lay in the cool sand in the shade at the edge of tall reed beds along the rivers. One leopard moved from one shady spot 50 m from the Sand River to another, perhaps cooler, spot within 10 m of the river as the day got hotter.

Leopards resting in trees changed positions frequently. Sometimes they only turned their heads to face another direction; at other times they changed their body's position. These slight but periodic movements of stationary leopards probably indicated a higher than actual activity rate when I monitored radio-collared leopards.

TRAVELING

Traveling leopards were probably searching for prey. But hunting leopards are difficult to see because they are usually concealed in cover. When hunting, leopards are seldom in a hurry; they frequently stop to look and listen and seldom travel great distances per hunt. During one week, M10 killed two impala within 1 kilometer, then rapidly moved more than 10 km to the opposite end of his home range. Females rearing cubs often killed their prey only several kilometers from their dens. Leopards probably did not have to travel far to encounter prey, especially in the SRSA where impala densities averaged 90/ km^2. However, despite this abundance of prey, leopards moved over large areas, apparently for reasons other than hunting.

Sometimes leopards traveled rapidly and gave little notice of nearby prey. This occurred when they traveled along roads, trails, riverbanks, or dry streambeds. That these traveling leopards periodically stopped to scent mark or vocalize suggested that they were patrolling their home ranges or making themselves conspicuous. In the darkness the approach of a vocalizing leopard was impressive. To remain stationary, listening while a vocalizing leopard approached, was difficult. Leopards did not vocalize while hunting. Only once, after F4 unsuccessfully attempted to stalk impala, did she vocalize.

Leopards also traveled in response to certain sounds, including the calls of other leopards. Once, several minutes after M14 began vocalizing, F11, who was nearby, joined him and they traveled together. Another time, after F4's vocalizations increased in frequency, she was joined by M3. Thus leopards were attracted to the calls of the opposite sex during the breeding periods, and movements then appeared to be solely for the purpose of locating potential mates.

Leopards may occasionally travel to drink water. Although I never observed a leopard drinking, tracks in the mud beside waterholes and along rivers suggested that leopards drank at such places. Leopards probably had little difficulty finding drinking water during wet seasons, but during the peak of the dry seasons leopards probably traveled long distances to drink. S. Joubert (personal communication) once observed M10 drinking at the Nwaswitshaka waterhole, probably the only source of drinking water in this leopard's home range during the dry seasons. On another occasion a male leopard I was observing left its kill and traveled to the nearby Sabie River, perhaps to drink, and returned several hours later.

Another reason leopards traveled was to explore new areas. This form of activity was characteristic of young leopards, although nearly every radio-collared leopard was located at least once in an area where it had never before been located. The daily locations of several exploring males (M1 and M7) suggested they traveled rapidly while exploring.

Leopards also traveled for reasons other than hunting, patrolling territories, drinking, and exploring. Females with cubs frequently traveled between their cubs' hiding place and kills; leopards fled from lions, hyenas, and wild dogs; and, as I discovered, they sometimes followed me, probably out of curiosity, as I traveled on foot through their home ranges. Reports of leopards inside fenced areas at Skukuza and other camps suggested that they periodically traveled to campgrounds, perhaps out of curiosity or perhaps to capture prey that had sought safety inside the fences.

Leopards traveled at variable speeds. Some traveled slowly, almost casually, whereas others traveled rapidly. I never saw a leopard run for more than about 50 m. I once followed an adult male traveling along the edge of a road at night. He scent marked three times along 1.2 km in twenty-four minutes, for an average speed of 2.9 kph. This was probably the average traveling speed of patrolling leopards. If a leopard traveled at this speed for only four hours, it could easily travel 12 km overnight. Turnbull-Kemp (1967) reported that a caged leopard paced back and forth 3.7 to 5.0 kph, a captive leopard traveled up to 5.8 kph, and wild leopards traveled 3.4 to 6.1 kph. He accurately timed an old male leopard eleven times that traveled comfortably at 4.8 kph. Hamilton (1976) reported that one male leopard traveled at least 10.9 km/day, and a translocated male moved up to 13 km/day.

Daily Activity Pattern

Leopards were most active after 5:00 P.M. but before 6:00 A.M. (see fig. 6.1). Activity of leopards sharply declined between 4:00 A.M. and 7:00 A.M., about sunrise; and leopards were least active between 7:00 A.M. and 8:00 A.M. Midday activity peaked between 10:00 A.M. and 11:00 A.M., then declined between

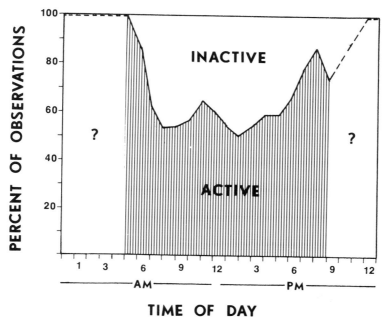

TIME OF DAY

FIGURE 6.1 Cumulative percentage of hourly observations of active and inactive leopards.

1:00 P.M. and 2:00 P.M., the hottest part of the day. Nighttime activity increased sharply between 6:00 P.M. and 7:00 P.M. and reached a peak between 7:00 P.M. and 8:00 P.M.

My own activity pattern probably biased the nocturnal activity information for leopards because I seldom monitored them between 10:00 P.M. and 5:00 A.M. However, some observations during the night suggested that leopards were often active, although a decline in activity occurred between 9:00 P.M. and 11:00 P.M.

One leopard, monitored for a twenty-four hour period and intermittently during other nights, provided information on its nocturnal activity pattern. When F11 fed on a kill on December 12 and 13, 1974, her active periods were between 4:00 A.M. and 5 A.M., 8:00 A.M. and 10:00 A.M., 3:00 P.M. and 4:00 P.M., and 10:00 P.M. and 1:00 A.M. Her major periods of inactivity were between 2:00 A.M. and 3:00 A.M., 5:00 A.M. and 7:00 A.M., 11:00 A.M. and noon, and 7:00 P.M. and 8:00 P.M. (see fig. 6.2). During the night she left the kill at midnight to attend her cubs, but returned at 4:30 A.M., forty-three minutes before sunrise. Another night she was inactive between 10:00 P.M. and 2:00 A.M., the same period she was most active only nine days earlier. When she fed on a kill at night, she was active 71% of the time. When she was not feeding, she was active only 23% of the time.

The observed feeding patterns of leopards also suggested that they were not

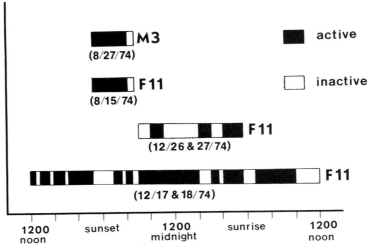

FIGURE 6.2 Hourly periods adult male M3 and adult female F11 were active.

active throughout the night. Leopards usually fed on carcasses as soon as it became dark. They usually fed once or twice before midnight, rested from midnight to early morning, and fed once again before sunrise. One leopard fed before 7:30 P.M. and after 1:00 A.M., but rested in between. Another fed in early evening, but rested on a nearby limb between 9:00 P.M. and 10:45 P.M. Male 3 rested near his kill most of the night and did not feed until 3:45 A.M. Aschoff (1966) reported that many nocturnal animals exhibited a double peak of activity, with the second peak being lower. In most species this pattern appears to be an independent property of the animal's behavior. Schaller (1972) reported that lions had two activity peaks during the night, one early in the night and another later in the morning, before dawn. In spotted hyenas the first nocturnal peak occurs between 10:00 P.M. and 11:00 P.M. and the second between 4:00 A.M. and 5:00 A.M. (Kruuk 1972). Hamilton (1976) reported a reduction in leopard activity and movements between 10:00 P.M. and 2:00 A.M.

Factors Influencing Activity

SEX, AGE, AND PHYSICAL CONDITION

Male leopards were more active than females during the night (see table 6.2). Differences in food habits and size, or competition with other nocturnal predators may have accounted for these differences. Male leopards were probably more capable than females of warding off hyenas and lions. Although a single hyena may take a kill from a female leopard, it may avoid adult male leopards. Hyenas and lions appropriated kills from female leopards in the study areas, but not from male leopards. I once saw a hyena following a safe distance (50 m)

TABLE 6.2 *Leopard Activity*

Sex	Day		Night	
	n	%	*n*	%
Male				
Active	762	55	154	82
Inactive	621	45	34	18
Female				
Active	635	62	67	70
Inactive	389	38	29	30
Total	2,407	—	284	—

Data based on radio observation.

behind M14, and M23 killed at least two hyenas near his kills. I could not detect any significant differences in daytime or nighttime activities of adult and sub-adult male leopards, and I observed no major differences between activity patterns of female leopards in good condition and those in poor condition. Physical condition had little influence on activity patterns of leopards more than 1.5 years old.

Reproductive status of female leopards influenced their nocturnal activity patterns. Females with cubs were active 83% of the time during nighttime observations, compared to 64% for those without cubs. Females with small cubs probably spent more time hunting and traveling between cached carcasses and their cubs' hiding places than females without cubs. During the day their activity patterns were similar. Females with cubs were active during 60% of the observations, compared to only 64% for those without cubs.

WEATHER

Temperature, cloud cover, precipitation, and wind velocity only slightly influenced leopard activity (see table 6.3). Although most movements of leopards occurred at night, when temperatures were lowest, neither very hot (31° to 41° C) nor very cold (0° to 10° C) temperatures caused significant changes in activity. Leopards sometimes moved about during midday when temperatures were extremely high. I saw one leopard moving through the brush at 1:44 P.M. when the temperature was at least 32.4° C, and Norton and Henley (1987) obtained an active radio signal from a leopard in the Cape Province, South Africa, when the air temperature in the shade was more than 30° C. Other observations, however, suggested that leopards avoided high temperatures in direct sunlight. During the hottest part of the day, leopards often lay in deep shade. Female 11 lay in deep shade panting 100 to 120 times/minute when the air temperature was 30.2° C. Hamilton (1976) reported that leopards in Tsavo National Park were more active from 6:00 A.M. to 9:00 A.M. and 4:00 P.M. to 7:00 P.M., than from 9:00 A.M. to 4:00 P.M., the hottest part of the day.

Although cloud cover appeared to have little influence on leopard activity,

TABLE 6.3 *Influence of Weather on Leopard Activity, 1973–75*

Weather Parameter	Active		Inactive	
	n	%	*n*	%
Temperature				
0–10° C	78	59	54	41
11–20° C	412	57	311	43
21–30° C	1,069	62	666	38
31–41° C	59	58	42	42
Cloud Cover				
Clear	673	58	492	42
Partly cloudy	443	64	253	36
Overcast	501	60	329	40
Precipitation				
None	1,534	60	1,015	40
Raining	84	59	58	41
Wind velocity				
0–5 kph	1,131	60	769	40
5–10 kph	301	60	204	40
10+ kph	186	65	100	35

few leopards were observed hunting on clear days. Of eight observed daytime hunts, on four the day was overcast, on three partly cloudy, and on one clear. Because cloud cover usually meant cooler daytime temperatures, it was difficult to separate their relative importance.

Moderate rainfall did not influence leopard activity. Several leopards were observed walking during light rainfalls, and movements of leopards indicated that they traveled considerable distances during nights when heavy rain occurred. Because intense rainfall was usually short, I had no information on leopard activity during such periods. However, just before a severe thunderstorm arrived, one leopard left a tree where he was feeding and retreated to cover. Winds had little influence on activity of leopards. High winds, which were uncommon, were usually associated with thunderstorms.

SEASON AND DENSITY OF COVER

Leopard activity during daytime was significantly related to season. During dry seasons, leopards were less active during the day, compared to wet seasons (see table 6.4). Density of stalking cover may have affected the diurnal seasonal activity patterns. During wet seasons, when cover and shade were abundant, leopards could move about, but still remain concealed during the day. During dry seasons, when cover density was greatly reduced, leopards would have little opportunity to move about unnoticed by prey or other predators.

Habitat significantly influenced the activity of leopards during dry seasons (see table 6.4). Leopards rested in riparian habitats more frequently than in thorn thicket habitats during the day. This difference may also have been related to stalking cover. Because riparian vegetation provided more cover and

TABLE 6.4 *Activity of Leopards*

	Dry season				Wet Season			
	Dense Riparian Cover		Sparse Thorn-Thickets		Dense Riparian Cover		Sparse Thorn-Thickets	
Activity	*n*	%	*n*	%	*n*	%	*n*	%
Active	325	54	440	61	329	61	523	64
Inactive	282	46	285	39	212	39	295	36

Dry season: riparian versus thorn-thicket: Chi-square = 6.91, *df* = 1, $P < 0.010$
Wet season: riparian versus thorn-thicket: Chi-square = 1.40, *df* = 1, $P < 0.010$
Dry versus wet season activity: Chi-square = 7.76, *df* = 1, $P < 0.010$

shade than the sparse thorn thickets during the dry seasons, leopards probably preferred riparian cover for resting.

The activity patterns of leopards may have also corresponded with daytime activity patterns of impala. The highest rates of activity were recorded from those leopards using thorn thickets during wet seasons, when impala also used them. Activity of leopards was low in riparian vegetation during wet seasons because those habitats were seldom used then by impala.

Habitat Availability

Because detailed vegetation maps were not available for the study areas, only two major habitats were assigned to each radio location of leopards: (1) riparian habitat adjacent to the major rivers and their tributaries and (2) *Acacia Combretum* habitat on the drier upland sites. Riparian habitat was characterized by an overstory of large evergreen trees and a dense shrub and grass understory. Dense reed beds (*Phragmites communis*) were also common along the banks of and on islands in the Sabie and Sand Rivers and in the dry, sandy riverbed of the Nwaswitshaka River. Regardless of season, shade and cover in riparian habitats were always available to leopards. I assumed that riparian habitat extended 100 m on each side of the Sabie and Sand Rivers, 30 m on each side of the Nwaswitshaka River and the Msimuku and Nhlanganeni tributaries, and 20 m from the remaining smaller tributaries.

The *Acacia-Combretum* habitat was characterized by smaller deciduous trees and an open understory. Because the *Acacias* and *Combretums* shed their leaves seasonally, less cover was available to leopards, especially during dry seasons. In the SRSA *Acacias* were more common than *Combretums*, but in the hilly uplands of the NRSA *Combretums* were dominant. I subtracted the area occupied by riparian habitat from the total study areas to estimate the amount of *Acacia-Combretum* habitat within each study area. Riparian and *Acacia-Combretum* habitats made up 16% and 84% of the SRSA and 4% and 96% of the NRSA. Although the NRSA had many small tributaries, riparian habitats made up proportionately less of its total area because of its larger size.

Leopard Habitat Use Patterns

The home ranges of female leopards encompassed almost twice as much riparian habitat as did the home ranges of male leopards. The percentage of riparian habitat within the home ranges of leopards averaged 9% (7% to 11%) for nine male leopards and 15% (12% to 19%) for four female leopards in the SRSA. But despite their proximity to rivers, female leopards were less apt than males to move across rivers.

In contrast to home ranges in the SRSA, those of four males in the NRSA averaged 5% (4% to 6%) riparian habitat and those of six female leopards 6% (4% to 8%). Although male and female leopards in the NRSA moved equal distances from riparian habitats, they seldom used *Acacia-Combretum* habitat far from the river.

Radio-collared leopards used riparian habitats significantly more often than expected. Data evaluation using the method outlined by Nicholls and Warner (1972), indicated all radio-collared leopards significantly favored riparian habitats (see table 6.5). No distinct pattern of habitat use by leopards of different sex or age classes was evident. However, the three highest-ranked leopards, those that strongly selected riparian habitats, were males from the SRSA. The three lowest-ranked leopards, those that selected the least riparian habitat, were two females from the NRSA and one male from the SRSA. The only comparative habitat data for leopards is that reported by Hamilton (1976) in Tsavo National Park, Kenya. Two of his most intensively monitored leopards favored woodland and brushland vegetation along rocky hillsides during the dry seasons. During wet seasons, this pattern of habitat selection was less evident. Hamilton speculated that the differences were due to the greater availability of shady resting places during the wet seasons. During dry seasons, leopards sought shady areas among the trees and brush of the rocky hillsides. Riparian habitats like those in my study areas were apparently uncommon in the Tsavo National Park leopard study area.

DAILY AND SEASONAL HABITAT USE

Leopards were seldom found far from dense cover, particularly during the daytime. Leopards usually avoided crossing large open areas, keeping near cover at the edge when traveling through such areas. When leopards crossed open areas in the daytime, they appeared nervous, frequently scanning their surroundings and quickly traveling to the nearest cover. During the night,

TABLE 6.5 *Location-Fixes by Habitat Type*

Leopard No.	Sex	Age	Number of Location-Fixes				Chi-Square Value [2]
			Riparian		Acacia-Combretum		
			Observed	Expected [1]	Observed	Expected [1]	
1	M	SA	147	17	121	251	1,061.5
2	M	SA	145	30	123	238	496.4
3	M	A	141	21	114	234	747.3
4	F	A	126	46	134	214	169.0
5	M	A	20	3	15	32	105.4
6	F	A	55	11	142	185	186.4
7	M	SA	60	18	88	130	111.6
9	M	A	11	3	34	42	22.9
10	M	A	32	6	85	111	118.8
11	F	A	44	9	160	195	142.4
12	F	SA	45	17	97	125	52.4
13	F	A	53	8	77	122	269.7
14	M	A	59	12	146	193	195.5
16	F	A	18	4	5	19	59.3
17	F	A	14	3	30	41	43.3
18	M	SA	30	3	36	63	254.6
19	M	SA	8	1	16	23	5.11
21	F	A	35	6	37	66	152.9
22	M	A	8	1	3	10	53.9
23	M	A	46	8	38	76	199.5
24	F	SA	17	2	6	21	123.2
25	F	A	6	1	10	15	26.7
26	M	SA	28	5	17	40	119.0

[1] Expected observations were determined by the method outlined by Nicholls and Warner (1972).
[2] All Chi-square values were significant at the $P < 0.050$ level.

however, leopards were more confident, and their tracks in the dusty roads and clearings indicated they frequently crossed open areas.

The leopard's use of riparian and *Acacia-Combretum* habitats also differed between day and night. Leopards in the SRSA used riparian habitats most often during the day (52%) and *Acacia-Combretum* habitats most often during the night (57%) (see table 6.6). These patterns suggested that leopards selected habitats with the densest cover during the day. Leopards in the NRSA also used *Acacia-Combretum* habitat more frequently at night (81%) than during the day (68%). The frequent use of *Acacia-Combretum* by leopards in the NRSA during the day (68%) compared to leopards in the SRSA (48%) was probably due to the greater

TABLE 6.6 *Location-Fixes During Day and Night*

Vegetation Type	Study Area			
	Sabie River		Nwaswitshaka River	
	Day	Night	Day	Night
Riparian	728	84	319	17
Acacia-combretum	685	111	675	72
Total	1,413	195	994	89

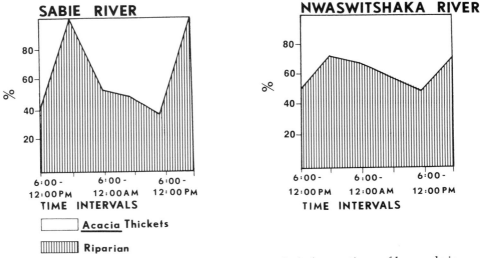

FIGURE 6.3 Cumulative percentage of time-period observations of leopards in *Acacia* thickets and riparian habitat types.

extent of this habitat in the NRSA. The sex of leopards did not influence their selection of habitat. Peak use of *Acacia-Combretum* habitats by leopards occurred between 6:00 P.M. and midnight (see fig. 6.3). This corresponded to a peak in leopard activity; apparently leopards initially selected open habitats in which to hunt. Lowest use of *Acacia-Combretum* habitat occurred in the early morning between midnight and 6:00 A.M., followed by a slow but gradual increase between 6:00 A.M. and noon. This suggested that the use of *Acacia-Combretum* habitats was correlated to leopard activity patterns. As leopard activity increased, use of *Acacia-Combretum* habitat also increased.

The pattern of daily habitat use by leopards indicated that they spent most of the day resting in dense riparian habitat. Leopards often cached carcasses of their large kills in riparian habitat in order to feed and rest in dense cover. As darkness fell, leopards became active, leaving protective cover and venturing into the more open *Acacia-Combretum* habitats to hunt impala. This probably occurred because impala selected open habitats at night as an antipredator response to leopards and lions. After hunting, and with the coming of daylight, leopards again returned to denser cover to spend the day.

Peak seasonal use of riparian habitats by leopards occurred when vegetative cover was minimal in the late dry and early wet seasons. Peak seasonal use of *Acacia-Combretum* habitats occurred during the middle of the wet season, when vegetative cover was most dense (see fig. 6.4). Increased use of riparian habitat at the end of the dry season demonstrated the importance of dense cover for resting. Similarly, the cover, shade, and cooler temperatures preferred by leopards during the wet seasons was not available in *Acacia-Combretum* habitats until after the rains arrived and vegetative growth reached a peak.

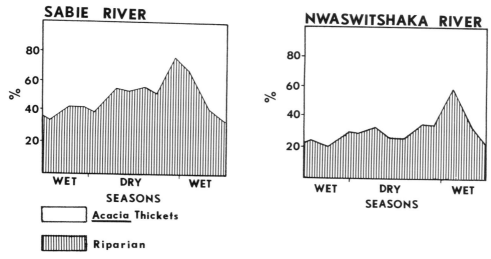

FIGURE 6.4 Cumulative percentage of seasonal observations of leopards in *Acacia* thickets and riparian habitat types.

A peak in riparian habitat use by SRSA leopards in September was probably a result of the atypical dry and wet seasons during the study. Most wet-season data for Sabie River leopards were collected in 1973–74, a year when early, extremely heavy rains fell after an exceptionally dry period. Most wet-season data for leopards in the NRSA were obtained in 1974–75. The rains that season arrived later and were less intense than the wet seasons of 1973–74. Because the rains arrived later, the response of vegetation was also late. Leopards in the NRSA delayed using the *Acacia-Combretum* habitats because of the delayed vegetation response.

Although leopards selected riparian habitat in their home ranges, they used habitats more uniformly during wet seasons, probably because cover was more widespread even in the *Acacia-Combretum* habitat. Leopards in the NRSA used *Acacia-Combretum* habitat more often during all seasons than leopards in the SRSA, so leopards were capable of adjusting to local environments. Thus, if riparian cover for resting was not available, leopards used the next best type of cover.

Leopards will frequently use rocky outcrops (koppies) if they are available. Hamilton (1976) reported that leopards in Tsavo National Park selected rocky hillsides and koppies, especially during dry seasons, and others (Turnbull-Kemp 1967; Pienaar 1969; Schaller 1972) noted use of koppies by leopards. Although koppies were scarce in my study areas, the NRSA encompassed several small ones in remote locations. Because of their remoteness, I was unable to monitor their use by leopards, but from reports elsewhere in the park I was aware that such koppies were frequently used by leopards.

INFLUENCE OF WEATHER ON HABITAT USE

Although habitat use was not significantly influenced by cloud cover, seasonal temperatures, wind velocities, or precipitation (see table 6.7), weather conditions had subtle influences on leopard habitat use. Leopards in the SRSA used *Acacia-Combretum* habitats more frequently on partly cloudy (54%) or overcast (55%) days compared to clear days (42%). Those in the NRSA used *Acacia-Combretum* habitat without regard to cloud cover. Because cloud cover and temperatures were related, use of these open habitats may have been associated with the cooler temperatures on partly cloudy and overcast days.

Leopards used open habitats (*Acacia-Combretum*) more frequently when temperatures were low; they used closed, shady riparian habitat when temperatures were high. Leopards were usually inactive during the hottest part of the day and often lay in deep shade in trees or on the ground. Once, on an extremely hot day, I approached within 5 m of F11, who lay resting under a shade tree. Her respiration rate was between 100 to 120/minute, even while resting. Resting leopards often shifted their positions to remain in the shade if the sun started to shine on them. Hamilton (1976) also reported that leopard activity was significantly related to temperature. These data suggest that leopards were intolerant to high temperatures and lay in direct sunlight less frequently than lions do (Schaller 1972).

Wind velocity had little influence on leopard habitat use. Although leopards in the SRSA used riparian habitat slightly more often on days with low (0–5 kph) wind (51%) compared to windy (10+ kph) days (48%), leopards in the NRSA exhibited opposite behavior. They used riparian habitat slightly more often on windy days (35%) than on days with low wind (31%).

Moderate rainfall did not significantly influence leopard habitat use. The

TABLE 6.7 *Location-Fixes Under Various Weather Parameters*

Weather Parameter	Riparian		Acacia-Combretum	
	n	%	n	%
Temperature				
0–10°C	49	37	83	63
11–20°C	330	46	393	54
21–30°C	719	41	1,016	59
31–40°C	50	50	51	50
Cloud cover				
Clear	544	47	618	53
Partly cloudy	277	40	419	60
Overcast	327	39	503	61
Precipitation				
None	1,095	43	1,455	57
Raining	53	38	88	62
Wind Velocity				
0–5 kph	817	43	1,084	57
5+ kph	217	43	286	57

only pattern observed was a slight increase of use of *Acacia-Combretum* habitats in both study areas during rainfall (62%), compared to precipitation-free days (57%). During three days of continuous rainfall, one male leopard stayed in a dense thicket of palms along the Sand River. After the rains stopped he resumed his normal movements. But leopards did not avoid water; they periodically swam rivers and they traveled during rains. On one occasion a leopard jumped from a tree into a waterhole to avoid me and swam away.

Summary

1. Signal patterns from radio-collared leopards indicated they were more active during the night than day. But most forms of activities observed at night were also observed during the day.
2. Resting was the dominant form of activity during the day. Leopards usually rested in shady, secluded locations, especially during the hot, wet seasons.
3. Traveling activities included hunting, feeding, patrolling territories, drinking, and exploring new areas. The average speed of one male leopard patrolling his home range was 2.9 kph.
4. Daily activity patterns indicated that leopards were most active in the evenings (7:00 to 8:00) and least active in the afternoons (1:00 to 2:00). Activity increased abruptly between 6:00 and 7:00 in the evening and declined sharply in the morning between 4:00 and 7:00. Leopards also periodically rested during the night.
5. Male leopards were more active than female leopards during the night. Age of leopards did not significantly influence activity patterns. Females with cubs were more active during the night than females without cubs.
6. Seasonal variations in temperature, cloud cover, precipitation, and wind velocity had little influence on activity patterns. However, leopards appeared intolerant to high temperatures and often rested in deep shade. Movements during daytime increased on overcast or partly cloudy days.
7. Leopards were more active during the day in wet seasons than they were in dry seasons. Seasonal differences in leopard activity patterns may have been related to vegetative cover and the activity patterns of impala.
8. Habitats in the leopard study areas were classified as riparian or *Acacia-Combretum*. Riparian habitat was composed of dense grasses and tall evergreen trees along watercourses. *Acacia-Combretum* habitat on drier upland sites was more open and composed of shorter deciduous trees and shrubs.
9. Sixteen percent of the habitat in the Sabie River and 4% in the Nwaswitshaka River study areas were riparian habitat; the remainder was *Acacia-Combretum* habitat.
10. Riparian habitat in home ranges of leopards in the Nwaswitshaka River Study area averaged 5% for males and 6% for females. In the SRSA, it averaged 9% for males and 15% for females.

11. Leopards used riparian habitat significantly more often than expected, considering the amount available in their home ranges.

12. Leopards used riparian habitat significantly more often during the day than night. Use of open *Acacia-Combretum* habitats increased during the night and peaked between 6:00 P.M. and midnight. Use of *Acacia-Combretum* habitats increased with increasing leopard activity.

13. Seasonal use of riparian habitats by leopards peaked in the late dry and early wet season when overall vegetative cover was minimal. Seasonal use of *Acacia-Combretum* habitat did not peak until the middle of the wet season, when vegetative cover was at a maximum.

14. Use of habitats by leopards was most uniform during the wet seasons, when vegetative cover was widespread and dense.

15. Although weather did not significantly influence habitat use, leopards used the open *Acacia-Combretum* habitat slightly more often when it was cooler, there was more cloud cover, or when it was raining.

7

Movements and Home Ranges of Leopards

ALTHOUGH some mammals move seasonally between distant feeding or breeding areas, the majority move about within a familiar home range, where they find their way with speed and assurance (Ewer 1968). However, a species will sometimes exhibit both types of movements. Some lions in the Serengeti remain in the woodlands year round, preying on resident game; others follow migrating prey to their seasonal ranges on the plains (Schaller 1972). Individuals may also exhibit different types of movements as they mature. The young individual becomes familiar with an area as it moves about with its mother. Later, as it sexually matures, this familiar home range is often abandoned and the individual seeks new areas. Once a suitable area is discovered, the individual eventually becomes familiar with it and the area becomes its new home range.

A study of the movement patterns of leopards can, therefore, reveal environmental adaptations. Movements can reveal how leopards are influenced by prey or other leopards. And important characteristics of leopard habitat can be assessed by examining home range use and dispersal patterns. I collected data on leopard movements with this in mind, realizing that because of leopards' adaptability, the patterns I documented might differ from those of leopards inhabiting other areas.

Daily Movements of Leopards

I measured the linear distance between consecutive daily radio locations to obtain an index of leopard movements. These were not the actual distances leopards moved between one day's location and another. The relationship between this distance and the distance leopards actually moved was quite variable. A female (F11) I once monitored for twenty-four hours left a suspected kill, traveled 4 km to her cubs, and returned. A once-daily check on her location might have indicated no net movement, when she actually moved at least 6.5 km. In contrast, another leopard (M23) that was located 4 km from his previous day's location actually traveled only 4 km as revealed by his tracks in a straight, dusty road. In Tsavo National Park, Hamilton (1976) reported that a male leopard monitored hourly actually traveled two to five times the distance between his daily resting locations.

The greater the interval between radio locations, the greater the variability among distance measurements. This applied equally to all radio-collared leopards (see table 7.1). Part of this variability was due to the leopards' circuitous movement pattern. Leopards often traveled in circuits, passing by their previous locations every four to eight days. Because of this variation, I limited my analysis to only those distances leopards moved on consecutive days and excluded days leopards were stationary. Data were obtained on twenty-four leopards, monitored eight days to sixteen months per leopard (see fig. 7.1).

The average distance between daily locations of all radio-collared leopards was 1.7 km/day with 40% of the locations less than 1 km/day (see table 7.2). Ninety-five percent were less than 5 km/day; only 4 of 1,510 distances exceeded 10 km/day; and the maximum daily distance (adult M10) was 13 km. The average distance between daily movements of individual leopards varied 0.6 to 3.1 km/day. Four of twenty-four individual leopards averaged less than 1 km/day; twelve leopards averaged 1 to 2 km/day; and seven leopards averaged 2 to 3 km/day. Only one leopard's average exceeded 3 km/day. Leopards seldom remained in the same place on consecutive days; if they did, they were usually feeding on a kill. During the study, radio-collared leopards were stationary on only 86 (5%) of 1,596 consecutive daily locations.

Leopards in Tsavo National Park averaged 2.6 km/day with a range of 0.9 to 4.2 km/day (Hamilton 1976), nearly twice the average daily distance I recorded for KNP leopards. In Wilpattu National Park, Sri Lanka, Muckenhirn and Eisenberg (1973) estimated a subadult leopard traveled 5.5 km in twenty-four hours. In the southwestern Cape Province of South Africa, the majority (54% to 85%) of daily distances moved by three male leopards were less than 3 km (Norton and Henley 1987), but another male moved 17 km in two days and 24 km in three days (Norton and Lawson 1985). Although Turnbull-Kemp (1967) considered it unlikely that leopards moved more than 24 km/night, male leopards in the southern Kalahari Desert moved a maximum of 27.3 km (nonmating period) to 33 km (mating period) per day (Bothma and Le Riche 1984). Male leopards in the interior dune areas of the Kalahari Desert averaged 14.3 km/day, and females with cubs 13.4 km/day. But in the Nossob riverbed, female leopards with cubs averaged only 3.8 km/day.

TABLE 7.1 *Distances Between Locations of Leopards One to Ten Days Apart*

Distance Between Locations (in Kilometers)

Interval Between Locations (in days)	Subadult M1		Adult M3		Adult F4		Adult F13		Adult M14		Subadult M18	
	x	n	x	n	x	n	x	n	x	n	x	n
1	1.5	(137)	2.7	(154)	2.1	(156)	1.4	(55)	3.0	(105)	2.9	(11)
2	2.2	(70)	3.5	(59)	2.9	(63)	1.9	(25)	3.6	(38)	2.5	(21)
3	2.6	(22)	2.9	(18)	3.8	(15)	1.7	(15)	3.3	(18)	1.7	(10)
4	2.1	(10)	3.8	(10)	2.4	(5)	1.6	(10)	2.6	(9)	1.7	(5)
5	2.6	(6)	3.4	(3)	3.4	(3)	1.6	(1)	3.2	(6)	2.6	(4)
6	3.1	(5)	3.3	(2)	4.2	(3)	1.5	(8)	2.2	(2)	2.4	(3)
7	2.2	(1)	6.8	(1)	0.6	(2)	2.7	(1)	2.8	(4)	—	
8	1.7	(2)	—		—		1.8	(2)	2.2	(1)	3.6	(2)
9	2.1	(1)	5.2	(1)	—		—		—		4.8	(1)
10	—		—		2.3	(2)	3.5	(2)	4.3	(2)	4.3	(1)
10+	7.7	(2)	4.0	(1)	5.4	(3)	1.6	(3)	4.2	(2)	2.5	(5)

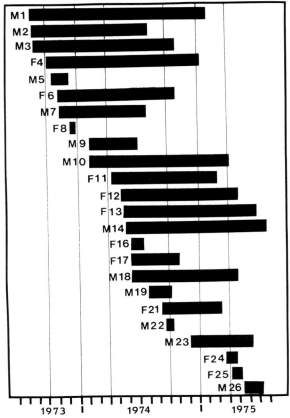

FIGURE 7.1 Periods that individual radio-collared leopards were monitored throughout the leopard study.

FACTORS INFLUENCING MOVEMENTS OF LEOPARDS

Sex.   Adult male leopards moved greater distances than adult females. Adult males averaged 2.8 km/day between locations, or 1.9 times farther than the average 1.5 km/day moved by adult females (see table 7.3). Frequency distributions of distances moved by adult males and females were also significantly different. Males moved more than 3 km/day 40% of the time, compared to only 8% for adult females. In contrast, adult female leopards moved less than 1 km/day 45% of the time compared to 21% for males. Why did male leopards move greater distances than females? If they hunted for the same prey, their movements should be similar. Might female leopards, like lionesses (Schaller 1972), be more successful hunters than males? Might males travel greater distances than females before making a kill? Although I had no data to confirm or refute this speculation, I believed that leopards spent only a fraction of their time actually hunting. Thus factors besides hunting, such as territory maintenance,

TABLE 7.2 *Daily Locations of Leopards*

Leopard No.	Sex	Age	Percentage of Observations per Distance Interval (in Kilometers)							Mean	
			(0–1.0)	(1.1–2.0)	(2.1–3.0)	(3.1–4.0)	(4.1–5.0)	(5.1–10.0)	(10+)	x	n
1	M	SA	39	36	18	2	5	0	0	1.5	(137)
2	M	SA	47	31	14	5	5	0	0	1.4	(193)
3	M	A	23	21	18	10	16	13	0	2.7	(154)
4	F	A	29	33	15	9	8	6	0	2.1	(156)
5	M	A	54	27	0	15	4	0	0	1.1	(26)
6	F	A	62	26	9	3	0	0	0	1.0	(115)
7	M	SA	39	27	13	6	8	3	3	2.1	(95)
8	F	SA	71	29	0	0	0	0	0	0.6	(7)
9	M	A	33	24	5	19	0	19	0	2.4	(21)
10	M	A	21	31	10	7	14	14	3	3.1	(29)
11	F	A	45	29	24	1	0	0	0	1.4	(141)
12	F	SA	51	40	6	3	0	0	0	1.1	(78)
13	F	A	38	38	22	2	0	0	0	1.4	(55)
14	M	A	17	23	15	20	10	15	0	3.0	(105)
16	F	A	56	22	11	11	0	0	0	1.3	(18)
17	F	A	63	13	25	0	0	0	0	1.3	(8)
18	M	SA	27	18	9	18	18	9	0	2.9	(11)
19	M	SA	33	56	11	0	0	0	0	1.4	(9)
21	F	A	73	19	8	0	0	0	0	0.9	(26)
22	M	A	60	0	20	20	0	0	0	1.4	(5)
23	M	A	58	33	5	3	0	0	0	1.0	(60)
24	F	SA	53	13	20	7	7	0	0	1.4	(15)
25	F	A	56	44	0	0	0	0	0	0.9	(9)
26	M	SA	27	32	16	11	3	11	0	2.2	(37)

Sex: M = male; F = female
Age: A = adult; SA = subadult

TABLE 7.3 *Daily Locations of Adult Leopards*

| Sex | Percentage of Observations per Distance Interval (in Kilometers) | | | | | | | Mean | |
	(0–1.0)	(1.1–2.0)	(2.1–3.0)	(3.1–4.0)	(4.1–5.0)	(5.1–10)	(10+)	x	n
Male	21	23	16	14	12	14	0	2.8	309
Female	45	30	17	4	2	2	0	1.5	501

Difference between means: Adult male and female leopards ($t = 8.26$, $df = 808$, $P < 0.001$).

locating mates, and rearing of young, probably accounted for the different movement patterns of male and female leopards.

Age.   Age significantly influenced the daily movements of leopards. Adult males averaged greater distances (2.8 km/day) than subadult males (1.6 km/day). By comparison, the frequency distribution of movements of subadult males was similar to that of adult females (see table 7.4), and there was no significant difference between their average daily movements. Adult female leopards averaged 1.5 km/day, but a subadult female (F12) averaged only 1.1 km/day. Subadult males moved significantly greater distances than subadult females.

Physical condition.   Leopards in poor physical condition (emaciated or infested with parasites) had different movement patterns from healthy leopards. The average distance moved per day by adult males in good condition was 2.6 times greater than males in poor condition. Healthy adult M10 moved farther (3.1 km/day) than any other radio-collared leopard. Emaciated male leopards averaged only 1.1 km/day, with most movements less than 1 km/day (see table 7.5). Like adults, subadult males in good condition moved farther each day than emaciated subadults.

Unlike males, the movements of female leopards in poor condition were similar to those of healthy females. However, an emaciated subadult female leopard (F8) moved only 0.6 km/day in eight days before she died. These data suggested that female leopards were moving minimum distances required to obtain prey, and that other factors were responsible for the greater movements of males.

TABLE 7.4 *Daily Locations of Adult and Subadult Leopards*

| Sex | Age | Percentage of Observations per Distance Interval (in Kilometers) | | | | | | | Mean | |
		(0–1.0)	(1.1–2.0)	(2.1–3.0)	(3.1–4.0)	(4.1–5.0)	(5.1–10)	(10+)	x	n
Male	Adult	21	23	16	14	12	14	0	2.8	309
	Subadult	41	31	15	5	5	2	1	1.6	173
Female	Adult	45	30	17	4	2	2	0	1.5	501
	Subadult	51	40	6	3	0	0	0	1.1	78

Difference between means: adult and subadult males ($t = 9.50$, $df = 780$ $P < 0.001$); adult and subadult females ($t = 2.39$, $df = 577$, $P < 0.050$).

TABLE 7.5 *Daily Locations of Leopards in Good and Poor Physical Condition*

Sex	Age	Condition	Percentage of Observations per Distance Interval (in Kilometers)							Mean	
			(0–1.0)	(1.1–2.0)	(2.1–3.0)	(3.1–4.0)	(4.1–5.0)	(5.1–10)	(10+)	x	n
Male	Adult	Good	21	23	16	14	12	14	0	2.8	309
		Poor[1]	55	32	8	4	1	0	0	1.1	100
	Subadult	Good	41	31	15	5	5	2	0	1.6	473
Female	Adult	Good	45	30	17	4	2	2	0	1.5	501
		Poor	57	24	10	6	2	0	0	1.2	49
	Subadult	Good	51	40	6	3	0	0	0	1.1	78
		Poor	57	24	10	6	2	0	0	1.2	49

[1] Poor condition = below average weight; ribs and pelvis conspicuous; pelage dull; often infested with numerous external parasites.

Difference between means: Adult males in good and poor physical condition ($t = 8.25$, $df = 407$, $P < 0.001$); subadult males in good and poor physical condition ($t = 3.67$, $df = 571$, $P < 0.001$); adult females in good and poor physical condition ($t = 1.80$, $df = 548$, $P < 0.100$); subadult females in good and poor physical condition ($t = 0.13$, $df = 125$, $P < 0.200$)

TABLE 7.6 *Daily Locations of Leopards*

Sex	Period	x	SD	n
		Mean Linear Distance Between Daily Locations (in Kilometers)		
Male	Peak breeding	3.0	2.1	97
	Nonbreeding	2.7	2.1	212
Female	Peak breeding	2.5	2.1	41
	Nonbreeding	1.8	1.5	178

Peak breeding period = July through September
Nonbreeding period = October through June
Difference between means: male leopards between peak breeding and nonbreeding period (t = 1.23, df = 307, $P <$ 0.400); female leopards between peak breeding and nonbreeding periods (t = 2.79, df = 217, $P <$ 0.010).

Reproductive condition.   Healthy adult male leopards and females without cubs moved greater distances during the breeding period (see table 7.6). Although males increased their movements between nonbreeding and breeding seasons from an average of 2.7 to only 3.0 km/day, females increased theirs 1.8 to 2.5 km/day. This suggested that female leopards moved greater distances than males during the breeding period. The behavior of at least one radio-collared female leopard supported this view. On the evening of July 29, 1974, adult female 4 called nine times between 5:30 P.M. and 6:42 P.M., outside her normal home range. Two days later, she moved even further outside and west of her normal range, and during the next four days she moved 4 km west along the Sabie River. She later returned to her home range and did not visit the outside area again. Her daily movements during the breeding season, up to 6.8 km/day, were highly erratic compared to her previous movements.

In the southern Kalahari Desert the maximum daily distance moved by male leopards was much greater during the mating season (33 km) than during the nonmating period (27.3 km) (Bothma and Le Riche 1984). Female tigers in estrous travel more widely than usual (Schaller 1967) and the movements of a female leopard appeared to increase during her estrous cycle (Hamilton 1976).

Female leopards with cubs moved significantly shorter distances each day than females without cubs. Three females without cubs averaged 1.9 km/day, whereas three females with cubs averaged 1.2 km/day (see table 7.7). Females with small cubs frequently moved less than 1 km, and they were seldom located more than 4 km, from their previous day's location. The average daily move-

TABLE 7.7 *Daily Locations of Female Leopards*

Status of Female	Percentage of Observations per Distance Interval (in Kilometers)							Mean	
	(0–1.0)	(1.1–2.0)	(2.1–3.0)	(3.1–4.0)	(4.1–5.0)	(5.1–10.0)	(10+)	x	n
Rearing young	55	27	16	2	0	0	0	1.2	282
Not rearing young	32	34	17	7	5	4	0	1.9	219

Difference between means: Female leopards rearing and not rearing young (t = 5.90, df = 499, $P <$ 0.001).

TABLE 7.8 *Daily Locations of Female Leopards*

Period	Mean Linear Distance Between Daily Locations (in Kilometers)		
	x	SD	*n*
Six months before parturition	1.4	0.9	105
Six months after parturition	1.1	0.8	117

Difference between means: Female leopards six months before and after parturition ($t = 3.00$, $df = 220$, $P < 0.005$).

ments of three radio-collared females (F6, F11, and F21) six months before parturition was 1.4 km/day, but only 1.1 km/day six months after parturition (see table 7.8). Young cubs apparently restricted the daily movements of female leopards. Females with cubs were usually located in the same places day after day. Female 11 hid her cubs in a patch of dense reeds along the Nwaswitshaka River, and when I approached, she refused to leave. Before her cubs were born, she immediately fled when I approached her. Sometimes, however, instead of staying with her cubs, she rested on a hillside 50 m away, where she could watch the patch of reeds from a distance.

Female F21 often stayed at a small koppie along the Nwaswitshaka River where her cubs were apparently hidden, and F6 hid her cubs in the same koppie used by F11 a year later. Female leopards usually left their cubs in order to hunt during the night. Radio-tracking confirmed that they were capable of traveling at least 4 km between their cubs and kills during the night. Lionesses are known to travel 5.6 to 8.0 km overnight before returning to their cubs (Schaller 1972). Female leopards frequently traveled between their cubs' hiding place and cached kills until the kills were consumed; other females remained near the kills between meals. This behavior probably varied with age of the cubs and distance to kills. Because cubs left behind might be killed if discovered by another predator, it would be to a female's advantage to hunt close to the cubs' hiding place so she could return to protect them when not feeding.

The behavior of one female (F11) was probably typical of a female's movements between her kill and her cubs more than 2 km away. Although she had a kill, I was unable to locate it because she fled when I was more than 100 m away. I monitored her activity and movements for twenty-four hours to determine where her cubs were concealed. She stayed near the kill from 10 A.M. to midnight, then left and traveled 4 km to the koppie where her cubs were apparently concealed. She remained with the cubs from 1:15 A.M. to 3:45 A.M., then, traveling a shorter 2.5 km route back to the kill, she returned to feed at 4:30 A.M., forty-five minutes before sunrise (see fig. 7.2).

Capture and encounters with humans. The movements of leopards one day after they had been immobilized was less than their preimmobilization movements. This difference was significant, however, only for adult males. Before

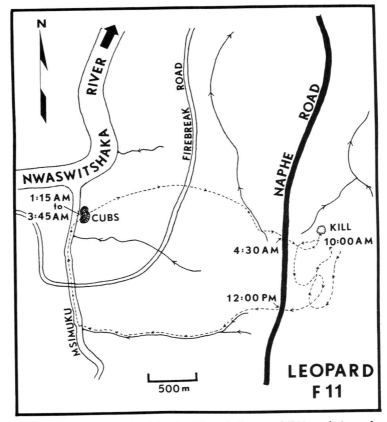

FIGURE 7.2 Movements (dashed line) of female leopard F11 and time she spent with her cubs and at a kill approximately 2.5 km from the koppie where her cubs were hidden.

capture, they moved an average of 2.0 km/day; one day after immobilization, they averaged only 1.5 km/day (see table 7.9).

Although immobilized male leopards moved less after release than nonimmobilized ones, this difference was not evident until two days after their re-

TABLE 7.9 *Mean Linear Distance (in Kilometers) Between Daily Locations of Immobilized and Nonimmobilized Leopards Before and After Capture*

| | | | | | After Capture | | | | | |
| | | Before Capture | | | Immobilized | | | Nonimmobilized | | |
Sex of Leopard	Number of Days	n	x	SD	n	x	SD	n	x	SD
Male	1	23	2.0	1.2	15	1.5	1.6	16	2.1	1.6
	2	17	2.2	2.1	10	1.1	1.0	5	3.6	2.6
	3	9	1.8	1.2	5	1.6	1.4	1	2.4	—
Female	1	9	1.7	0.9	9	0.9	1.1	8	1.9	1.9
	2	4	1.4	1.3	9	1.4	0.7	2	2.3	0.8
	3	2	1.1	1.2	7	1.1	0.6	1	3.1	—

lease. Then, nonimmobilized males averaged 3.6/day, compared to 1.1 km/day for immobilized males. Thus apparently immobilized and nonimmobilized males made an equal effort to leave the capture site the first day after their release. However, after the second day, nonimmobilized males continued their normal movements, whereas immobilized males traveled less, perhaps responding to the side effects of immobilization.

Several immobilized males stayed in the same location two to three days after their release. These locations were usually in dense riparian vegetation where they could rest undisturbed. Immobilized leopards were darted in the hindquarters, and the dart's impact may have stiffened their legs and restricted their movements. I did not know why female leopards behaved differently from males following immobilization.

Encounters between leopards and humans on foot had little impact on the leopards' movements, at least up to three days after such encounters. Usually the leopards saw me and fled before I saw them. Although male leopards traveled slightly greater distances the day after such encounters compared to the pre-encounter distance, the difference was not significant. The movements of female leopards were not influenced by such encounters.

Spacing Between Kills

Only adult male leopards moved significantly greater distances after leaving a kill compared to the pre-kill distances. Twenty-four hours before making a kill, adult males moved 1.5 km/day, but after leaving a kill site, they moved 4.7 km/day. This was more than four times the distance they moved the day before making a kill, and twice as great a distance they normally moved.

These data suggested that adult male leopards moved to more distant parts of their home range as quickly as possible after leaving a kill. A similar, but not significant, difference was also evident among movements of subadult male and adult female leopards. After consuming a carcass, leopards apparently left the site, regardless if other prey were nearby. Only a few exceptions to this behavioral pattern were observed. Male 10 once fed on an impala for two days and then killed another a day later only 300 m away. Male 26 once fed on two different impala carcasses only 700 m apart in four days.

Consecutive kills of adult male leopards were farther apart than those of other leopards. In addition, kills of all leopards were further apart in dry seasons than in wet seasons. The average distance between consecutive kills was 3.9 km ($n = 34$) for adult males, 3.7 km ($n = 63$) for subadult males, and 2.7 km ($n = 11$) for emaciated males. The distance between kills of female leopards averaged 2.5 km ($n = 37$) for adults and 1.8 km ($n = 17$) for subadults. Kills of adult male leopards were thus spaced 1.5 times further apart than those of adult females, and those of subadult males were 2.1 times farther apart than those of subadult females. The distance between dry-season kills was 19% to 75% greater than that of wet-season kills (see table 7.10). Because the greatest

TABLE 7.10 *Kills of Leopards*

Leopard		Season								
		Wet			Dry			Combined		
Sex	Age or Condition	n	x	SD	n	x	SD	n	x	SD
Male	Adult	19	3.6	2.1	15	4.3	3.4	34	3.9	2.7
	Subadult	32	3.3	2.5	32	4.0	2.7	64	3.7	2.6
	Emaciated	11	2.7	1.2						
Female	Adult	21	2.2	1.8	16	3.0	2.6	37	2.5	2.2
	Subadult	7	1.2	0.7	10	2.1	1.4	17	1.8	1.2

Linear distances (in kilometers) between consecutive confirmed and suspected kills determined by radio-tracking.

seasonal difference between kills of leopards was for subadult females, season may have had a greater impact on the hunting success of female leopards than male leopards.

The kills of older adult male leopards in the SRSA were randomly distributed. Most of resident adult M3's kills were randomly spaced along the Sabie River or the Sand River and its tributaries. Adult M23's kills were also randomly distributed (see table 7.11), but subadults M1 and M2 had randomly distributed kills only during the wet seasons. Male 2's kills were aggregated north of the Sabie River, near the center of his home range, during the dry seasons. The kills of adult F4 were aggregated near the confluence of the Sand and Sabie Rivers, and those of subadult F12 were in the eastern half of her home range during the wet seasons. The kills of the adult females were almost maximally spaced. Those of the subadult female were randomly spaced during the dry seasons.

The factors affecting the seasonal spacing of kills of leopards were difficult to determine. One factor may have been the availability of trees in which to cache kills. Leopards usually selected trees that provided maximum concealment of

TABLE 7.11 *Values of Spacing of Confirmed and Suspected Kills of Leopards*

Leopard No.	Sex	Age	Study Area	Season			
				Wet		Dry	
				n	R	n	R
3	M	A	Sabie River	10	1.09	—	—
23	M	A	Sabie River	10	0.74	—	—
4	F	A	Sabie River	17	0.74[1]	7	2.09[1]
1	M	SA	Sabie River	13	1.18	6	0.76
2	M	SA	Sabie River	21	0.96	14	0.69[2]
12	F	SA	Sabie River	8	0.63[2]	10	0.77
14	M	A	Nwaswitshaka River	5	0.53[2]	9	1.12

R = measure of spacing: in a random distribution R = 1; with maximum aggregation R = 0; with maximum spacing R = 2.15 (Clark and Evans 1954).
[1] Level of significance $P < 0.010$.
[2] Level of significance $P < 0.050$.

kills. During dry seasons, evergreen trees, which provided the best conceal-ment, were restricted to the narrow riparian zones, and leopards sometimes dragged carcasses as much as 0.5 km to such trees. However, spacing of kills was obviously influenced by other factors. Indeed, kills of leopards were often aggregated during wet seasons when cache trees were most widespread and available. The distribution of impala and stalking cover may also have con-tributed to the aggregated kills of leopards. Few impala were killed by leopards in the short grass, brackish flats regions of the SRSA. Only six (8%) of seventy-two known and suspected kills in the SRSA were in sparse stalking cover, but four (66%) of six wet-season kills occurred after impala concentrated on the flats. In the NRSA adult M14's kills were concentrated east of the Naphe Road, where impala were abundant. However, he seldom visited the area until impala moved in to feed on the new vegetation. Because of the lack of drinking water, few impala used this region during dry seasons.

PREY ABUNDANCE AND MOVEMENTS

Prey abundance has a profound influence on the movements of large African felids. In Serengeti National Park woodlands, where prey of lions was abundant year round, resident lions traveled only four to five km overnight and had small home ranges. But on the plains where prey was seasonally scarce, lions had large home ranges and few prides lived there permanently (Schaller 1972). In Kalahari-Gemsbok National Park, a semidesert where game is also scarce, lions travel up to 35 km/day and hunt areas of at least 274 km^2 (Eloff 1973).

The movements and home ranges of leopards also appeared to be affected by the abundance of prey. In the NRSA where average prey biomass was 15% lower than that in the SRSA, the average distance between daily locations of male leopards was 12% to 78% greater than for males in the SRSA. Similarly, subadult males in the NRSA traveled at least 2.9 km/day, whereas those in the SRSA traveled only 1.6 km/day.

The average distance between daily locations of female leopards without cubs was 52% greater in the SRSA than in the NRSA. The most likely explana-tion was a difference in habitat and prey population densities. The home ranges of most females in the NRSA encompassed prey-abundant riparian zones; those of females in the SRSA encompassed short grass areas where stalking cover was sparse.

Female leopards did not appear to respond to changes in prey abundance as rapidly as male leopards. After the September 1973 rains, when impala left riparian habitats along the Sabie and Sand Rivers, male leopards followed them northward, but F4 remained in her old home range south of the Sand River.

Exploratory Movements

Subadult leopards, especially males, made frequent exploratory trips outside their home ranges. Subadult M1 from the SRSA left his home range often, perhaps eleven times, in nine months; was monitored outside his home range up to twenty-four days; and traveled at least 9.8 km outside his western boundary. Subadult M2, also from the SRSA, left his area at least eight times during a six-month period.Unlike M1, however, he was only gone up to eight days per exploratory trip. Two of his exploratory trips, like those of subadult M1, were south of the Sabie River in an area abundant in prey. Another subadult (M7) explored outside his home range at least six times in ten months, was gone up to fifty days per exploratory trip, and traveled a straight line distance of at least 24 km. In contrast, subadult F12 left her home range to explore only three times while monitored and never remained gone longer than four days. Most subadult leopards that were monitored either died after a brief period, dispersed, or disappeared from the study area.

Subadult M7 died from an unknown cause outside his previous home range. Small pieces of decomposed skin were found near the radio collar, so he may have met a violent death. Subadult F12 was in poor condition but still inside her home range when I removed her radio collar.

Each subadult leopard that was monitored provided information on the frequency and duration of exploratory movements. Subadult M1 was captured August 18, 1973, and was located each day within a well-defined area for eight months. In late April 1974 he suddenly disappeared from the area and I was unable to locate him nearby. A month later, on May 20, he reappeared within his old home range but left again on June 19. I located him south of the Sabie River on July 2 several kilometers outside his home range (see fig. 7.3), where he remained at least four days, feeding on a suspected kill. He then left the suspected kill and explored an area south of the Sabie River. I monitored him regularly until July 24 when I again lost radio contact. On July 26 he returned to his old home range northeast of Skukuza. He remained there between July 26 and August 5, but from August 5 through January 23 frequent loss of radio contact suggested that he left his old home range at least nine times. At no time did his exploratory trips outside his home range exceed nine days.

I observed M1 for the last time on January 3, 1975. He was in excellent condition and feeding on an impala carcass within his old home range. I lost radio contact with M1 in February 1975, and because he was never again recaptured, I assumed that he left the area. Excluding his twenty-four-day exploratory trip, this male inhabited an area of at least 25.1 km^2. If the area included in the exploratory movements is considered, he used an area at least 74.1 km^2, three times larger than his previous home range.

The movements of subadult M7 were also characteristic of maturing male

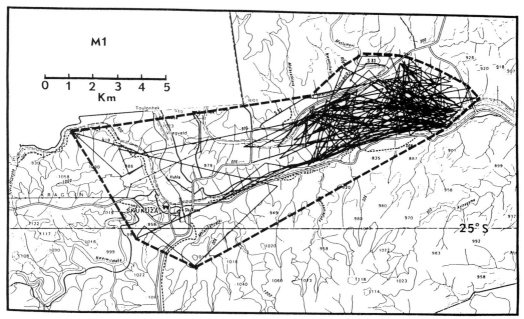

FIGURE 7.3 Exploratory movements of subadult male leopard M1 southwest of his usual home range between the Sabie and Sand rivers northeast of Skukuza, August 1973 to February 1975.

leopards. After being fitted with a radio collar in November 1973, he used an area of 44.6 km^2 west of Skukuza south of the Sabie River (see fig. 7.4). He left this area in December and spent seven days east of Skukuza before returning to his familiar range. After staying nineteen days, he left again and traveled east for five days, crossing the flood-swollen Sabie River on the evening of January 12. The following day, he also swam across the flood-swollen Sand River near the confluence of the Mutlumuvi. Three days later he recrossed the Sand River but stayed north of the Sabie River. I located him just across the Sabie River from his familiar home range where he remained from January 25 to February 13.

On February 15, M7 was again located south of the Sabie River but east of his former home range. He then returned and remained within his familiar home range from February 26 to March 12, 1974. On March 13, he again left his familiar home range, apparently retraced his previous route eastward, and over a fifty-day period moved as far as the Nkabeni Waterhole 24 km east of his home range. He remained near Nkabeni at least five days and then moved westward, reversing his direction of travel. I lost radio contact with him near the Mkhulu Picnic Spot on May 2, but on May 28 recaptured him inside his old home range about 30 km from Mkhulu.

Five days after his release with a new radio collar, he left again and spent at

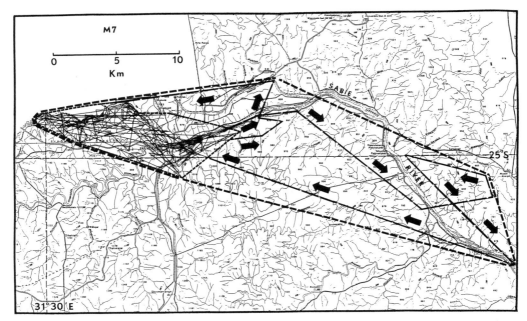

FIGURE 7.4 Exploratory movements (arrows) of subadult male leopard M7 east of his usual home range west of Skukuza, November 1973 to August 1974.

least thirty-two days 11 km east of his range, returning to his home range for five days, then leaving again on another exploratory trip. He stayed away seven days, returned for one day, and left again for at least twelve days. During the final seven days before his death, he crossed the Sabie River, probably on the night of July 30, and traveled west along the river until he was on the side opposite his familiar home range. He died or was killed on August 9 while he was still on the north side of the river. Thus M7 left his familiar home range to explore new areas at least six times, but returned to his familiar home range from at least 24 km away on five occasions.

The exploratory movements of subadult leopards had several common features: (1) leopards were at least one year old before they began to explore new areas, (2) the distances leopards traveled and the time they spent away from their familiar home ranges increased with age, (3) subadult male leopards traveled more extensively than subadult females, and (4) during the initial phases of exploratory movements and up to nine months thereafter, leopards periodically returned to their familiar home ranges.

Subadult M1 was probably twenty-two to twenty-eight months old and subadults M7 and M2 were probably eighteen to twenty-four months old when they first began exploratory movements. Male 7 and M1 were, therefore, about thirty-two and thirty-six months old, respectively, when they died or radio contact was lost. Male 2 was about twenty-eight to thirty-two months old, but still within his home range, at the time of his death. Female 12 was thirty-six to forty-two months old and also within her home range at the time of her last capture. These data suggested that some subadult male leopards in the study areas did not disperse from their natal ranges until they were about three years old, and at least one female did not disperse even after she was 3 to 3.5 years old.

The age of dispersing leopards appears quite variable and is probably related to the availability of resources and competition with resident leopards. Although this and other studies (Schaller 1972; Sunquist 1983; Le Roux and Skinner 1989) suggest leopards may become independent at 11.6 to 18 months of age, they may also remain in their natal range an additional two to twenty-three months before dispersing. The estimated ages of dispersing leopards range from 14.2 months (Le Roux and Skinner 1989) to 36 months (Muckenhirn and Eisenberg 1973; Bertram 1978). In prey-abundant areas such as my study areas, dispersal may be delayed, especially if adjacent habitats are already occupied by resident leopards. The presence of resident leopards may also explain why exploring leopards frequently returned to their natal areas. In the Londolozi Game Reserve near KNP, two litters of leopard cubs were last seen with their mother at 11.6 and 13.8 months, and a male and female sibling remained together an additional eleven weeks, until 14.2 months of age, before the male dispersed (Le Roux and Skinner 1989). He returned five months later, remained two days and left again. Female cubs from two litters did not disperse, remaining in the female's area for at least thirty months.

The distances explored and the time spent exploring areas outside natal home ranges increased with the increasing age of leopards. The dispersal process was apparently gradual, and older subadults roamed farther and spent more time in unfamiliar areas than younger leopards during each exploratory trip. Young subadult M2 stayed a maximum of only eight days outside his home range while exploring, but older M7 stayed at least fifty days. Similarly, M2's exploratory trips took him only 2.5 km outside his natal home range, but M7 traveled at least 24 km. Older M7's first documented exploratory trip lasted at least seven days, the second twenty-six days, and the third fifty days. The last two trips before his death lasted thirty-two and seven days, and immediately before his death he had been gone at least twelve days. Young subadult M2 spent at least 15 (4%) of 350 days exploring outside his familiar range; older subadult M1 spent at least 43 (8%) of 537 days; and oldest subadult M7 spent at least 137 (49%) of 274 days. In contrast, subadult F12 spent only 5 (2%) of 331 days exploring outside her home range. Exploring leopards also appeared to spend less time in their natal home ranges after each successive exploratory trip. The six consecutive periods M7 spent within his natal home range decreased from forty days to nineteen, fourteen, six, four, and one during his nine-month monitoring period.

Although the Sabie River usually functioned as a natural barrier, exploring

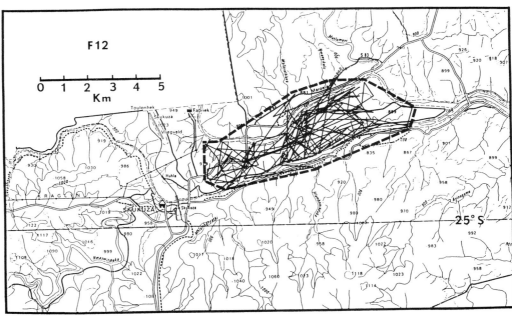

FIGURE 7.5 Exploratory and routine movements of subadult female leopard F12 in the Sabie River study area, May 1974 to May 1975.

subadults swam across it regularly, even during high water levels associated with floods. Male 7 swam the Sabie River at least five times and the Sand River twice. Three of the Sabie River crossings and both of the Sand River crossings occurred during the wet season when the river was very high. Resident leopards, however, crossed the river only during low water levels in dry seasons or on the concrete causeway spanning the river at Skukuza. Radio-collared resident leopards did not cross the Sabie River during this period. Male 2 did not swim across, but he used the Sabie River causeway to cross the flood-swollen river in December 1973.

The exploratory movements of the subadult female differed significantly from those of the males. Her exploratory trips occurred less often and did not take her far outside her home range. She crossed the Sabie River only on one occasion and then during the dry season when the river was low. However, she crossed the shallower Sand River at least four times but never went more than 1 km beyond her familiar home range. She also explored previously unvisited areas west of her home range instead of the regions across the Sand and Sabie Rivers (see fig. 7.5). Subadult female 12's home range overlapped that of adult F4, but she moved north and south between the Sabie and Sand Rivers, whereas F4 moved east and west. Later during the study, F12 also used an area seldom used by F4. This shift may have been caused by a vacancy left after the death of adult F16. Female 16's movements suggested that she once inhabited the area later used by subadult F12. Thus F12 may have avoided exploring unfamiliar distant regions because a vacant area was available nearby. This contrasted with the situation faced by subadult males in the SRSA. Two adult males, M3 and M23, already occupied the entire study area when subadult males M1, M2, and M7 began exploring distant areas.

Radio-collared subadult leopards frequently returned to their familiar home ranges after long absences. Because dispersal is usually an abrupt process, with individuals seldom returning to their natal ranges, one might ask if the exploratory trips of subadults were merely indicative of a local food shortage rather than a prelude to dispersal? However, if subadults left because of a local food shortage, why didn't radio-collared residents also leave? Adults M3 and F4 remained in the SRSA, but subadult M1 left to explore. Subadult M7 also left the same area occupied by adult M5 and passed through areas occupied by M9 and M14. The data suggested that the exploratory movements of subadult leopards were indeed a prelude to eventual dispersal. However, home range familiarity was an extremely attractive force opposing dispersal, as evidenced by leopards periodically returning to their familiar areas.

Movement Patterns and Home Ranges

ADULTS

Males.   Excluding exploratory movements of subadults and the occasional wandering movements ($n = 7$) of resident leopards, the movement patterns and home ranges of resident male leopards were significantly different from those of adult females and subadult leopards. Resident males had the largest areas and moved rapidly from one end of their home range to the other. In the SRSA M3 used a 37.5 km^2 area (see fig. 7.6). Although his home range was bounded on the south by the Sabie River, the narrower Sand River to the north did not obstruct his movements. He crossed the flooded Sand River during the wet season, so apparently the Sabie River did not prevent him from moving south.

The northern boundary of M3's home range was not well known because it lay beyond the range of ground radio reception. The eastern boundary of his range was defined by a tarmac tourist road, but the road itself was not a barrier.He frequently crossed a similar but more intensively traveled tourist road paralleling the Sand River. Although no physical barrier existed on the west side of his home range, he spent little time there. Instead, he spent most of his time in the eastern portion of his range and little time in its northern section.

Male 3's usual movements were in a clockwise direction, west along the Sabie River, north to the Sand River, east along the Sand River to its confluence with the Sabie River, and then west again along the Sabie River. Occasionally

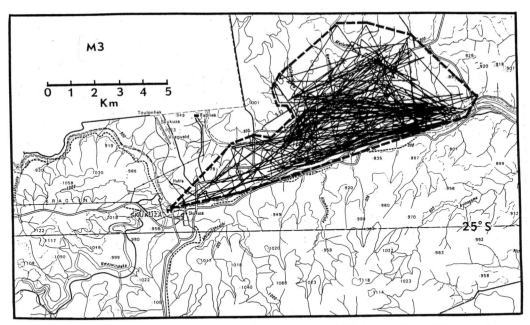

FIGURE 7.6 Movement pattern of adult male leopard M3 in the Sabie River study area, August 1973 to November 1974.

during the wet seasons he crossed the Sand River, instead of moving parallel to it, and spent several days north of the river before returning.

The rapidity of M3's movements in his home range was striking. He usually spent two to three days traveling a 6- to 7-km section of the Sabie River and two to three days along the Sand River before completing a circuit. During the dry season he would pass a given point within his home range about once a week. During the wet season the interval between circuits increased and became more variable because he preyed on impala north of the Sand River. During the wet seasons, M3 was often found at random locations along the Sabie River, but during dry seasons his movements from the eastern to western boundary of his home range could be monitored and his location accurately predicted daily.

The only other resident male leopard known to inhabit the SRSA during 1973–75, M23, had a home range of 16.4 km² between the Sand and Sabie Rivers west of M3's home range (see fig. 7.7). Although monitored for only seven months, this distinctive male was observed and photographed early in the study, and his presence was periodically reported to me by tourists and park staff. His movements were restricted to the south by the Sabie River, to the northwest by a game-proof fence between KNP and the Sabi-Sand Game Reserve, and to the north by the Sand River. No physical barrier prevented his movement to the east, where his range adjoined that of M3. The few times he was located north of the Sand River or south of the Sabie River he had probably

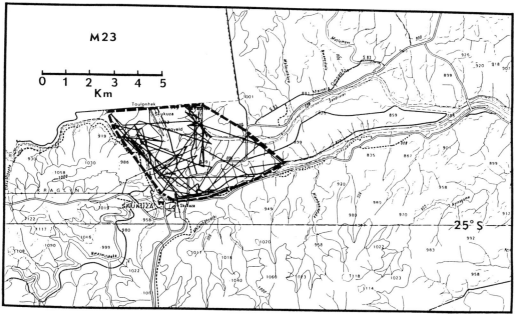

FIGURE 7.7 Movement pattern of adult male leopard M23 in the Sabie River study area, December 1974 to June 1975.

crossed via causeways. I once saw him trotting north on the causeway over the Sand River, followed by a vehicle, and received reliable reports of his crossing this causeway and the one over the Sabie River near Skukuza. The game-proof fence along the northwest boundary of his range, however, may have been a barrier; he was once unable to scale a similar fence surrounding the airport runway and had to be forced to leave through an open gate (N. de Beer, personal communication).

Male 23's movements along the north side of the Sabie River often visually exposed him to tourists at the Skukuza Rest Camp on the south side of the river. I periodically received reports of him as he passed by this particular section of his range because he made little attempt to conceal himself. Reports by park staff suggested that he inhabited this area at least several years prior to my study. His distinctive dark color, large size, and cropped ears left little doubt that most people were observing the same leopard.

The movements of M23 suggested that he spent more time in the eastern than in the western half of his home range. On several occasions he may have been attracted to refuse that had been uncovered in a landfill by hyenas. His presence nearby suggested that he may have scavenged for food more frequently than other leopards. Male 23 was an extremely bold and cunning leopard compared to the other study area leopards. He killed and ate spotted hyenas attracted to his kills and escaped from livetraps by sliding the door open

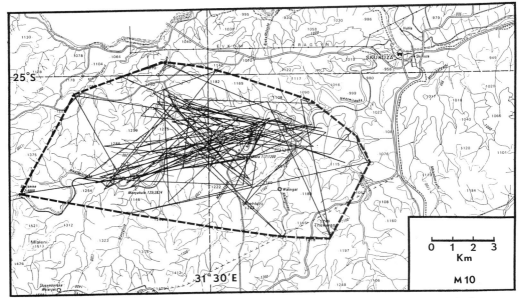

FIGURE 7.8 Movement pattern of adult male leopard M10 in the Nwaswitshaka River study area, February 1974 to April 1975.

from the inside. He was conditioned to the presence of humans and vehicles and was frequently reported on roads in the park near Skukuza. My admiration and respect for this leopard, as well as my occasional fear when I approached him on foot, were unmatched by my feelings for other leopards.

The movements of leopards in the NRSA were not obstructed by physical barriers. Adult M10 frequently moved across the Nwaswitshaka River, which was usually dry, and used an area of at least 96.1 km^2. This included most of the study area north of the Nwaswitshaka River, and was the largest home range used by a resident leopard (see fig. 7.8). Although he spent most of his time along the Nwaswitshaka River, he periodically traveled up its tributaries such as the Msimuku, Nhlanganeni, and Manyahule. He was once observed north of the Doispan Road stalking an impala and was located there on another occasion. Male 10 traveled a circuit similar to that traveled by M3, but in a counterclockwise direction.

Because M10's home range was much larger than that of M3, he took ten to fourteen days to complete a circuit. Normally, he traveled parallel to the river from east to west in two to three days. Once, however, he traveled 13 km, almost the entire length of his range, overnight.

Resident M14's home range, which lay east of M10's home range, was 56.4 km^2 and encompassed the remainder of the NRSA, plus an area south of the Sabie River east of Skukuza (see fig. 7.9). No natural physiographical features

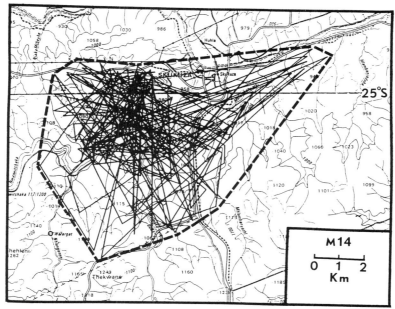

FIGURE 7.9 Movement pattern of adult male leopard M14 in the Nwaswitshaka River study area, May 1974 to July 1975.

influenced M14's home range other than the Sabie River at its northeastern corner. He seldom crossed north of the Kruger Gate road but frequently crossed the Naphe road. When first monitored he spent most of his time near the Nwaswitshaka River; later he enlarged his home range to the southeast. This expansion of his home range occurred during the wet season of 1974–75. Although he did not have predictable movement patterns like other resident males, he usually visited the northwestern boundary of his range once a week.

Limited data were also obtained for several adult male leopards outside the principal study areas. Male 5, captured south of the Sabie River west of Skukuza, had a home range of at least 15.2 km^2. During the brief period he was monitored, he traveled parallel to the Sabie River. On one occasion he completed a circuit by traveling east along the Sabie River, turned south, then moved west again, passing his starting point twelve days later.

Male 9, also captured south of the Sabie River east of Skukuza, stayed in an area of 29.3 km^2 before suddenly leaving. Later his radio collar was recovered more than 14 km away in a citrus estate outside the Park boundary. Because M9 was often located along the Lower Sabie tourist road, he was probably the leopard I frequently observed there early in the study. I initially suspected that M9 used the eastern half of the NRSA. Subsequent captures and monitoring revealed that the area was used by adult M9 and M14. Male 9 usually stayed within 2 km of the Sabie River, but occasionally moved 4.5 km south before

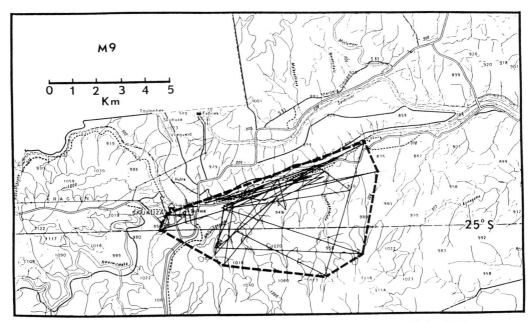

FIGURE 7.10 Movement pattern of adult male leopard M9 south of the Sabie River study area, February 1974 to July 1974.

returning (see fig. 7.10). Data on this male's movements were probably biased because he occasionally traveled south beyond range of radio reception and then could only be located by aircraft.

Another old male leopard (M22) was captured within the SRSA in the dry season of 1974. Upon his release, however, he immediately crossed the river, which was at its lowest level, and completed a circuit of a 3.7-km² area south of the Sabie River before radio contact was lost. This old, decrepit male apparently inhabited the area west of M9 and was probably the same decrepit male I had previously observed courting a female on the Lower Sabie River road near his ultimate capture location.

Females. The movements of adult female leopards were dependent on the mobility of their cubs. Females with small cubs usually moved outward in a radial pattern from a central location, harboring their cubs. During the day they stayed near their cubs, but in the evening and at night they left the cubs to hunt. Adult females also had smaller home ranges and less predictable movement patterns; they repeatedly used certain localities within their home ranges more often than males.

Most data from adult female leopards were obtained in the NRSA. Of six captured adult females, three (F6, F11, and F21) gave birth to cubs during their radio-tracking periods, and another (F25) probably had cubs two years earlier.

Female 13 did not rear cubs during the study period. In the SRSA F4 probably gave birth to cubs after her radio ceased to function; older F16 died before her entire home range could be determined; and F27, F28, and F29 were only ear-tagged because they were captured when the study was ending. Female 28 probably inhabited the northwestern corner of the study area; she was once seen and photographed north of the Sand River near her capture site. Female 29 was probably the mother of radio-collared subadult F12 and inhabited the central part of the study area. Old F27 apparently inhabited the area south of the Sabie River, outside the study area boundary, and was probably the old female I observed courting old M22. In 1976, more than a year after my study, F27 was seen south of the Sabie River along the Lower Sabie road (S. Joubert, personal communication).

Female 6 was pregnant when first captured in the NRSA in November 1973. Shortly after her release she moved 3 km northeast, where she gave birth to cubs. Although I once approached her on foot, I was unable to observe her. But a large termite mound nearby with several large aardvark burrows at its base could have served as a den. After this disturbance, she moved her cubs near the Msimuku where she stayed for six to seven months. I did not disturb her at the new den site.

After F6 left her original den area, I discovered that she was probably using a small koppie to hide her cubs. The same koppie was later used by another female (F11) with cubs. When F6 was using the koppie, I seldom located her more than 3 km away. Although I saw F6 on several occasions when her cubs were less than six months old, she was always alone. Finally, one day as I approached an impala she had cached in a tree, I saw one of her cubs climb a nearby tree (see photo 7.1). During the eleven-month period F6 was monitored, she used an area of only 12.3 km^2 near the koppie that concealed her cubs. If her first unlocated den was included in her home range, she used a 20.8-km^2 area (see fig. 7.11).

Another female leopard in the NRSA (F11) shared most of the area used by F6. While monitored from May through November 1974, F11 used an area adjacent to but south of the Nwaswitshaka River. In July and August she associated several times with adult M14 and on at least one occasion, vocalizations suggested that they copulated. In November her movements changed abruptly; she stayed in a dense patch of reeds along the Nwaswitshaka River. Her movements and response as I approached her on foot indicated that she had given birth to cubs and hidden them in the dense vegetation.

After she had cubs, F11, like F6, seldom moved more than 3 km from the patch of reeds (see fig. 7.12). But during January 1975 her movements suddenly changed again, and she was seldom found near her cubs' suspected hiding place. Because her cubs were too small to follow her, and because two other females (F6 and F21) seldom moved far when their cubs were of comparable age, I assumed that F11 had lost her litter. In February 1975, F11 and M14 were

7.1. A small cub of a female leopard (F6) climbs a tree near a kill to escape my presence.

located together again. Before this time, when she still had her cubs, M14 was never located near her, and only once in her absence was he located near the cubs' hiding place. Female F11's total home range, excluding wandering movements, was 13.9 km^2.

Female 21 in the NRSA had a home range southwest of F6's and F11's home ranges. F21, monitored from September 1974 to March 1975, spent more time in the southern portion of her range than fig. 7.13 depicts because she was often out of range of radio reception. From September to December 1974 she used a 5.6 km^2 home range; after she had cubs she used an area of only 1.8 km^2. After her cubs were born she spent most of her time hunting along the river's riparian vegetation zone.

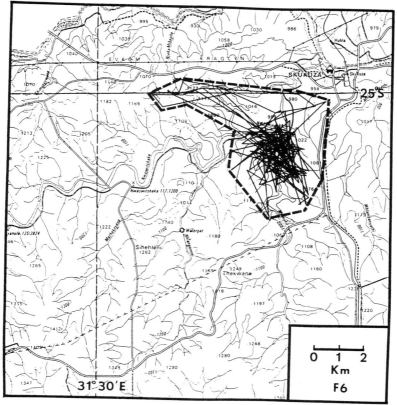

FIGURE 7.11 Movement pattern of adult female leopard F6 in the Nwaswitshaka River study area, November 1973 and November 1974.

Several females in the NRSA did not rear young cubs during the study. Female 13 inhabited a 10.4 km² area north of the Nwaswitshaka River (see fig. 7.14) and was neither lactating nor pregnant when captured. She followed a circuitous path, like a male leopard, but over a smaller area. She visited the southwestern and northeastern portions of her home range every ten days or less, stayed within 1 km of the Nwaswitshaka River, and often visited previously used resting places. She rarely traveled north of an intensively used tourist road that formed the northern perimeter of her home range.

Another lone female in the NRSA was F17. She usually traveled back and forth along the Nhlanganeni, a tributary of the Nwaswitshaka River. She probably spent more time in the center of her home range, near several koppies, than my data indicated. The eastern and western boundaries of her home range may have also extended farther than indicated (see fig. 7.15). Her 14.2-km² home range was the largest I documented for a female in the NRSA and probably reflected the relative scarcity of prey in the water-scarce Nhlanganeni region.

Female 17 apparently had cubs before my study began, because I later cap-

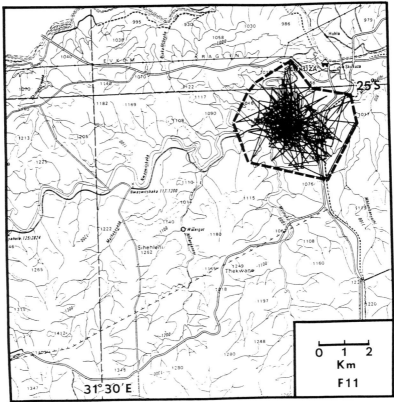

FIGURE 7.12 Movement pattern of adult female leopard F11 in the Nwaswitshaka River study area, April 1974 to March 1975.

tured a 1.5-year-old female leopard (F15) within F17's home range. Furthermore, the tracks of a young leopard were periodically seen in the riverbed of the Nhlanganeni. It was already traveling alone when captured and was probably born during the 1972–73 wet season.

Another adult female (F25) captured in the NRSA was monitored only briefly before she died. Her movements were confined to a 3.0-km^2 area around the Nwaswitshaka River waterhole, and she was never located more than 1 km away from the Nwaswitshaka River. She was probably the female with two small cubs who left tracks there during the wet season of 1973–74. If her cubs survived, they would have been at least two years old and traveling independently when I first captured F25.

Adult female leopard F4 was intensively monitored in the SRSA. She used an 18.0-km^2 area between the Sabie and Sand Rivers from October 1973 to January 1975. Although she never crossed the Sabie River during this period, she occasionally crossed the Sand River. She had predictable movements and passed the Sand River confluence with the Sabie River about every seven days.

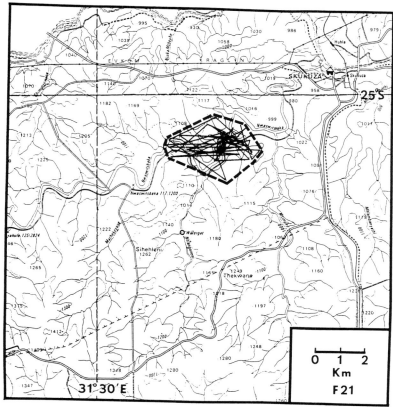

FIGURE 7.13 Movement pattern of adult female leopard F21 in the Nwaswitshaka River study area, September 1974 to March 1975.

Her movements were similar to M3, who used the same area. She typically traveled west along the Sabie River, north through thorn thickets to the Sand River, and then east along the Sand River to her starting point. Sometimes she reversed her direction of travel, covering the same route counterclockwise.

Female 4's movements changed during the dry season of 1974. She began to use another 11.9-km² area west of her former range—an area she had rarely visited. She also began to vocalize and associate with M3. On at least one occasion vocalizations indicated that she and M3 were copulating. Radio contact was lost three months later, and despite intensive trapping I was unable to recapture her. Females with small cubs avoided traps, so she may have had cubs.

An extremely emaciated old female leopard (F16) was captured and monitored only two months before she died near the Sabie River. During this period she stayed within 1 km of the north bank of the Sabie River. Just before she died, however, she crossed the Sabie River causeway near Skukuza. She died nearby, on the south side of the river. Large cub (F20) and older subadult

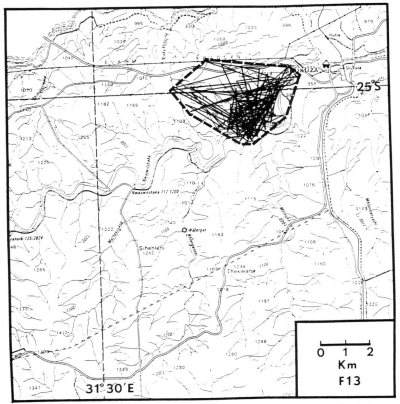

FIGURE 7.14 Movement pattern of adult female leopard F13 in the Nwaswitshaka River study area, May 1974 to July 1975.

female (F24) were also captured in F16's 4.5-km^2 area. Both were emaciated and had mange when first captured. One, probably F20, may have been F16's offspring.

SUBADULTS

The movements and home ranges of subadult leopards varied with age. Young, partially independent leopards stayed within the home ranges of their mothers. But as they matured they explored new areas adjacent to, as well as distant from, their mothers' home ranges. The frequency, duration, and distance of these exploratory movements have been discussed. In this section, additional movement data are presented.

Males.   Leopard M2, one of the youngest subadult males captured, was radio-collared in August 1973 and monitored until his death a year later. During this period he stayed primarily within a 17.2-km^2 portion of his 29.9-km^2 area in the

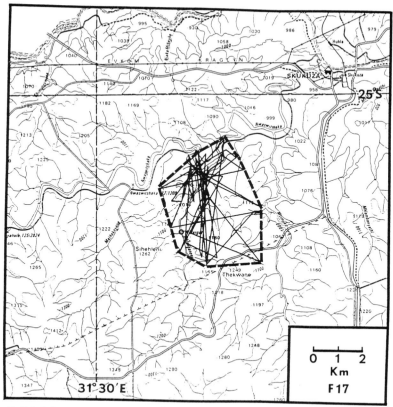

FIGURE 7.15 Movement pattern of adult female leopard F17 in the Nwaswitshaka
River study area, May 1974 to July 1975.

western half of the SRSA (see fig. 7.16). On at least eight occasions he explored
at least 2.5 km outside his natal area, but still displayed a predictable pattern.
He spent most of his time along the Sabie River, seldom spent more than one
or two days along the Sand River, and he moved in a clockwise pattern. He
typically moved west along the Sabie River, north to the Sand River, and east
along the Sand River until he returned to his starting point. Sometimes his
movements were interrupted by a trip 1 to 2 km north of the Sabie River, but
he always returned to the Sabie River to resume the circuit. Male 2's primary
home range was similar in size to those of several female leopards (F6, F22, and
F24).

Nine months after M2's death, an older subadult male (M26) was captured in
the same area. Although I monitored M26 only two months, I located him
almost daily. He used a slightly smaller area (12.4 km²) than M2 but traveled a
similar pattern. His movement pattern differed from M2's in only two respects:
he spent more time near the Sand River, and most movements occurred along
the east-west axis of his home range (see fig. 7.17). Although M26 was never

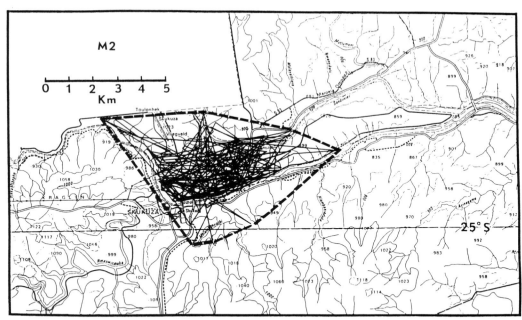

FIGURE 7.16 Movement pattern of subadult male leopard M2 west of the Sabie River study area, August 1973 to August 1974.

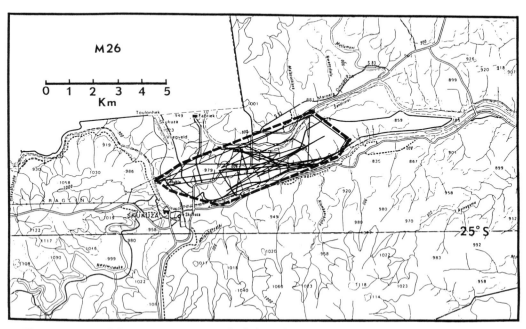

FIGURE 7.17 Movement pattern of adult male leopard M26 in the Sabie River study area, May 1975 to June 1975.

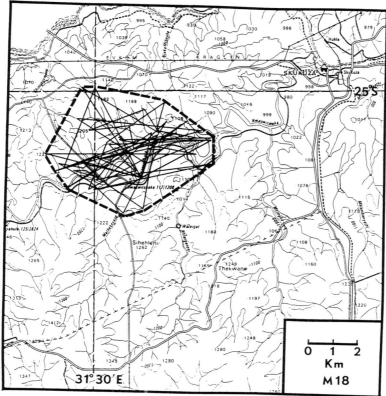

FIGURE 7.18 Movement pattern of subadult male leopard M18 in the Nwaswit-shaka River study area, June 1974 to May 1975.

located south of the Sabie River nor north of the Sand River, his monitoring period was probably too brief to reveal any exploratory movements.

Two subadult male leopards (M8 and M9) were captured and monitored in the NRSA. Male 18 used a 22.2-km² area in the southwestern portion of the study area. Although he usually stayed within several kilometers of the river, he periodically visited areas 3 km away. Male 18's home range was slightly larger than the home ranges of adult female leopards in the same area, but three times smaller than those of adult males. Most of M18's movements were along the Nwaswitshaka River (see fig. 7.18), and he made at least two exploratory trips up the Mathekeyane tributary.

Subadult M19 was monitored only briefly before his collar was recovered from the bottom of the Sabie River. After being released from the trap, this severely emaciated, mange-infested male wandered erratically over a 33-km² area north of the Nwaswitshaka River. His erratic movements were unusual and suggested that he may have come from another region, perhaps south of the Sabie River east of Skukuza. Therefore, during the period M19 was monitored, he may have been on an exploratory trip outside his natal home range.

Females. Two subadult female leopards (in addition to F12) provided information on movements. Female 8 was extremely emaciated when captured along the Msimuku in the NRSA in December 1973. She slowly traveled in a northwesterly direction along the Nwaswitshaka River after release, but later died in a heavy rainstorm. Female 24 was also emaciated and infested with mange when captured along the Sabie River. She moved parallel to the Sabie River over a 19.6-km² area within the river's riparian vegetation zone. The only characteristics common to the movements of these two subadult females were their exclusive use of the rivers' riparian vegetation zones and their travel parallel to the rivers.

Home Range Characteristics

Home ranges of leopards were strongly influenced by habitat quality and by the leopards' sex and age. Adult male leopards had larger home ranges than adult female leopards. In the Sabie River region three adult males (M3, M9, and M23) had an average home range of 27.7 km², which was 54% larger than the 18-km² primary range used by an adult female (F4) in the same habitat. The average home range of two adult male leopards (M10 and M14) in the NRSA was 76.2 km², or about 400% larger than the 14.8-km² average for four adult female leopards (F6, F11, F13, and F17) living in the same habitat.

The home ranges of adult male leopards in the NRSA were 175% larger than those of males in the SRSA. Female F4's primary home range in the SRSA was 22% larger than the average home range for four adult females in the Nwaswitshaka River region. These data suggest that adult female leopards were capable of surviving in smaller areas than male leopards. Considering the male's greater physical size, a male's home range was probably determined by factors other than available prey. Because there was less difference in the size of home ranges of adult female leopards (both within and between study areas) than in home ranges of male leopards, apparently the home ranges of females were determined by the food supply and were the minimal areas needed for survival and reproduction.

Several characteristics of the home ranges of adult female leopards suggested that they selected higher quality habitats than male leopards. In the NRSA, the home ranges of six females included more prey-rich riparian vegetation area (79%) than did the home ranges of two males (63%). Furthermore, the major axes of home ranges of female leopards were usually oriented along the Nwaswitshaka River or its major tributaries. This suggested that the home ranges of females were oriented to take maximum advantage of the edge effect along the riparian vegetation zone. Four of six home ranges of female leopards also encompassed at least one rocky outcrop, or koppie, that could be used to conceal cubs. Finally, the home ranges of two female leopards encompassed permanent waterholes.

In contrast, the home ranges of adult male leopards in the NRSA included more dry, prey-scarce, thorn-thicket vegetation zones than did those of female leopards. Male 10 and M14 were located more frequently than females on the drier ridges separating watersheds. These areas of sparse ground cover and *Combretum* supported less wildlife than riparian vegetation zones (Pienaar et al. 1966a).

I compared home ranges of KNP leopards to those in other areas. In Zimbabwe's Matopos (or Matobo) National Park, leopards used areas averaging at least 18 km² and fed primarily on rock hyraxes (Smith 1977). Although prey biomass data were not available for Matopos National Park, a leopard home range averaging 18 km² is less than the 24.7 km² average for thirteen adult leopards in the game-rich SRSA and NRSA. The methods used to assess home range size probably accounted for most of the apparent differences. Because home ranges of leopards in Matopos National Park were determined by spoor and other signs along preestablished routes, their sizes were probably underestimated.

In Serengeti National Park, Schaller (1972) estimated the minimum areas occupied by two female leopards were 40 and 60 km². But later in Serengeti National Park, Bertram (1982) found that a radio-collared female and cub used an area of only 15.9 km² during a five-month period, and a young male, sometimes seen with its mother, used an area of 17.8 km². In Wilpattu National Park, Eisenberg and Lockhart (1972) did not believe home ranges of leopards exceeded 10 km². In Tsavo National Park, where leopard home ranges were also determined by radio telemetry (Hamilton 1976), males had an average minimum home range size of 36.3 km², with a range between 17.9 and 63.4 km². One adult female leopard's home range was 14.4 km², and a subadult male's home range was 9.1 km². The average home range size for Tsavo males (36.3 km²) was between the average for males in the SRSA (27.7 km²) and NRSA (76.2 km²) study areas. This probably represents a home range size common to male leopards throughout the drier *Acacia*-savanna woodlands, where prey densities are similar. In Nepal an adult female used an 8-km² area (Seidensticker 1976a), and her male offspring used areas up to 8 km² before dispersing (Sunquist 1983).

In areas of low prey densities leopards have much larger home ranges. In the mountainous areas of the southwestern Cape Province in South Africa, three adult radio-collared males had home ranges between 53.5 and 127.7 km² in the Cedarberg Wilderness Area (Norton and Henley 1987). A male and female leopard in another mountainous area of the Cape Province had home ranges of 388 and 487 km², respectively (Norton and Lawson 1985). In the southern

Kalahari Desert two adult males and three females with cubs used an 800-km² area in the interior dunes (Bothma and Le Riche 1984).

Home Range Occupancy

In order to measure differential use of home ranges by leopards, I divided the study areas into 1 km x 1 km quadrats. The frequency with which leopards were located in the quadrats indicated their intensity of use. The total number of quadrats used by leopards was also a crude estimate of home range size. Quadrat occupancy analysis revealed that leopards did not use their home ranges uniformly. Some parts were used more frequently than others; other areas were avoided. Usually, one to three quadrats within leopard home ranges were used more intensively than others (see table 7.12). In the SRSA five of seven radio-collared leopards were located 8% to 20% of the time in a single 1-km² quadrat, and two leopards were located 13% to 20% of the time in only two quadrats. Two other leopards near the Sabie River were located 8% to 24% of the time within a single quadrat. In the NRSA 8% to 46% of the locations of seven leopards were within one quadrat, and one leopard had 31% of its locations equally distributed in six quadrats. Overall, 17% of the radio locations of fourteen leopards were in a single quadrat that made up only 4% of their home ranges.

Female leopards with cubs used single quadrats most intensively. These quadrats usually contained sites capable of concealing small cubs. From September 1974 to March 1975, 46% of F21's seventy-two locations were within a quadrat containing a koppie where she concealed her cubs (see fig. 7.19). She used this quadrat intensively three months after her cubs were born in Decem-

TABLE 7.12 *Use Intensity of Home Ranges By Radio-collared Leopards*

Number of Locations per 1 km² Quadrat	Number of Quadrats																
	Sabie River Study Area									Nwaswitshaka River Study Area							
	Leopard Identification Number									Leopard Identification Number							
	M1	M2	M3	F4	M7	M9	F12	M23	M26	F6	M10	F11	F13	M14	F17	M18	F21
1	12	11	9	6	23	10	4	9	6	9	26	5	2	17	6	13	3
2	12	2	3	5	9	3	3	3	8	4	7	5	2	10	3	5	3
3	1	2	8	3	7	3	4	1	2	2	6	1	3	7	4	6	3
4	3	2	5	2	4	2	5	5	0	0	3	0	1	6	1	0	0
5	4	1	4	0	5	0	1	0	0	1	2	0	1	2	1	0	0
6	2	0	2	1	1	0	3	0	0	1	6	1	1	1	0	0	1
7	0	0	2	2	2	0	1	2	0	1	0	1	0	0	0	2	0
8	1	2	2	1	0	0	1	0	1	0	0	2	0	2	0	0	0
9	1	0	5	2	0	0	3	2	1	0	0	0	1	3	0	0	0
10	0	1	3	3	0	0	1	0	0	0	0	1	1	1	0	0	0
10+	10	12	4	10	2	0	2	1	0	7	0	3	4	4	1	1	2
Total	46	33	47	35	53	19	28	23	18	25	50	22	16	53	16	27	12

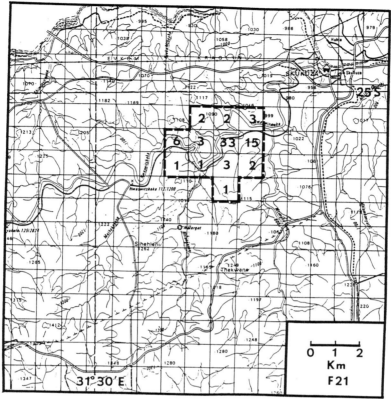

FIGURE 7.19 Number of radio locations per 1 km² quadrat for adult female leopard F21 in the Nwaswitshaka River study area.

ber. The second most intensively used quadrat (21%) was east and adjacent to the quadrat containing the koppie.

Another female with cubs, F6, was located 87 (44%) of 197 times in two quadrats on the Msimuku tributary (see fig. 7.20). This quadrat contained a narrow ravine and a small koppie where her cubs were concealed from December 1973 to February 1974. Most of F6's other locations were within five quadrats surrounding the quadrat containing the koppie, and these were used for hunting. Two impala killed by the female were in these quadrats.

Twenty percent of the locations of F11 were in a single quadrat containing a reed bed where she concealed her cubs (see fig. 7.21). She later moved her cubs to the quadrat containing the small koppie previously used by F6. When her cubs were a month old, F11 stayed with them during the day. She also used some of the quadrats formerly used by F6, but only after F6's cubs were more than a year old and F6 herself no longer used the area.

Because females rear their cubs alone, as much as 66% of their time may be spent with the cubs (Seidensticker 1977), especially when the cubs are small

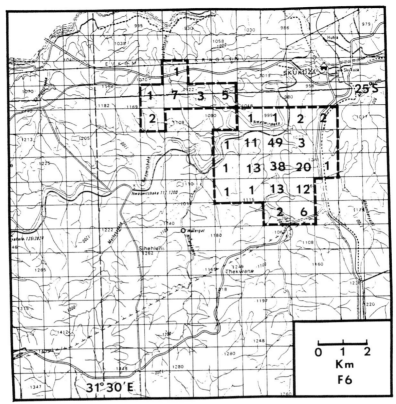

FIGURE 7.20 Number of radio locations per 1 km² quadrat for adult female leopard F6 in the Nwaswitshaka River study area.

and helpless. When their cubs were less than one month old, 50% to 60% of the daily radio location of females I monitored were within the quadrats harboring the cubs. When F21's cubs were one to two months old, 60% of her locations were still in the same quadrat harboring her cubs. Comparable data for females with older cubs were not available.

Female leopards without dependent cubs also intensively used one to three quadrats within their home ranges. Nineteen percent of F13's locations were in a single quadrat by the Nwaswitshaka River. Two of the other intensively used quadrats (32% of locations) were also by the river (see fig. 7.22). Such intensive use of quadrats in riparian vegetation was common among all radio-collared female leopards.

Twenty-five percent of F17's locations were in a quadrat at the confluence of the Nhlanganeni and Nwaswitshaka Rivers. Most of her other locations were also in quadrats bordering the Nhlanganeni (see fig. 7.23). In the SRSA 25% of F4's radio locations were in four quadrats encompassing riparian vegetation adjacent to the Sand River 3 km upstream from its confluence with the Sabie

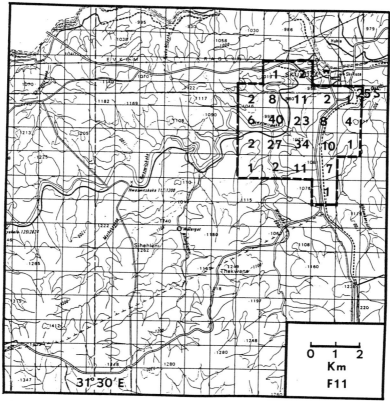

FIGURE 7.21 Number of radio locations per 1 km² quadrat for adult female leopard F11 in the Nwaswitshaka River study area.

River. Female 4 seldom used quadrats in the open brackish flats habitat, but occasionally she used quadrats on their boundaries (see fig. 7.24). Subadult F12 used quadrats west of the brackish flats but, unlike F4, she also used one quadrat in the flats (see fig. 7.25).

A conspicuous feature of the terrain or of certain vegetation was often associated with quadrats intensively used by leopards. Two leopards in the SRSA (M3 and F4) frequently used quadrats that contained large evergreen trees for shade or cover, especially during dry seasons. Other intensively used quadrats contained dry streambeds that leopards used as resting places. Areas where streams joined together were intensively used by F6, F11, F13, and F17. Male 9 was captured in, and later frequently located in, a quadrat at the confluence of the Mhlambanyathi and the Sabie River (see fig. 7.26).

Male leopards in the SRSA used their home ranges similarly, regardless of age. Subadult M1 and adult M3, who shared the eastern half of the SRSA, often used quadrats bordering the Sand and Sabie Rivers (see figs. 7.27 and 7.28) but seldom used those in the open brackish flats at the center of their home ranges.

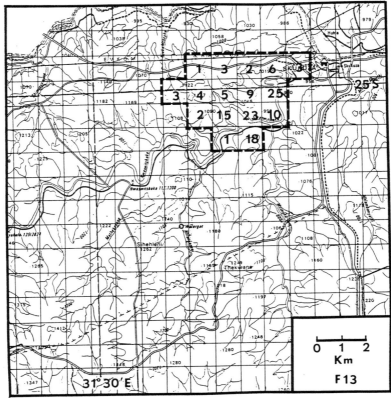

FIGURE 7.22 Number of radio locations per 1 km² quadrat for adult female leopard F13 in the Nwaswitshaka River study area.

As M1 matured, he visited scattered quadrats southwest of his normal range. Male 3 used quadrats north of the Sand River bordering tributary streams such as the Mutlumuvi and Malayathlava, but quadrats more than 1 km away were seldom used. The intensity of quadrat use by leopards in the SRSA decreased with increasing distance from riparian vegetation zones. Male 23 used quadrats along the Sand and Sabie Rivers more often than those located between the rivers (see fig. 7.29).

Adult M10's pattern of home range use in the NRSA was different from comparable-aged males in the SRSA. Instead of using a single quadrat intensively, he used quadrats uniformly along the Nwaswitshaka River. Two quadrats encompassed waterholes along the river, three were adjacent to the river, and one was more than 1 km away (see fig. 7.30). Although M10 favored those parts of his home ranges along the river, he also used an impala-abundant area north of the river. He was seldom located in impala-scarce regions of his home range south of the Nwaswitshaka River.

Subadult M18 used those parts of his home range bordering the Nwaswit-

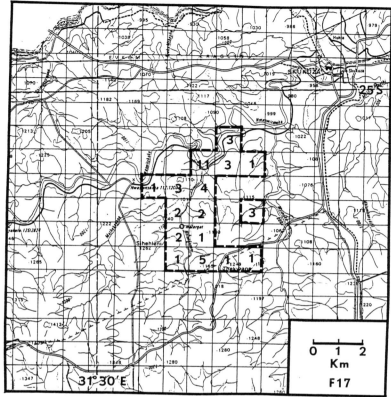

FIGURE 7.23 Number of radio locations per 1 km² quadrat for adult female leopard F17 in the Nwaswitshaka River study area.

shaka River more frequently than those 1 km or more away. The most frequently used quadrat was one near the Nwaswitshaka waterhole where he was periodically recaptured (see fig. 7.31).

Some of the factors biasing the observed home range use patterns of leopards were baits in traps, the predominance of daytime observations, and leopard activity patterns. Quadrats with baited traps may have been avoided by females, especially if they were previously captured there. In contrast, two males (M14, M18) and two emaciated leopards (M2, F16) appeared to be attracted to quadrats with traps because the bait provided an easy meal.

Quadrats encompassing daytime resting places of leopards may have been overrepresented because of the predominance of daytime observations. A comparable number of radio locations taken during the night might have revealed a different home range use pattern. During the night leopards may have used areas with less vegetative cover, taking advantage of the darkness to stalk prey. Although leopards were seldom located in the open brackish flats during the day, tracks in dusty firebreak roads suggested that they at least passed through

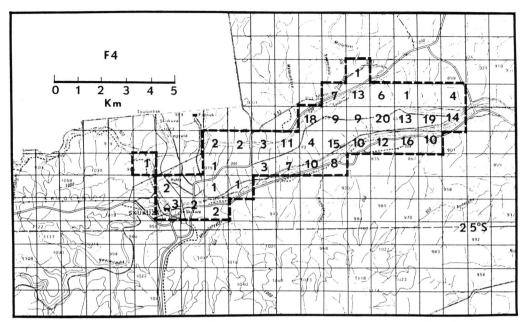

FIGURE 7.24 Number of radio locations per 1 km² quadrat for adult female leopard F4 in the Sabie River study area.

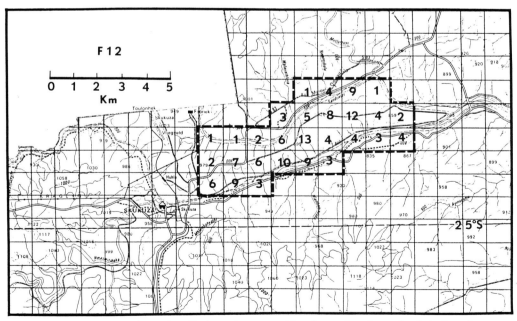

FIGURE 7.25 Number of radio locations per 1 km² quadrat for subadult female leopard F12 in the Sabie River study area.

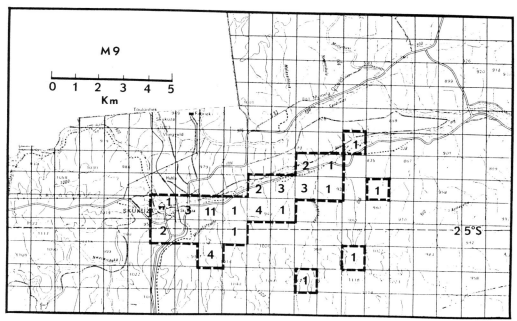

FIGURE 7.26 Number of radio locations per 1 km² quadrat for adult male leopard M9 south of the Sabie River study area.

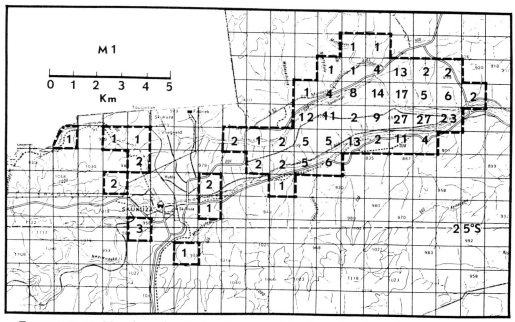

FIGURE 7.27 Number of radio locations per 1 km² quadrat for subadult male leopard M1 in the Sabie River study area.

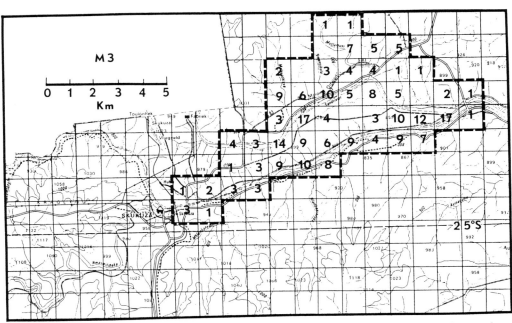

FIGURE 7.28 Number of radio locations per 1 km² quadrat for adult male leopard M3 in the Sabie River study area.

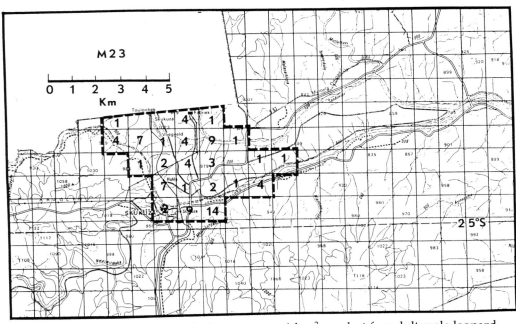

FIGURE 7.29 Number of radio locations per 1 km² quadrat for adult male leopard M23 in the Sabie River study area.

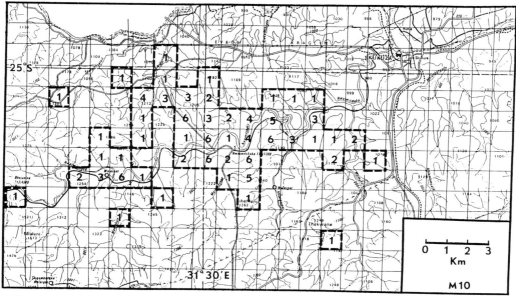

FIGURE 7.30 Number of radio locations per 1 km² quadrat for adult male leopard M10 in the Nwaswitshaka River study area.

the areas at night. Leopards may also have used quadrats near their home range boundaries more frequently than daytime observations indicated. The activity patterns of leopards may also have biased home range use pattern data because of the long intervals between observations. Because leopards were more active during the night, more quadrats may have been used at night than during the day.

Despite daytime biases the observed home range use patterns of leopards indicated areas that were more important to them; they spent at least half of each day in the observed quadrats. However, the observed use patterns probably did not portray the nocturnal use of leopard home ranges. Hunting areas were probably underrepresented, but kill and feeding sites were adequately represented because their use was less influenced by time of day. Night observations would probably reveal more use of quadrats in drier, upland *Combretum* habitats.

As a measure of home range size, the number of quadrats leopards used overestimated their size compared to the polygon method. Twelve of sixteen leopard home ranges were overestimated, using the quadrat method. Overestimates varied from 10% to 114%, with an average of 37%. Four underestimated home ranges averaged 31% less compared to the polygon method. Most home ranges were overestimated using the quadrat method because only a single observation in a 1-km² quadrat added 1 km² to the home range size. In the

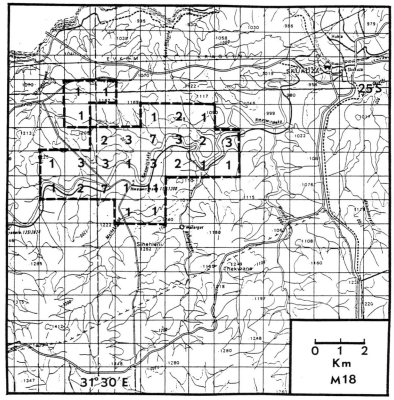

FIGURE 7.31 Number of radio locations per 1 km² quadrat for subadult male leopard M18 in the Nwaswitshaka River study area.

polygon method only the area within the polygon was included in the estimate of size.

The home ranges of individuals that were difficult to monitor were usually underestimated. And the exploratory movements of younger leopards raised the question of whether the areas they used could properly be called home ranges. Male 7's home range was grossly overestimated by the polygon method, because he seldom used quadrats between several widely separated quadrats of intensive use.

Seasonal Movement Patterns

Leopards in the study areas did not have distinct seasonal home ranges. But the seasons did induce changes in home range size, geometric centers of activity, mean activity radii, and home range use patterns. Thus some leopard home

TABLE 7.13 *Minimum Sizes of Observed Seasonal Home Ranges*

		Size of Observed Seasonal Home Ranges (km²)[1]									
		Sabie River Study Area					Nwaswitshaka River Study Area					
		Leopard Identification Number					Leopard Identification Number					
Year	Season	M1	M2	M3	F4	F12	F6	M10	F11	F13	M14	M18
1973	Dry	11.0	10.6	11.6								
1973–74	Wet	19.9	22.7	30.8	14.9		18.8					
1974	Dry	49.3	14.3	27.4	28.0	16.9	7.1	75.5	11.9	9.6	21.9	19.5
1974–75	Wet	21.2		17.4	15.6	15.4		52.3	12.2	5.3	40.2	19.5
1975	Dry										50.2	

[1]Minimum number of observations per leopard per season = 17; mean = 71.6.

ranges seasonally expanded and contracted. Others changed little between seasons.

The size of leopards' home ranges sometimes varied seasonally, but such changes varied with individual leopards, the study area, and the season. In the SRSA changes in home ranges of individual leopards varied from 1.5 to 29.4 km² between consecutive seasons (see table 7.13). The greatest home range size difference occurred among male leopards. Subadult M1's home range varied between 11.0 and 49.3 km² during four seasons. Those of adult M3 varied between 11.6 and 30.8 km². The home ranges of two female leopards (F4 and F12), an adult and a subadult, varied only from 1.5 to 13.1 km² between seasons, considerably less than that recorded for males.

In the NRSA the seasonal home ranges of adult male leopards also varied more than those of adult females. Between the 1974 dry season and the 1974–75 wet season, M10's home range decreased 23.2 km²; M14's increased 18.3 km². During the same period the home range of one female leopard (F11) increased only 0.3 km², whereas that of another female (F13) decreased 4.3 km².

The home ranges of subadult leopards appeared less influenced by seasonal changes than those of adults. The size of adult male home ranges varied an average of +33% between seasons. Those of adult females averaged only −11%. By comparison, the home range variation of one intensively radio-tracked subadult male (M18) remained unchanged, and that of subadult F12 varied only 1.5%.

Some seasons had a greater influence on the home range size of certain leopards than others. In the SRSA the greatest change in average home range size (112%) occurred between the 1973 dry season and the 1973–74 wet season. Between the 1973–74 wet and the 1974 dry season, seasonal home ranges averaged another 14% increase. This trend was eventually reversed between the 1974 dry and 1974–75 wet season, when the average decrease in seasonal home ranges was −34%. Between the 1973–74 wet and 1974 dry season,

TABLE 7.14 *Seasonal Use of Home Range by Leopards in the Sabie River Study Area*

Number of Locations per Quadrat	Number of Quadrats										
	Leopard M1				Leopard M3				Leopard F4		
	1973 Dry	1973–74 Wet	1974 Dry	1974–75 Wet	1973 Dry	1973–74 Wet	1974 Dry	1974–75 Wet	1973–74 Wet	1974 Dry	1974–75 Wet
1	6	7	11	10	8	9	9	10	5	6	7
2	4	7	4	7	4	5	5	2	2	10	5
3	3	3	2	3	1	6	9	1	1	2	3
4	2	2	1	1	2	5	3		4	5	1
5		1	3			4	3		2	4	
6			1			1			2	1	
7	1					1	2		1	3	
8		1				2			3		
9			1			1			1	1	
10+							1		3		
Total	16	2	2	2	15	34	32	13	24	32	16

the home range size of four of five monitored leopards decreased an average of 38%.

The abundance, distribution, and behavior of impala influenced the size of seasonal home ranges of leopards. When impala moved away from the Sabie and Sand Rivers after the rains in 1973, several leopards shifted their home ranges north of the Sand River to follow them. Although leopards continued to use the area south of the Sand River, they spent less time there. In the NRSA the seasonal home ranges of leopards changed as they apparently sought impala that had moved away from drying waterholes. As impala moved away from permanent waterholes in the wet seasons and toward them in dry seasons, the leopards followed. The distribution of impala also changed seasonally. During the wet seasons impala avoided dense cover and favored short grass areas. Some leopards altered their home ranges during wet seasons to include more of the short grass habitats preferred by impala.

Seasonally, leopards used some parts of their home ranges more frequently than others (see tables 7.14 and 7.15). In the SRSA, M1, who was frequently located between the Sand and Sabie Rivers during the 1973 dry season, focused his activity north of the Sand River during the 1973–74 wet season, then explored the area to the west during the 1974 dry season (see fig. 7.32). Although he visited the area between the rivers regardless of seasons, he spent more time in other areas. In other words, he did not abandon use of one area in favor of another; he merely reduced use of one portion while increasing it in another. Male 3, who spent most of his time between the Sand and Sabie Rivers during the 1973 dry season, used areas north of the Sand River during the 1973–74 wet season, then reduced the size of his home range the following dry season (see fig. 7.33). During the 1974 dry season, M3 also extended his home range about 4 km west. Adult F4 exhibited similar seasonal movement patterns (see fig. 7.34).

TABLE 7.15 *Seasonal Use of Home range by Leopards in the Nwaswitshaka River Study Area*

Number of Locations per Quadrat	Number of Quadrats								
	Leopard F6		Leopard F11		Leopard F13		Leopard M14		
	1973–74 Wet	1974 Dry	1974 Dry	1974–75 Wet	1974 Dry	1974–75 Wet	1974 Dry	1974–75 Wet	1975 Dry
1	8	6	5	6	2	4	11	19	19
2	3	1	3	1	5		7	6	4
3	1	2	2	1	3	1	1	6	6
4		1		1	2	1	2	2	6
5	1	1		1		1	4	2	2
6			1	1			1		1
7	1			1			1	1	
8	1		2	2	1		1	1	1
9	1	1						1	
10+	4	1	4	3	1	2			
Total	20	13	17	17	14	9	27	32	33

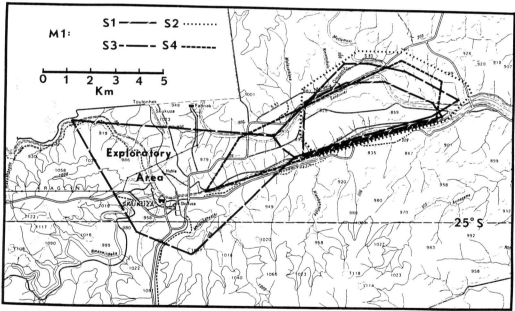

FIGURE 7.32 Polygons encompassing the seasonal radio locations of subadult male leopard M1 in the Sabie River study area. Seasons: S1 = Dry 1973 (August to September); S2 = Wet 1973–74 (October to March); S3 = Dry 1974 (April to September); S4 = Wet 1974–75 (October to March).

In the NRSA F6's home range was not significantly influenced by season after she moved her cubs (see fig. 7.35), nor was that of F11 with cubs (see fig. 7.36). These data suggested that the family rearing responsibilities of female leopards prevented them from making seasonal home range adjustments. In

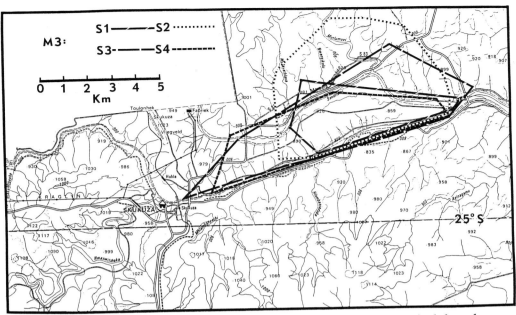

FIGURE 7.33 Polygons encompassing the seasonal radio locations of adult male leopard M3 in the Sabie River study area. Seasons: S1 = Dry 1973 (August to September); S2 = Wet 1973–74 (October to March); S3 = Dry 1974 (April to September); S4 = Wet 1974–75 (October to March).

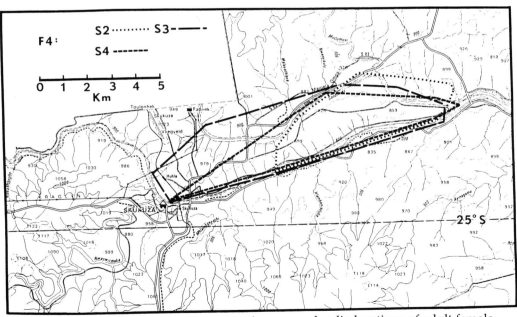

FIGURE 7.34 Polygons encompassing the seasonal radio locations of adult female leopard F4 in the Sabie River study area. Seasons: S2 = Wet 1973–74 (October to March); S3 = Dry 1974 (April to September); S4 = Wet 1974–75 (October to March).

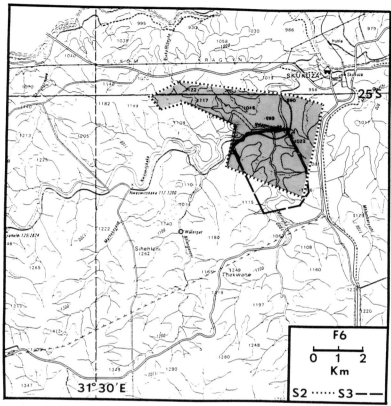

FIGURE 7.35 Polygons encompassing the seasonal radio locations of adult female leopard F6 in the Nwaswitshaka River study area. Seasons: S2 = Wet 1973–74 (October to March); S3 = Dry 1974 (April to September).

contrast, the home range of a female without cubs (F13) varied considerably between seasons. She moved extensively north of the Nwaswitshaka River during the 1974 dry season, but during the wet season she used a smaller area adjacent to the Nwaswitshaka River (see fig. 7.37). Adult M14, who used a small area during the 1974 dry season, began using a new area southeast of his former range during the 1974–75 wet season. During the 1975 dry season, he enlarged his home range again but remained in the same region (see fig. 7.38).

Geometric centers of activity of leopards also changed with the seasons. Of seventeen radio-collared leopards, differences between seasonal centers of activity exceeded 1 km for eleven of them. Of these, distances were between 2–3 km for three leopards, between 3–4 km for two leopards, and greater than 4 km for one leopard. Seasonal differences were greatest between the 1973–74 wet season and the subsequent dry season (see table 7.16). This period of high rainfall and increased vegetation growth caused impala populations to increase. The least seasonal difference occurred between the 1974 dry season and

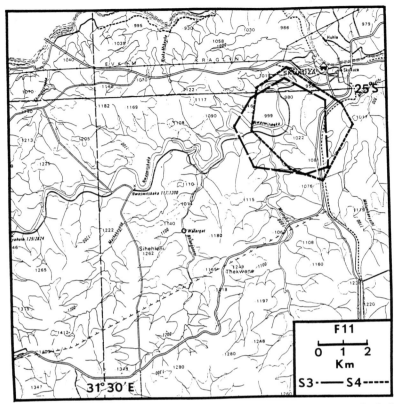

FIGURE 7.36 Polygons encompassing the seasonal radio locations of adult female leopard F11 in the Nwaswitshaka River study area. Seasons: S3 = Dry 1974 (April to September); S4 = Wet 1974–75 (October to March).

the subsequent wet season, when changes in rainfall, vegetation growth, and prey abundance were minimal.

Changing seasonal centers of activity also revealed directional shifts in seasonal home ranges of leopards. In the SRSA, following the 1973 rains, leopards focused their activities northeast of the previous season's activities. Thereafter, centers of activity shifted northward and westward until the leopards returned to previously used areas. Although similar seasonal patterns were exhibited by three male leopards in the NRSA, two males did not reduce the size of their home ranges nor exhibit previous home range use patterns between 1973 and 1975. Instead, M14 slowly expanded his home range, and the size of M18's remained unchanged.

Seasonal changes in mean activity radii also revealed how leopards responded to the seasons. Nine of twenty-two between-season activity radii were significantly different. Of these, four increased between wet and dry seasons and three decreased between dry and wet seasons (see table 7.17). No single

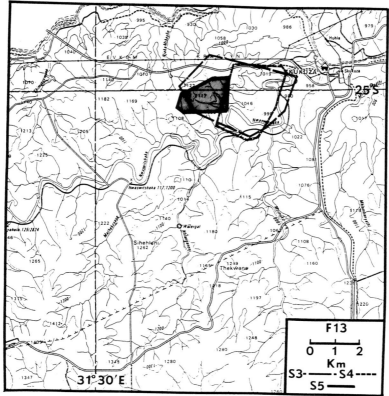

FIGURE 7.37 Polygons encompassing the seasonal radio locations of adult female leopard F13 in the Nwaswitshaka River study area. Seasons: S3 = Dry 1974 (April to September); S4 = Wet 1974–75 (October to March); S5 = Dry 1975 (April to July).

factor could explain these observed variations. Male 1's activity radii were 1.8, 1.6, 3.5, and 2.5 km between the 1973 dry and 1975 dry seasons. Those of M3 were 2.3, 2.5, 2.9, and 2.7 km for the same period.

The daily movements of leopards along the Sabie River abruptly declined between November and December when most impala fawns were born. Between June and September, after impala dispersed from permanent waterholes, leopard daily movements increased. Greater daily movements suggested that leopards were either spending less time at kills or they killed prey less often. Because it was unlikely that they spent less time feeding, it appeared that the kill rate declined during the dry season, particularly from May through August. The increased trapping success of leopards during the dry season also supported the view that, during dry seasons, their prey was more difficult to capture.

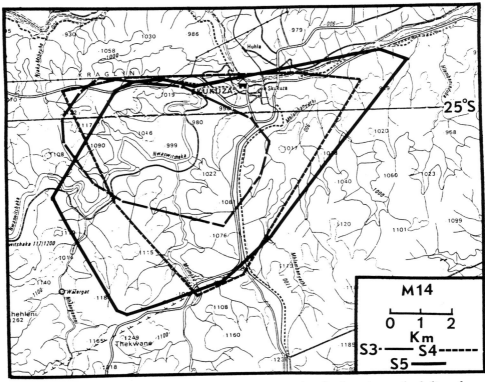

FIGURE 7.38 Polygons encompassing the seasonal radio locations of adult male leopard M14 in the Nwaswitshaka River study area. Seasons: S3 = Dry 1974 (April to September); S4 = Wet 1974–75 (October to March); S5 = Dry 1975 (April to July).

TABLE 7.16 *Distance Between Seasonal Geometric Centers of Activity of Radio-collared Leopards*

			Distance Between Geometric Centers of Activity (in Kilometers)			
Study Area	Leopard No.	Sex	1973 Dry and 1973–74 Wet	1973–74 Wet and 1974 Dry	1974 Dry and 1974–75 Wet	1974–75 Wet and 1975 Dry
Sabie River	1	M	1.4	3.3	0.6	
	2	M	0.8	0.1	0.6	
	3	M	1.5	1.6	0.6	
	4	F		1.4	0.7	
	7	M		4.5		
	9	M		2.4		
	12	F			2.7	
	23	M				1.1
	24	F				3.9
Nwaswitshaka River	6	F		1.2	1.7	
	10	M		2.2	0.1	0.7
	11	F			0.5	
	13	F			0.6	
	14	M			1.1	0.4
	17	F			0.9	
	18	M			0.9	
	21	F			0.4	

TABLE 7.17 *Mean Activity Radii in Kilometers of Radio-collared Leopards*

Study Area	Leopard No.	Sex	1973 Dry			1973–74 Wet			1974 Dry			1974–75 Wet			1975 Dry		
			n	x	SD	n	x	SD	n	x	SD	n	x	SD	n	x	SD
Sabie River	1	M	38	1.8	0.8	119	1.6	1.0	70	3.5	2.7	41	2.5	1.5			
	2	M	38	2.0	0.7	140	1.6	0.9	90	1.5	0.8						
	3	M	27	2.3	0.9	115	2.5	0.9	97	2.9	1.6	17	2.7	1.7			
	4	F				122	2.0	1.1	108	2.8	1.8	30	2.4	1.4			
	7	M				93	5.2	4.8	55	4.2	6.4						
	12	F							87	1.8	0.9	51	2.0	1.0			
	23	M										74	1.8	0.7	10	1.4	0.7
Nwaswitshaka River	6	F				100	1.6	1.1	95	0.9	0.5						
	10	M				12	3.4	1.7	62	3.1	1.8	39	3.1	1.8			
	11	F							109	1.0	0.5	95	1.2	0.6			
	13	M							52	1.4	0.6	65	1.0	0.4	13	1.0	0.2
	14	M							71	1.8	0.9	69	2.3	1.2	65	3.2	1.7
	18	M							33	1.9	0.9	31	1.6	0.8			

Underlined value differs significantly ($P < 0.050$) from previous seasonal value.

Summary

1. The average distance between daily locations of radio-collared leopards was 1.7 km. Forty percent of 1,510 daily movements were less than 1 km, 90% were less than 5 km, and the greatest recorded daily movement was 13 km.

2. Adult male leopards traveled nearly twice the distance (2.8 km) between days as adult female leopards (1.5 km). Adult male leopards also traveled significantly greater daily distances (2.8 km) than subadult males (1.6 km), but subadult male and adult female leopards traveled nearly equal distances between days.

3. The physical condition of leopards affected their movements. Adults in good condition moved greater distances between days than adults in poor condition. Adult males in good condition traveled 2.6 times as far each day as adult males in poor condition.

4. Adult female leopards traveled significantly greater distances between days during the peak breeding period (July through September) than at other times. The average distance traveled daily by adult males did not increase significantly between nonbreeding and breeding periods.

5. Young cubs restricted the daily movements of female leopards, especially during their first six months. Female leopards without cubs moved an average of 1.9 km/day, compared to 1.2 km/day for females with cubs.

6. The immobilization procedure significantly reduced the distances that male leopards traveled one day after release. Encounters of leopards with humans on foot did not significantly influence the distances they moved daily.

7. Adult male leopards traveled three times farther (4.7 km) the day after leaving a kill than the distance traveled (1.5 km) the day before they made a kill.

8. The kills of adult male leopards were spaced farther apart (3.9 km) than kills of subadult males (3.7 km), adult females (2.5 km), or subadult females (1.8 km).

9. The distribution of impala appeared to affect the spacing of kills of leopards because kills were more widely spaced during dry than wet seasons. Kills of adult male leopards were randomly spaced during the dry seasons, whereas those of an adult female were aggregated one wet season but maximally spaced during a dry season.

10. The daily movements of leopards were related to prey abundance. In the Nwaswitshaka River study area, where prey biomass was 15% lower than in the Sabie River study area, the daily movements of leopards were correspondingly greater, especially those of male leopards.

11. Female leopards were less responsive to seasonal movements of impala than males. When impala moved north of the Sand River during the wet season, several male leopards apparently followed them. However, a female leopard used her area in the same manner regardless of season and distribution of impala.

12. Subadult leopards eighteen to twenty-eight months old periodically explored new areas outside their normal home ranges. These exploratory trips lasted at least fifty days; exploratory behavior lasted at least ten months; and leopards explored at least 24 km outside their normal ranges.

13. Subadult male leopards made more and longer exploratory trips than a subadult female. Although exploring leopards frequently returned to their natal home ranges, most eventually dispersed. Unlike resident leopards, exploring subadults frequently crossed flood-swollen rivers.

14. The home ranges of adult male leopards varied between 16.4 and 96.1 km^2 and were much larger than those of adult females. Unlike females, adult males moved rapidly from one end of their home range to the other in predictable patterns.

15. Adult female leopards had smaller home ranges (5.6 to 29.9 km^2) than those of adult males. Females' home ranges changed with their age and the mobility of their cubs. Their movement patterns were less predictable than those of adult males.

16. The home ranges of subadult leopards, excluding exploratory movements, were similar in size to those of adult females but smaller than those of most adult males. Subadults' movement patterns were less predictable than those of adult leopards.

17. Habitat quality, sex, age, and reproductive status influenced home range sizes of leopards. Home ranges of adult female leopards appeared to include higher quality habitat than those of adult males.

18. Female leopards, especially those with small cubs, intensively used relatively few quadrats in their home ranges. Male leopards used more quadrats in their home ranges than females. Intensively used quadrats were characterized by dry streambeds, koppies, and dense stands of evergreen trees along streambeds.

19. Although leopards did not maintain distinct seasonal home ranges, size of home ranges, home range use patterns, and length of seasonal mean activity radii changed between seasons.
20. Seasonal changes in leopard home range use appeared to be related to the density, distribution, and behavior of impala. In the Sabie River study area, leopard home ranges increased in size between the 1973 dry and 1973–74 wet seasons. Most of those in the Nwaswitshaka River study area decreased in size or remained unchanged between the 1974 dry and 1974–75 wet seasons.

8

The Feeding Ecology of Leopards

THE leopard's feeding ecology is not as well known as those of the larger, more visible African carnivores such as the lion, spotted hyena, wild dog, and cheetah (Eltringham 1979). Leopards in KNP are known to prey on at least thirty-one species (Pienaar 1969) and leopards in Serengeti National Park fed on at least twenty-four species (Schaller 1972). Only a few of these species, however, mainly medium-size ungulates such as Thompson's gazelle and impala, constitute the majority of the leopards' diet (Mitchell et al. 1965; Kruuk and Turner 1967; Pienaar 1969; Hirst 1969b; Graupner and Graupner 1971; Schaller 1972; Hamilton 1976; Smith 1977). Because most data on leopard food habits are based on kill remains, the information is undoubtedly biased in favor of large prey. The significance of smaller prey in the leopard's diet is debatable. Several food habits studies of leopards based on fecal analysis suggest that in some areas small prey such as rock hyraxes form a significant portion of the leopard's diet (Grobler and Wilson 1972; Hamilton 1976; Smith 1977; Norton et al. 1986). Regardless of study techniques, data indicate that leopards are highly adaptable in their feeding ecology and are often opportunistic feeders.

Because much was already known about the occurrence of larger prey in the leopard's diet, especially in KNP (Pienaar 1969), I concentrated on those aspects of leopard feeding ecology that were less known or difficult to determine because of the leopard's nocturnal habits and preference for dense cover. These included leopard hunting and feeding strategies, scavenging, use of small prey, and kill and consumption rates.

Hunting Strategies

The untainted condition of the flesh of most leopard kills I discovered suggested leopards killed most of their prey late in the evenings or at night. Uneaten carcasses of prey killed by leopards were discovered only in the late evenings. By morning most kills of leopards had already been partially consumed. Because leopards usually ate part of their prey within several hours of killing it, my observations suggested that the uneaten carcasses were less than three hours old and those partially eaten less than twelve hours old. No recently killed, uneaten carcasses of prey were discovered between 7:00 A.M. and 5:30 P.M.; apparently few large prey were killed by leopards during the day in my study areas.

For a predator that depends on concealment for hunting success, the advantages of hunting at night are obvious. First, the chances of prey detecting a stalking leopard or one lying in ambush is less under the cover of darkness. Second, many of the ungulates such as bushbuck and impala are active at night because they must periodically feed to keep their digestive systems functioning (Leuthold 1977). Movement and activity of prey at night would increase their chances of being detected by hunting leopards. Third, although some antelopes such as impala select open cover when feeding at night, they cannot completely avoid feeding in areas near cover where leopards may be concealed. Thus, rather than hunt during the day when they are easily detected by prey, leopards frequently wait until evenings or darkness to take advantage of maximum concealment, a strategy also used by female lions when cover is a constraint (Elliot et al. 1977).

Although leopards in my study areas appeared to hunt primarily at night, they occasionally hunted during the day. I observed leopards stalking and pursuing prey at 12:15 P.M. and attempting to ambush prey at 11:20 A.M., 11:30 A.M., and 1:57 P.M. On four of these days skies were overcast, and cloud cover may have contributed to the leopards' increased activity or influenced the alertness of their prey (Leuthold 1977). During the observed daytime hunts temperatures ranged from 19 to 28°C and averaged 22.8°C. No observed daytime hunts were successful. Leopards primarily used sound or vision to locate prey. I never observed leopards using scent to track down prey, and their apparent difficulty in detecting baits hung in trees suggested that their sense of smell may have been inferior to their other senses. Leopards may, however, take wind into consideration when stalking prey (Tehsin 1980). Hunting leopards spent much time looking and listening for prey.

May 13, 1975, Naphe Road, 3:50 P.M. to 5:45 P.M.: After walking through the grass beside the road a female leopard stops, sits on a rock and licks itself for ten minutes. After cleaning, she walks, sits in the grass and scans her surroundings for three minutes. She gets up and walks to a nearby tree where she alternately watches and cleans herself for one hour. During this time she ignores lions roaring 1 km away but instantly becomes alert when a young impala begins bleating. She becomes tense and alert, staring in the direction of the impala, but after it stops bleating, she relaxes and does not investigate.

At night, sound played an important role in leopards' detecting prey.

July 4, 1975, Lower Sabie Road, 6:00 P.M. to 6:55 P.M.: A leopard sits on the edge of the road looking into the darkness. After three minutes the leopard gets up and walks by my vehicle and sits again, looking in all directions and turning its ears to detect any sounds. Three more times the leopard gets up, walks 20 to 30 m, sits down, and stares into the darkness. Suddenly at 6:15 the leopard jumps up, looks behind, and crosses the road. Slowly it stalks into the vegetation and within two minutes the night is filled with piercing squeals and rattling sounds. Driving to where the leopard disappeared, I focus my spotlight on the leopard wrestling a ratel in its mouth and front paws. The back of the ratel's scalp has been bitten off. There are deep bleeding wounds on the leopard's chest and front legs. In the struggle the leopard releases the ratel, which runs away squealing loudly and running erratically in circles. The leopard does not pursue the ratel, but spends the next ten minutes licking its wounds. While cleaning itself, the leopard becomes alert again and stalks into the darkness in the opposite direction taken by the ratel. Several minutes later I see the leopard standing over a freshly killed civet. After three minutes of cleaning itself the leopard plucks large mouthfuls of fur from the civet and then eats from the civet's hindquarters. After watching for ten minutes I leave. When I return an hour later, the leopard is gone and the civet is consumed except for the plucked fur, tail, and feet.

Leopards appear to be opportunistic predators, taking the most abundant or easily captured prey. Some prey, such as impala in KNP, may be more vulnerable to leopard predation than others because of prey habitat preferences, relative abundance, size, and behavior. Because the leopard's hunting success depends on cover, prey, such as bushbuck, that favor dense cover are more likely to be captured by leopards than prey using open habitats. In KNP nursing young of species that favor grassland, such as zebra and wildebeest, are seldom killed by leopards. But the young of species such as kudu and waterbuck, which favor bush, are regularly killed, despite their lower relative abundance (Pienaar 1969). However, abundant prey are probably killed by leopards in dense cover merely because they are frequently encountered relative to other prey. From an energy conservation viewpoint, smaller prey are probably killed less often by leopards in areas where larger prey are readily available. Behavior of prey may also play an important role in prey selection. Aggressive or potentially dangerous prey, such as warthogs or baboons, are probably avoided if less dangerous, alternative prey are available.

The persistence of hunting leopards is probably dependent on their degree of hunger and past hunting success. Leopards seemed to be aware that their chances of success were sometimes poor.

July 4, 1975, Lower Sabie Road, 5:40 P.M.: A leopard is sitting in the middle of the road as the sun sets. After five minutes of scanning nearby cover, the leopard rises and slowly walks along the edge of the road. Suddenly a genet appears on the road 60 m ahead of the leopard. The leopard stops and remains motionless, watching the genet. After several seconds, the leopard swiftly darts forward in a low crouch, alternately spurting ahead and remaining motionless as it closes the distance between itself and the unwary genet. When 15 to 20 m away the leopard bursts into a rush for the genet, but the genet quickly jumps into the dense undergrowth beside the road and disappears from view when the leopard is only 6 to 10 m away. The leopard stops but remains on the road and looks intently into the vegetation where the genet disappeared. After thirty seconds the leopard resumes its slow walk down the road.

Small elusive prey are probably seldom pursued great distances by adult leopards, especially if larger prey is available. But if smaller prey is abundant and larger prey scarce, leopards apparently become adept at capturing small prey. In Matopos National Park where impala are less abundant than in KNP, leopards frequently fed on rock hyraxes or dassies, small rodentlike mammals living in rocky habitat (Grobler and Wilson 1972).

Although adult ungulates may be pursued only a short distance, young unexperienced ungulates may be chased considerable distances. I never observed leopards pursuing adult impala long distances but once observed a leopard pursue an impala fawn at least 100 m.

December 12, 1973, Kruger Road, 12:15 P.M.: All impala in a herd are alert and many are giving alarm calls. Suddenly a lone two- to three-week-old impala fawn runs toward my vehicle with a leopard 20 m behind in pursuit. The fawn deviates from the game trail it is on, but doubles back and crosses the road in front of me. The leopard continues to pursue the fawn to the edge of the road but hesitates to cross, thus giving the fawn a thirty- to forty-five-second advantage to run ahead and out of view. Eventually the leopard also crosses the road, stops at the opposite side to look for the fawn, and then slinks off after the fawn. Although the fawn is no longer in view, the leopard courses back and forth through the brush searching for the fawn. After several minutes the leopard sits down and looks in the direction taken by the fawn. A single adult female impala, apparently the fawn's mother, exits the brush and comes to the edge of the road nervously looking in the direction taken by the fawn. The remainder of the impala resume feeding less than 100 m away. Twenty minutes later I am unable to see the leopard as it stalks off through the brush.

By comparing photographs, I identified the leopard as an adult female that was later captured in the vicinity. Apparently the leopard considered it advantageous to pursue the impala fawn. The fawn's mother and other impala made no attempt to protect the fawn nor to divert the leopard's attack. Only a cursory attempt was made by the presumed mother to locate her missing fawn. The fawn did not seek refuge among the impala herd and its ultimate fate was unknown.

Stalking and ambushing appeared to be the leopard's most frequent hunting strategy. Of thirteen observed hunting attempts by leopards, they stalked prey seven times, attempted to ambush prey five times, and pursued prey only once. Leopards took advantage of cover when stalking prey and appeared to anticipate likely locations where attacks could be made. One leopard watched an impala in an open area but did not stalk it until it moved into cover. The leopard then swiftly trotted down a game trail the impala was approaching from an angle. The leopard kept a screen of vegetation between it and the impala, and both eventually disappeared from view. Another adult male leopard also used cover to screen its approach to a herd of impala.

October 4, 1974, Msimuku Spruit, 5:40 P.M. to 5:50 P.M.: Male 14 crosses the road and climbs to the top of a 2-m bank to watch a group of impala crossing the streambed about 100 m away. After two minutes he jumps down into the dry streambed and approaches the impala by keeping to the tall reeds growing in the streambed. When 50 m from the impala he stops at the edge of the reeds to watch the impala. They are unaware of the leopard. After three minutes the leopard leaves the reeds and climbs the opposite bank, stalks along the edge of the stream, and remains hidden behind a narrow belt of tamboti trees. I last see the leopard crouched motionless behind some trees about 20 m from the impala. He disappears from view moving slowly toward the unsuspecting impala.

Cover is important to leopards stalking prey. I never observed leopards stalking prey across the short grass, open areas during the day, despite the fact that both impalas and leopards regularly used such areas. In several observed attempts by leopards to capture impala in open areas, the leopards remained concealed in small patches of cover within the open areas, waiting for impala to come close enough for a rush. The locations of leopard kills suggested that most were made in dense vegetation. Of 50 kills, 46% were found in riparian vegetation, 44% in medium to dense thornbush thickets, and 10% in open habitats. Leopards also took advantage of features of the terrain to approach prey. The frequency that leopard tracks were seen in sandy dry streambeds suggested that they frequently used them to approach prey. Prey would have difficulty seeing leopards stalking them along streambanks. Once, while searching on foot for a radio-collared leopard, I saw it watching me 15 m away and lying in a dry streambed. The leopard would have been invisible had I not been above it. When the leopard left it slinked along the streambed, remaining close to the grass-covered bank.

Leopards may have been especially successful ambushing prey. I once followed the tracks of a warthog in a muddy road to where adult M23 was lying in grass 70 cm high. The signs of the struggle indicated the warthog walked within 3 m of the hidden leopard before the leopard attacked and killed it. Twenty-eight percent of the kills of leopards I found were near game trails; leopards may have been waiting to ambush prey, primarily impala, passing by on their way to waterholes or resting areas. Ambushes near waterholes may have been particularly successful.

June 20, 1974, Nwaswitshaka Windmill, 10:53 A.M. to 12:15 P.M.: A leopard crosses a firebreak road and open area to lie beneath a lone tree at a waterhole. The leopard lies in the shade but occasionally gets up to survey its surroundings. At 11:20 a herd of impala approach the waterhole and the leopard immediately takes cover by crouching behind the tree and several small saplings. The impala see my vehicle and are reluctant to come closer. However, they eventually approach the waterhole, unaware that the leopard is watching. The impala pass within 20 m of the concealed leopard, but the leopard does not launch an attack. For thirty minutes the leopard watches the impala, which slowly move away. Finally the impala move from view and the leopard lies down again in the shade. At 12:15 the leopard gets up and slowly walks away toward the dense vegetation along the Nwaswitshaka River.

The leopard was never seen by the impala apparently because it blended well into the background vegetation, despite the lack of dense cover. This observation also revealed how close leopards must be to their prey before launching an attack. It also portrayed the seemingly indifferent attitude leopards have following unsuccessful hunts. Leopards were just as quick to abandon a hunting attempt as they were to initiate one.

October 6, 1974, Naphe Road, 11:30 A.M. to 11:50 A.M.: A leopard sits looking intently in one direction only. After several minutes it gets up and stalks through the grass toward an *Acacia* tree and jumps to a fork in the lower branches 1.5 m above ground. The leopard, now motionless, continues to look intently at something nearby. After several minutes a feeding female steenbuck approaches the tree. The leopard becomes tense when the steenbuck is 30 m away and remains rigid as the steenbuck comes within 5 m of the crouched leopard. When the steenbuck is 3 m away and the leopard appears about to spring, the steenbuck suddenly bolts away. The leopard remains in the tree but faces the direction taken by the steenbuck. It watches the steenbuck for another five minutes, jumps from the tree, and slowly walks toward the steenbuck. The leopard is last seen moving in the direction taken by the steenbuck.

Leopards generally ignored prey in trees, apparently because trees provided secure refuge for partially arboreal prey such as baboon and vervet monkeys. Leopards also ignored vervet monkeys giving alarm calls and jumping in branches above the leopard. But given certain habitat conditions and perhaps individual leopard hunting experience, some leopards apparently can become significant predators on vervet monkeys (Isbell 1990). The response of baboons to leopards usually was to flee to the nearest tall tree. If the leopard approached closely, baboons and monkeys climbed to the ends of small tree limbs for protection. Such small limbs could not support the weight of leopards but could support the smaller baboons and monkeys. Chimpanzees, another partially arboreal potential prey of leopards in some habitats, usually react to leopards by giving alarm calls and going to or remaining in trees (Izawa and Itani 1966; Lawick-Goodall 1968; Nishada 1968). Guinea fowl also roosted on the very ends of small branches at night for protection from predators.

At least one leopard in my study areas sometimes killed scavengers that were attracted to his kills. Adult M23 killed and ate at least two spotted hyenas that were apparently attracted to his kills. On January 15, 1975, three days after he made an unidentified kill, I cautiously approached the leopard on foot, heard him feeding, but was unable to see him because of the dense vegetation. The following day I found the remains of a juvenile spotted hyena in a nearby tree. All except the head and feet of the hyena had been eaten by the leopard. On February 2, 1975, M23 was flushed from the remains of another hyena of undetermined age upon which he had been feeding. On March 3, 1975, I monitored M23 as he approached two hyenas feeding on a warthog he had killed. The leopard rushed them when they were 15 m away, but both hyenas escaped.

Hunting Success

Leopards were not highly successful at capturing prey. If only visual observations of stalking leopards, leopards waiting in ambush, and leopards pursuing prey are considered, leopards were successful on only two of thirteen (16%) occasions. All observed daytime hunting attempts were unsuccessful, but two attempts during darkness were both successful. Observations of other cats also suggest higher hunting success at night. Of 513 attempts by lions to capture prey, 33% were successful during the night and 21% successful during the day

(Schaller 1972). Like lions, leopards are probably able to conceal themselves better at night and thus closely approach unsuspecting prey.

Little comparative information is available on leopard hunting success in different habitats. In Serengeti National Park, only one of nine (11%) (Schaller 1972) and three of sixty-four (5%) (Bertram 1978) attempts by leopards to capture prey in the daytime were successful. Male leopards and female leopards with cubs in the Kalahari Desert killed about 12% and 23%, respectively, of the prey they contacted. Females with cubs were also more successful (28%) than males (8%) when prey was chased, but were less successful (4%) than males (7%) when prey was stalked (Bothma and Le Riche 1984).

Factors influencing hunting success of lions probably also apply to leopards. Lions were more successful hunting during the night than day, in thickets rather than in open habitats, and during upwind rather than downwind approaches. Lion hunting success was also greater during unexpected hunts than during stalks, and running after prey was highly unsuccessful. Stalking by single lions was as successful as ambushing prey, but hunting success generally increased as the number of lions hunting together increased (Schaller 1972; Elliot et al. 1977). Single lions were successful on only 15% (Schaller 1972) to 29% (Elliot et al. 1977) of their attempts. The hunting success of single lions was similar to or slightly higher than that of leopards.

Comparative information on the hunting success of other solitary felids is limited. Of forty-five obvious attempts on elk and mule deer, cougars tracked in the snow were successful thirty-seven times (82%) (Hornocker 1970). Of forty-three similar attempts by lynx to capture prey, eighteen (42%) were successful (Saunders 1963). Over a period of ten years, winter hunting success of lynx in Alberta, as determined by snow-tracking, varied from 0% to 67% with a mean of 21% (Brand et al. 1976). Eleven of twenty-two attempts (50%) by adult bobcats to capture black-tailed jackrabbits and cottontail rabbits during winter in Idaho were also successful (Bailey, unpublished data). However, as Hornocker (1970) pointed out, when felids are hunting the approach to prey is most crucial, and snow-tracking does not reveal whether an animal failed to place itself within actual attacking distance. Hunting success determined from snow-tracking is, therefore, a measure of success only after a stalk has successfully placed the animal within threshold distance. Obviously this excludes those instances when the prey fled or detected the predator before an attack was launched.

Method of Killing Prey

Leopards appeared to kill most larger, impala-size prey by biting the neck. Although I was unable to observe leopards actually killing prey, of fifty kills I examined, nineteen had teeth marks on multiple locations on their necks.

Thirteen of these (68%) had teeth holes in their throats; five (26%) had been bitten on the top of the neck behind the head; two (10%) had been bitten on the nape of the neck midway between the head and shoulders; and two (10%) had broken necks. I could not determine if the prey had been killed by a specific bite or whether the teeth marks were made as the prey was dragged by the leopards. Most bite marks on impala suggested the prey had been strangled, but a warthog had been killed by several bites around the nape of its neck. Another leopard grasped a ratel by its neck and shoulders, but the ratel twisted within the leopard's grip and bit the leopard's chest and forelegs before the leopard released it. Schaller (1967) observed a leopard grasp a tethered goat by its throat and reported that an axis deer fawn had been bitten through the nape of the neck. Kruuk and Turner (1967) and Turnbull-Kemp (1967) also reported that leopards killed prey by biting the neck or throat.

When felids kill prey, their bite frequently damages the spinal cord (Leyhausen 1965). Their jaws and the shape of their canine teeth apparently enable felids to penetrate the intervertebral space in the spinal column, and by forcing the two vertebrae apart, they partially or completely sever the spinal cord without damaging the vertebrae themselves. Leyhausen contended that the canine teeth of felids have little use in opening the carotid arteries of the prey's neck and that bleeding is not common from a cat's kill. Although I found large amounts of blood on the throats of some impala killed by leopards, I was unable to determine if loss of blood was the cause of death.

Felids employ different methods to kill small prey (Leyhausen 1965). Birds are usually killed by a bite between the shoulders and base of the neck. Struggling smaller prey are often shaken to confuse the normal functions of the inner ear for a few seconds, allowing a better-placed killing bite. The neck hold is an apparently innate killing technique, but killing must be reinforced in young, developing felids by stimuli through competition from siblings or mother. The forelimbs of felids are used solely for seizing prey and play no direct role in actual killing.

Behavior at Kills

Leopards usually drag larger kills into cover before eating, but small prey is eaten immediately. I once followed the trail of a leopard that dragged an impala more than 500 m and trailed another 300 m as it dragged an impala to dense cover beside the Sand River. Most kills, however, were moved less than 100 m, and kills that could be eaten entirely within a short period were eaten on the spot. In the sparsely vegetated southern Kalahari Desert, female leopards with cubs and males dragged their prey an average distance of 742 m and 410 m, respectively, to suitable cover in order to feed (Bothma and Le Riche 1984). In

TABLE 8.1 *Scavengers Observed at Kills of Leopards*

Scavenger	Number of Kills	Percent of Total
Spotted hyenas	28	52
Vultures	4	5
Tawny eagles	3	6
Lions	2	4
Crocodile	1	2
No observed scavenger	16	30

Sri Lanka leopards commonly dragged kills 11 to 12 m to the forest's edge before feeding (Eisenberg and Lockhart 1972). The man-eating leopard of Rudraprayag sometimes dragged human prey 6.4 km before eating them (Corbett 1947). A stock-killing leopard once dragged an adult sheep more than 3 km to feed (Stuart 1986). Smith (1977) reported that leopards moved prey farther during wet (260 m) than dry (120 m) seasons in the Matopos Hills of Zimbabwe. The longest distance he recorded a leopard dragging prey, a sable antelope calf, was 1,000 m. Kills were usually moved to secluded spots so leopards could feed undisturbed without attracting other predators or scavengers. If leopards are disturbed while feeding it is not uncommon for them to move their kills to a more secluded location.

Leopards in the study areas cached most of their large kills in trees. Of fifty-five recent and old kills and scavenged carcasses, forty-six (84%) were found in trees. Of nine carcasses leopards left on the ground, three were impala that died during severe rainstorms and were being scavenged by leopards, three probably had been cached in trees before I discovered them, and two were being dragged by leopards to trees when I inadvertently disturbed them. A civet killed by one leopard was not cached in a tree but eaten on the ground immediately.

Leopards cached kills in trees primarily to protect them from scavengers. Of fifty-four recent kills I discovered, thirty-one (57%) had already attracted spotted hyenas, lions, vultures, tawny eagles, or crocodiles. Spotted hyenas were present at twenty-eight (52%) of the kills; other scavengers were present at fewer kills (see table 8.1). At least two species of scavengers were at seven kills (13%) and three species at one kill (2%). Although black-backed jackals were common in the study areas, I did not observe them at or near kills of leopards.

The frequency of leopards caching kills in trees is probably a function of the number and species of scavengers in an area. In Tsavo National Park, where spotted hyenas seemed to be uncommon, Hamilton (1976) reported that only seven of eighteen larger kills of leopards were cached in trees. In Serengeti National Park, where hyenas and lions were more common, leopards cached most of their kills in trees (Schaller 1972). In the Kalahari Desert, leopards usually dragged their kills underneath a bush or tree and fed on the ground,

TABLE 8.2 *Cache Trees Used by Leopards*

Species of Tree	Number	% of Total	Height of Kill in Tree (in Meters)		Diameter of Tree (in Centimeters)	
			x	SD	x	SD
Spirostachys africana	10	22	4.8	1.5	41.1	8.2
Diospyros mespiliformis	8	17	6.6	3.2	76.6	34.8
Sclerocarya caffra	6	13	5.6	1.3	41.8	15.0
Acacia nigrescens	6	13	6.7	2.6	48.2	13.2
Schotia brachypetala	5	11	6.4	3.6	50.0	12.2
Kigelia africana	4	9	6.1	0.7	67.0	12.7
Unidentified	3	7	10.9			
Ficus sycomorus	1	2				
Lonchocarpus capassa	1	2	3.0		57.0	
Combretum spp.	1	2	4.3		25.8	
Lannea spp.	1	2	6.4		31.3	
Average			5.9	2.4	53.7	24.9

but if disturbed they moved them into trees (Bothma and Le Riche 1984). Where scavengers are uncommon, leopards may not cache any kills in trees (Eisenberg and Lockhart 1972); they may also drag kills to secluded caves (Brain 1981).

A comparison of trees selected by leopards to trees available indicated that large trees with dense foliage were favored (see table 8.2). During dry seasons leopards dragged kills considerable distances in order to use large evergreen trees that frequently grew near water. During wet seasons deciduous trees were used more often. Many trees with dense foliage, such as *kigelia* trees and *schotia*, grew primarily along riverbanks, and many kills I discovered were cached near rivers. Certain specific trees were particularly attractive to leopards and were used repeatedly by them to cache their kills in. Two kills were once cached in the same *kigelia* tree within nine days, and other kigelia trees had many claw scars on their trunks and impala hair underneath their thick branches, which suggested repeated use.

Cache tree selectivity by leopards appeared to be related to the intensity of competition with scavengers. If harassed by scavengers after killing prey, leopards were less selective and often cached their kills in the nearest large tree regardless of its suitability. I once discovered a kill that a hyena had appropriated from a young female leopard. After I inadvertently frightened the hyena away, the leopard returned and cached the impala in the nearest knobthorn *Acacia*. Another kill of the same leopard was discovered in a very small marula tree with two spotted hyenas sleeping underneath. Stevenson-Hamilton (1947) described how a hyena attempted to dislodge a leopard's kill by chewing at the base of a small tree.

Most cache trees used by leopards had diameters of at least 25 cm, but less than 132 cm. Trees with either smaller or larger diameters were probably too difficult for leopards to climb with heavy kills. Sycamore fig trees were seldom used as cache trees by leopards, perhaps because their slippery bark and large

8.1. The carcass of an impala cached in a tree by a leopard who merely draped the carcass over a large limb.

trunks made climbing with carcasses of prey difficult. The average diameter of cache trees used by leopards was 53.7 cm, but size varied with tree species (see table 8.2). No kills were cached in the numerous *Acacia grandicornuta* trees in the study areas, probably because of their small diameter and short height.

The size, slope, and configuration of the tree's branches determined the height at which carcasses of prey were cached in trees by leopards. Often the lowest and largest major branches of a tree were used to hold a carcass. Average height of kills above ground was 5.9 m with a range of 1.4 to 12.8 m (see table 8.2). Sometimes kills were merely laid over the main trunk of the tree (see photo 8.1), but more frequently they were tightly wedged into forks of limbs. Leopards often changed the positions of carcasses in trees while feeding to prevent the carcass from falling to the ground. Kills cached in trees near the ground could be reached by scavengers. One carcass of an impala cached by a leopard only 1.4 m off the ground had two of its hind feet bitten off by hyenas. They also clawed and chewed bark from the tree under the kill. Leopards did not cache kills in trees whose trunks sloped less than 45 degrees because lions and sometimes hyenas could climb such trees. Lions were also able to appropriate kills of leopards if they could be reached from ground level. Kills cached out on limbs were probably safer from lions than kills cached near the trunk of the tree. One leopard I observed took its kill to the uppermost branches of a tree when a lion appeared nearby.

I did not observe leopards actually climbing trees to cache their kills. Scratches, hair, and blood from prey suggested that carcasses were carried up tree trunks between the leopard's legs. I once watched a leopard move the carcass of an impala already cached in a tree to another branch. The leopard grasped the impala around the neck and straddled it with its forelegs. Short jumps up the trunk were then made with the front claws greatly extended. Claw marks 1- to 2-cm deep in the bark of *kigelia* trees indicated the tremendous strength of leopards climbing trees with kills.

Feeding Behavior

I found little evidence to indicate that leopards always disemboweled large prey before feeding. All freshly killed impala I observed, except one, still had their entrails intact. Even after leopards began feeding I never saw any attempt to disembowel the carcass before eating and cover the entrails as reported by Stevenson-Hamilton (1947). Leopards usually opened the body cavity while feeding and the entrails were either pulled or fell out. Perhaps the reported disemboweling of prey is dependent on the position of the carcass when leopards feed. Carcasses eaten on the ground may be more easily disemboweled than carcasses lodged in trees.

Leopards frequently fed on the groin or anal region first, then fed in the following sequence: hindquarters, abdomen, chest cavity, shoulders, forelegs, and lastly the neck and head. Depending on the position of the carcass in the tree, one side of the carcass was sometimes entirely eaten before the carcass was turned over. Regardless of which side was first eaten, the eating sequence was invariably the same—from the rear of the carcass forward. Of thirty-four partially eaten carcasses I observed, leopards had already eaten the groin on all, the hindquarters on twenty-two (65%), the hindquarters and abdominal region of seven (21%), and the entire carcass posterior to the shoulders on two (6%). Only once did I observe a leopard eat the shoulders and ribs of an impala before eating the hindquarters; a behavior I could not explain.

The behavior of one adult male leopard at an impala carcass was typical of the feeding behavior of leopards I observed. The impala was killed in the late evening and I arrived at dark. The carcass, 6 m up in a marula tree, had not yet been eaten.

December 12, 1974, 3 km south of Tshokwane, 6:40 P.M. to 11:00 P.M.: Only one hyena is under the tree when I arrive, but it soon runs off where it can still see the carcass. At 7:43, after rushing suddenly through the grass, the leopard approaches the tree and jumps 3 m in one leap up into the tree on a limb by the impala carcass. After pausing a minute, the leopard begins feeding by biting and chewing the groin until the tail is bitten off and eaten. Next, the leopard separates the rectum from the tail with its carnassals and pulls it and part of the intestine out and eats them.

8.2. A large male leopard feeds at night on the carcass of an impala it has cached in a tree.

After eating a small amount of flesh around the anus, the leopard vigorously plucks out some hair from one hindquarter and eats the flesh between the hindquarters and the spinal column. After feeding intermittently for ninety minutes the leopard stops and changes the position of the carcass in the tree (see photo 8.2). The leopard struggles to maintain a grip on the carcass to prevent it from falling to the ground where the hyena is waiting. After the carcass is lodged in another limb of the tree, the leopard retains its grip on the impala's neck for three to five minutes before letting the weight of the carcass settle into a fork on the limb. After releasing its grip, the leopard pants deeply for five to eight minutes as if exhausted. At 8:42 the leopard feeds again at the anal region of the carcass and one hindquarter, eating most of the flesh down to the knee joint. While the leopard is feeding, the rumen is ruptured as the leopard pulls on nearby flesh. Blood spurts from the carcass and runs down the tree trunk with the rumen contents.

After feeding eighteen minutes the leopard stops and carefully cleans its forelegs and paws, chest and lower neck. It licks its paws then washes its face with wet paws. After cleaning itself the leopard climbs out on a nearby limb, stretches out, closes its eyes, and apparently goes to sleep. The leopard's tail and legs hang limply and sway, as winds pick up velocity from an approaching thunderstorm. The leopard changes its resting position on the limb at 9:28 and three more times in the next seventeen minutes. At 10:45 the leopard leaves its resting place and returns to the carcass where it merely licks blood off the neck hairs. After licking blood for five minutes, the leopard jumps to a lower branch and carefully cleans itself again. Once, the leopard watches a hyena only 15 m from the tree. After cleaning for eight minutes, the leopard slowly stands upright on the limb, turns toward the tree trunk, and jumps to the ground. The leopard slowly walks away into the tall grass. I leave ten minutes later amid the thunder and lightning of a storm.

Carcasses that took leopards several days to eat, especially during the hot wet seasons, decayed rapidly. But by first eating parts of the carcass with the most flesh, leopards had less putrefied meat to eat later. Although flies were abundant on most carcasses of impala kills, only three were riddled with maggots while being eaten by leopards.

Rumina and their contents were not eaten by leopards. The rumina of seven of thirty-four partially eaten carcasses of impala were discovered on the ground or hanging from limbs beneath the carcasses. Other rumina were eaten by hyenas as the contents were scattered on the ground under the cache trees. Uneaten entrails were found at only one carcass after it had been fed upon by a leopard. I found no evidence that leopards first ate the entrails as reported by Schaller (1972), but leopards usually ate the entrails within twenty-four hours after they began feeding. The stomachs of two civets were also left by leopards. Leopards did not bury or cover the stomachs of civets or the rumina of antelope.

Leopards often plucked some of the longer hair of impala before feeding. I observed leopards plucking hair from the hindquarters of impala and discovered impala carcasses with hair plucked from their hindquarters, shoulders, and chest. Large amounts of plucked hair were usually found on the ground under carcasses leopards had fed upon. Leopards also plucked large amounts of hair from civets. I observed one leopard plucking hair from a civet with its incisor teeth, and then dislodging the hair with its tongue. Despite plucking hair, leopards still ingested considerable amounts of hair while feeding; feces of leopards contained hair of prey such as impala, civets, vervet monkeys, cane rats, and porcupines.

Once kills were safely cached in trees, leopards fed leisurely. Smaller civet-size kills were eaten rapidly, but larger impala-size kills were sometimes fed upon for as long as six days. A several-week-old impala was eaten by a leopard overnight, but the same leopard fed for two days on an adult impala. The average number of days leopards fed from forty kills was 2.4 days (range 1 to 6 days). No significant differences were apparent between the average number of days males (2.5 days) and females (2.3 days) fed at a kill. Older, mange-infested male leopards spent more time feeding than healthy male leopards, and females with cubs spent more days feeding at kills than females without young (see

TABLE 8.3 *Days Leopards Fed at Large Kills*

Status of Leopard	Number of Observations	Number of Days	
		x	*SD*
Old male	5	3.6	1.5
Males with mange	2	3.5	2.1
Healthy prime males	16	2.0	0.5
Female with cubs	6	2.5	0.8
Females without cubs	10	2.1	0.6

table 8.3). Leopards that fed for several days utilized the carcasses of prey more efficiently than those that fed rapidly. Older and mange-infested leopards consumed more skin and bones from the carcass than did healthy leopards.

Leopards fed irregularly throughout the day and night. Although thirteen of nineteen observations (68%) of feeding leopards were made during the day, daytime feeding intervals were brief. Most feeding by leopards occurred during the night. Of 3,230 min I watched over kills of leopards, leopards actually ate only 358 minutes (11%) of the time. The average uninterrupted feeding interval was twenty-four minutes (range three to eighty-four minutes). The average interval between feeding periods was 164 min (range 32 to 324 min). After a period of feeding, leopards left the cache tree (79%), remained in the tree to rest (8%), or remained in the tree to clean themselves (13%).

The major feeding periods of leopards were in the evenings after sunset, early in the night, during the middle of the night, and early in the morning before dawn. Leopards fed, but sometimes only briefly, at all hours during the day except between 1:00 P.M. and 3:00 P.M., the hottest part of the day. Leopards often returned to feed after sunset or during the first hours of darkness, and feeding occurred at irregular intervals during the night. One leopard fed twenty minutes at 8:30 P.M., twenty minutes at 10:55 P.M., and three minutes at 4:20 A.M. Another fed eighteen minutes at 7:35 P.M and forty-six minutes at 12:17 A.M. Although probably capable of eating large amounts of food, leopards did not appear to gorge themselves. Although leopards were observed, captured, and weighed with filled stomachs, I did not see leopards with greatly distended stomachs, even after watching one feed for eighty-four minutes.

Leopards often concealed themselves near a kill if they were not feeding. Sometimes leopards were seen lying, apparently asleep, over a limb in the tree near their kill. Other times leopards lay at the base of the tree or in nearby thickets where the kill was visible. If cover was scarce nearby, leopards rested in the nearest thicket, sometimes more than 100 m away, especially in the open short grass regions of the study areas. A female leopard once left her kill, traveled more than 1 km to a small koppie to her cubs, and after several hours returned to the kill. Another leopard left its kill, probably to drink at a nearby river, and returned several hours later.

Leopards fed on kills until all edible portions were consumed. Judging from the remains left behind, I concluded that leopards ate everything from adult impala carcasses except the leg bones, hooves, about 25% to 50% of the combined skull, vertebral column, and pelvis. Ribs were often consumed from the brisket about halfway to the vertebral column. On younger impala most of the ribs, part of the skull (nasal bones), and sometimes even the hooves were consumed. Only the stomachs and feet remained after leopards ate two civets. Remains falling from carcasses in trees were quickly carried away or consumed by hyenas.

I estimated that 16 to 20 kg of inedible portions of impala carcasses were left by leopards. Of this, 7 to 9 kg was vegetable material from the rumen and intestines. If adult impalas weighed from 44 to 60 kg (Hirst 1975), leopards consumed an average of 35% (27% to 45%) of each adult impala they killed. A greater proportion of flesh was probably eaten from younger impala. Small hooves in some leopard feces suggested several-week-old impala may have been entirely consumed. Utilization of carcasses of other large prey was probably comparable to that for impala, but smaller prey such as cane rats and genets were probably eaten entirely.

Scavenging

Given the opportunity, leopards will feed on carrion. To explain why the famous leopard of Rudraprayag became a man-eater, Corbett (1947) noted: "Leopards are scavengers, and when driven by hunger will eat any dead thing that they may find in the jungle." He speculated that man-eating began after this particular leopard scavenged bodies of humans killed by an influenza epidemic. Stock-killing leopards in the southwestern Cape Province of South Africa sometimes return to their kills even after the carcasses have become putrefied (Stuart 1986). Leopards are also readily attracted to baits in various stages of decay. Even decomposed baits riddled with maggots were occasionally attractive to leopards in my study areas. Leopards in KNP have also been seen feeding from a variety of carcasses, including an elephant. During an anthrax epizootic, four leopards died after feeding from disease-ridden carcasses (Pienaar 1969).

I witnessed scavenging by leopards on a major scale after many impala died following heavy rains on September 27–29, 1973. After the prolonged, intensive rainfall, carcasses of weather-killed impala were scattered throughout the SRSA and adjacent regions (see photo 8.3). The first evidence of scavenging of these carcasses by leopards occurred on September 30 when a bloated, uneaten carcass of an impala was found cached in a tamboti tree along the Sabie River (see photo 8.4). The carcass was not fed upon for at least two days. Then the leopard returned and moved the carcass to a more secluded location, away from

8.3. Carcass of one of the many impala that died during a prolonged period of intense rainfall.

8.4. An impala carcass found by a leopard and cached in a Tamboti tree.

the tourist road. Also on September 30, 25 km away, I approached radio-collared M2, a subadult male and discovered he had also fed on a decomposed carcass of an impala, which he dragged under a sicklebush. Only the hindquarters were eaten; and the leopard did not return to feed again on the carcass.

On October 5, as I approached a radio-collared leopard I suspected was feeding on a kill because of the cessation of his movements, I discovered drag marks and tracks in the mud. Backtracking, I discovered a ravine where several impala had died. The ravine was crossed by many drag marks and trails of leopards. A partially eaten and decomposed carcass of an impala was found about 1 m above ground on a fallen tree. After following the freshest drag marks several hundred meters to the nearby Sand River, I accidentally startled an uncollared leopard from the decomposed carcass of an impala. I also found an older, partially-eaten, decomposed impala carcass nearby. Ten meters from this carcass was yet another uneaten carcass surrounded by leopard tracks. Although I did not observe the radio-collared leopard I had initially approached, he was nearby and also suspected of feeding on impala carcasses.

These scavenging leopards exhibited several characteristics that were not observed again. One leopard cached a carcass for at least two days before feeding upon it. Also, the leopards left carcasses lying on the ground within reach of hyenas. This behavior, as well as abandoning uneaten carcasses, suggested that food was plentiful, and leopards could risk losing carcasses to hyenas. Interestingly, during this period many of the impala carcasses were not eaten by hyenas or vultures, although hyenas passed by within several meters of carcasses. Perhaps hyenas were also satiated from eating other carcasses. This unusual behavior was not observed again in hyenas.

Leopards occasionally appropriate carcasses of prey from other predators. Leopards take kills from hyenas (Kruuk 1972), cheetahs (Schaller 1972), and, as I discovered, other leopards.

June 9, 1975, SRSA, 9:00 A.M.: While approaching a radio-collared subadult male leopard on foot, I discover an uncollared adult male leopard, feeding in a tree. The radio signal indicates the radio-collared leopard is only 10 m from the older adult male, but I am unable to see it. The feeding leopard eats from 9:00 to 10:21, tearing flesh from the head and neck of an almost completely eaten impala carcass. The head of the impala nearly falls from the tree four times, but each time the leopard grabs it with a paw just in time (see photo 8.5). At least two hyenas circle the base of the tree, looking up at the kill. After eating the nose of the impala, the leopard stops and cleans himself for three minutes, then jumps to the ground. The radio-collared subadult male is still not visible. When 3 m from me, the uncollared adult male snarls threateningly five times, presumably at the hidden leopard, walks in front of the bush I'm hiding behind, and calmly walks away. I then hear scratching on the tree trunk

8.5. An uncollared adult male leopard feeds on an impala carcass in a tree. Hidden nearby is a radio-collared subadult male waiting for an opportunity to scavenge from the kill.

and see the subadult male in the tree reaching for the remains of the impala (see photo 8.6). The subadult sees me, freezes, and then jumps 3 m to the ground and flees. Ten minutes later I locate him 50 m from the tree. When I return to the tree the next morning the remains of the impala are gone.

The large male had apparently killed the impala the previous day because most of the carcass had been consumed. The radio-collared male was not nearby the previous day, so apparently he discovered the carcass sometime during the night. Several weeks later I captured an unmarked adult male leopard near the kill and through photographs identified it as the one observed on June 9. The subadult (M26) was obviously scavenging from the kill of the older male, and on three other occasions I located him near kills of other leopards. On two of these occasions he scavenged from the carcasses.

On August 8, 1974, one leopard attempted to take the carcass of a bushbuck being consumed by another adult male leopard along the Lower Sabie tourist road. The leopards exchanged blows but the adult male retained control. The agonistic feeding behavior of these leopards and of the radio-collared subadult M26 contrasted with the peaceful feeding behavior of a then-unmarked adult female and her offspring (F12).

As a feeding strategy, scavenging may contribute to the leopard's success as

8.6. A radio-collared subadult male leopard begins to feed on an impala carcass in a tree minutes after the kill is abandoned by an uncollared adult male leopard.

a predator. Leopards will apparently feed on carrion, even when live prey is abundant, if the carrion can be easily obtained. Scavenging may also be an important feeding strategy of leopards temporarily suffering from physical handicaps that could prevent them from capturing prey. Scavenging may also be important to young leopards that lack hunting experience. Scavenging may help leopards survive periods when they are unable to capture prey.

Diet of Leopards

I collected food habits data by observing kills located after stalking radio-collared leopards on foot and by examining recently deposited leopard feces. Although fewer feces were found during wet seasons, I saw no significant difference between the types of prey remains found in leopard feces collected during wet and dry seasons.

COMPOSITION OF DIET

Ungulates were the most frequently eaten prey of leopards in both study areas. Ninety-three percent of the leopard kills and 64% of the feces were ungulates or contained ungulate remains (see table 8.4). Of the ungulates, impala were

TABLE 8.4 *Diet of Leopards in Study Areas*

	Diet Determination Technique			
	Kills (n = 55)		Feces (n = 94)	
Prey	n	Occurrence %	n	Occurrence %
Large ungulates	51	93	60	64
Large carnivores	2	4		
Smaller mammals	2	4	27	29
Birds			2	2
Unidentified mammal			7	7

the most common prey of leopards. Impala remains were found in 60% of ninety-four leopard feces. And of fifty-five leopard kills, forty-eight (87.3%) were impala and three (5.7%) were bushbuck, warthog, and steenbuck; the remainder were not ungulates.

Food habits of leopards throughout KNP and other similar habitats in South Africa indicate that ungulates form the bulk of the leopard's diet (see table 8.5). Impala are apparently also an important ungulate prey of leopards throughout southern Africa. Seventy-seven percent of all leopard kills in KNP were impala as well as 83% to 92% in similar habitats (see table 8.5). Only in one area, the Umfolozi Game Reserve, was another ungulate, the bushbuck, taken more frequently by leopards than were impala (P. Brooks, personal communication).

The high frequency of impala taken by leopards is probably related to impala abundance. Impala are often the most abundant prey in many African parks and reserves including KNP. In the dense vegetation of Tsavo National Park, leopards frequently killed impala (Hamilton 1976). But in the savanna region of East Africa, where Thompson's gazelle are abundant, they were the leopard's most frequently taken prey (Kruuk and Turner 1967). In Zambia leopards frequently killed reedbuck, puku, and waterbuck (Mitchell et al. 1965). Axis deer and wild pig were the principal prey of Asian leopards in Wilpattu National Park, Sri Lanka (Muckenhirn and Eisenberg 1973).

In my leopard study areas, small mammals ranked second in importance to ungulates in the leopard's diet. Although small mammals made up only 4% of the kills I located, at least 29% of the leopard feces examined contained remains of small mammals. Smaller mammals eaten by leopards included vervet monkeys, cane rats, porcupines, and genets. Leopards were also known to have killed two civets, but civet remains were not identified in feces. Although I could not identify other species of small mammals eaten by leopards, the remains suggested hares and mongoose (see table 8.6).

Vervet monkeys were an important small prey of leopards in both study areas. These agile primates were often seen on the ground, but I did not know where leopards captured them. Their intense alarm response to leopards also

TABLE 8.5 *Food Habits of Leopards in the Republic of South Africa*

Prey	Kruger N. P. (1)	Kruger N. P. (2)	Sabi-Sand G. R. (3)	Sabi-Sand G. R. (4)	Timbavati G. R. (5)	Umfolozi G. R. (6)
Impala	87.3	77.7	83.4	83.0	92.3	10.5
Bushbuck	1.8	3.9		2.4		21.5
Waterbuck		3.9	3.3	0.7	0.6	7.9
Warthog	1.8	1.4	0.5	5.4	0.6	15.8
Gray duiker				4.3		7.9
Steenbuck	1.8			0.9		
Common reedbuck		2.3				
Greater kudu		2.9	3.0	1.9	4.8	
Nyala		0.4				18.4
Mountain reedbuck		0.1				2.6
Wildebeest		1.3	2.1	0.9	1.2	2.6
Zebra		1.2	0.2	0.1		
Tsessebe		0.2				
Eland		0.2				
Sable antelope		0.1				
Roan antelope		trace				
Buffalo		0.1				
Giraffe					0.6	
Vervet monkey						7.9
Civet	3.6					
Chacma baboon				0.4		5.3
Black-backed jackal				0.1		
Spotted hyena	3.6					
Others[1]		4.5	7.4			
Total kills	55	5,501	1,135	882	168	38

Heading: Occurrence in Diet of Leopards (%)

SOURCES OF DATA: (1) = This study, (2) = Pienaar (1969), (3) = Graupner and Graupner (1971), (4) = Crabtree (1973, 1974), (5) = Hirst (1969b), (6) = P. Brooks (personal communication).
  [1] = gray and red duiker, steenbuck, Sharpe's grysbuck, klipspringer, sumi, bushpig, ostrich, bushbuck, and reedbuck.

TABLE 8.6 *Prey Remains in Leopard Feces from Leopard Study Areas*

Prey	N	Occurrence %
Impala	56	60
Gray duiker	1	1
Unidentified ungulate	3	3
Vervet monkey	8	9
Porcupine	2	2
Cane rat	5	5
Genet	3	3
Unidentified small mammal	9	10
Unidentified mammal unknown size	7	7
Guinea fowl	1	1
Unidentified bird	1	1

suggested leopards were an effective predator on vervets. Although porcupines and cane rats were probably more vulnerable than vervet monkeys, Turnbull-Kemp (1967) noted that porcupines occasionally injure and sometimes even kill leopards with their long, sharp, pointed quills. Small mammals appear to form an important part of certain leopards' diet and may be especially important to young or injured leopards that are unable to capture larger prey.

Where small mammals are particularly abundant, they may become important prey of leopards. In Matopos National Park, Zimbabwe, Grobler and Wilson (1972) discovered that hyraxes, vlie rats, cane rats, springhares, and hares were important leopard prey. In Tsavo National Park, Hamilton (1976) reported that rodents, hyraxes, and hares were frequently eaten by leopards despite the fact that kill data suggested that impala were the most frequently eaten prey. Hyraxes also appear to be an important prey of leopards in the mountainous areas of the southwestern Cape Province in South Africa (Norton et al. 1986). Remains of rodents and hares were also found in the feces of Asian leopards in Wilpattu National Park (Eisenberg and Lockhart 1972).

Birds ranked third in importance in the leopard's diet in the KNP study areas. Two percent of the leopard feces contained remains of crowned guinea fowl and an unidentified bird, perhaps a hammerkop. The frequency of occurrence of bird remains in the feces of leopards from other regions vary from none to as high as 27% (Hamilton 1976).

FACTORS INFLUENCING DIET OF LEOPARDS

Size.   Most ungulates killed by leopards, such as impala, Thompson's gazelle, bushbuck, reedbuck, and axis deer, weigh about 30 kg, or slightly less than the average weight of leopards. Smaller ungulates, such as duiker, puku, steenbuck, and klipspringer, appear to be frequently killed by leopards only in areas where medium-size ungulates such as impala are not abundant. Energy expenditure and behavior probably play a significant role in determining size of prey killed by leopards. Small ungulates such as duiker and steenbuck are solitary, well dispersed, and probably difficult to locate; therefore leopards would have to expend much energy locating them. Larger ungulates, such as adult kudu, are seldom killed by single leopards, perhaps because of the danger involved. Considering the amount of energy gained from small prey and the fact that leopards cannot risk killing large, potentially dangerous prey, impala and similar-size ungulates probably provide the greatest return of energy for that expended in locating and killing them.

Leopard predation on larger ungulates is primarily on young individuals that are about the same size as impala and bushbuck. Of fifty kills of kudu, zebra, wildebeest, waterbuck, and buffalo in KNP, the majority were nursing young (52%) or subadults (38%) (Pienaar 1969). In nearby Sabi-Sand Game Reserve,

93% of the kills of larger antelopes were also young animals (Crabtree 1973, 1974).

Although leopards are capable of killing ungulates two to three times their own weight, apparently they seldom do so. The largest ungulates killed by leopards have been adult kudu and hartebeest (Mitchell et al. 1965), topi (Schaller 1972), and wildebeest (Pienaar 1969; Kruuk and Turner 1967), which weigh 109 to 130 kg. A giraffe calf weighing not less than 91 kg was once observed 3.9 m in a tree, where it had been cached by a leopard (Stevenson-Hamilton 1947).

Extremely small prey are also outside the leopard's normal size range of prey, although leopards may occasionally eat small prey such as passerine birds, chameleon, scorpion, and insects (Grobler and Wilson 1972). Hamilton (1976) found the remains of grasshoppers, centipedes, and a scorpion in the feces of leopards, but these food items are probably opportunistically or accidentally eaten.

Vulnerability.   The vulnerability of prey to leopard predation probably varies with prey density and availability as well as the leopard's preferences. The population density of prey is significant because abundant prey will be encountered more frequently than scarce prey. Availability depends on prey population density, behavior, and habitat preferences. Baboons may be abundant in an area, but their aggressiveness and their ability to escape into trees may prevent leopards from capturing them. Young wildebeest and zebra may be seasonally abundant but unavailable to leopards because of their preference for open plains, where stalking is difficult. In Matopos National Park, where klipspringers were abundant, their remains occurred in 14% of the leopard feces; however, the remains of steenbuck, a similar-size but less abundant species, occurred in only 2% (Grobler and Wilson 1972).

The leopard is believed to be the most important predator of small ungulates in KNP. Fifty to 85% of all kills of klipspringer, bushbuck, Sharpe's grysbuck, steenbuck, oribi, and mountain reedbuck were attributed to leopards, and leopards killed more gray duiker, nyala, reedbuck, and impala than any other predator (Pienaar 1969). Seventy-seven percent of all baboon kills were also attributed to leopards. Most smaller ungulates are vulnerable to leopard predation because they are easily killed and often live in habitats favored by leopards. Klipspringers favor rocky outcrops, Sharpe's grysbuck and mountain reedbuck favor some rocky terrain, and gray duiker, bushbuck, and nyala favor riparian habitats, all of which are also favored by leopards. Smaller ungulates are of minor importance to other major predators in KNP because they are too small to be regularly preyed upon by large predators (lions) or social predators (lions, hyenas, and wild dogs). These smaller ungulates are also too large to be taken by smaller predators such as jackals, caracals, and servals, although these smaller predators sometimes kill and eat them.

Based on their low relative abundance and high kill frequency, bushbuck

TABLE 8.7 *Food of Leopards*

	Numbers of Occurrences					
	Sabie River Study Area			Nwaswitshaka River Study Area		
Prey	Kills	Feces	Total	Kills	Feces	Total
Ungulates	26	27	53	24	19	43
Carnivores	2		2			
Small mammals	2	15	17		6	6
Birds		2	2			
Unidentified mammals		2	2		5	5
Total	30	46	76	24	30	54

Chi-square test: Chi-square = 8.08, *df* = 4, *P* < 0.100 (Sabie River versus Nwaswitshaka River distribution of prey items in combined kills and feces).

appear especially vulnerable to leopard predation in Kruger Park, followed by reedbuck and waterbuck (Pienaar 1969). Impala are often killed, probably because they are most abundant and are of the size preferred by leopards. Impala are probably less vulnerable to leopard predation than bushbuck because impala favor more open habitats and live in herds. Even if impala are more difficult to stalk than bushbuck, the leopard may have greater success in taking impala because there are more opportunities. Although bushbuck may be killed by leopards more often than impala, their low relative abundance may reduce their current overall importance to leopards. In my leopard study areas, the smaller solitary ungulates were less abundant than impala and were killed less frequently by leopards, despite their vulnerability. Because of their greater numerical abundance, impala are probably currently serving as a buffer species, reducing predation on less abundant, smaller antelopes. Under observed leopard population levels, if the impala population in the park declined, an increased but temporary amount of predation on the smaller ungulates might occur. However, the leopard population would eventually decline because there are fewer of these smaller ungulates to support leopards, compared to impala.

The difference in the composition of the diet of leopards in the SRSA and NRSA was probably due to differences in prey vulnerability. Examination of kills and leopard feces revealed that leopards in the SRSA ate fewer ungulates and more small mammals and birds than leopards in the NRSA (see table 8.7). Small mammal remains were found in only 21% of the feces of leopards in the NRSA, compared to 33% in the SRSA. Differences in population density and distribution of small mammals and birds probably contributed to this difference in food habits. Ecological densities of smaller mammals and birds were two to three times greater in the SRSA. Furthermore, riparian habitat, favored by baboons, vervet monkeys, cane rats, and genets, was more widely distributed in the NRSA, thus increasing the leopard's hunting area. Leopards in the NRSA had larger home ranges and traveled greater distances than did leopards in the

TABLE 8.8 *Food of Leopards in the Study Areas*

| Prey | Number of Occurrences | | | | | |
| | Dry Seasons | | | Wet Seasons | | |
	Kills	Feces	Total	Kills	Feces	Total
Ungulates	21	26	47	30	2	32
Carnivores				2		2
Small mammals	2	12	14		1	1
Birds		1	1		1	1
Unidentified mammals		1	1			
Total	23	40	63	32	4	36

Chi-square test: Chi-square = 10.50, *df* = 4, *P* < 0.050 (dry season versus wet season distribution of prey items in combined kills and feces).

SRSA. However, leopards in the NRSA occasionally ate cane rats, so apparently they were capable of locating these relatively rare prey in that study area.

The diet of leopards varies considerably, even within KNP. Impala are apparently more vulnerable to leopard predation in the park's northern and central districts than in the southern district (Pienaar 1969). Vulnerability of impala to leopard predation in these districts is probably related to habitat and impala distribution. In the central and northern districts most impala occur along thickets near watercourses, because the remaining habitat is open grassland (central district) or mopane shrub and tall grassland (northern district). Thus, despite their lower overall abundance in these districts, impala are forced by their food habits and water requirements to live in riparian habitats frequented by leopards. In the park's southern district, where thorn thickets and other dense cover are more widespread, more habitat is available for impala. This may decrease impalas' chances of encountering leopards, relative to the northern and central districts. In those districts, where impala are less numerous, waterbuck and bushbuck, riparian-loving species, were utilized more intensively by leopards. This relationship also exemplifies the buffering influence that large numbers of impala have on other prey species within the park.

Wildebeest and zebra, both varying in relative abundance from 1.3% to 15.4% in the park, were not vulnerable to leopards, probably because they live in open habitats avoided by leopards. Reedbuck, which favor tall grass, were especially vulnerable to leopard predation in the park's central district, despite the fact that reedbuck habitat there is limited.

A significant difference between the diets of study area leopards during dry and wet seasons was also noted. Most leopard kills were located during wet seasons and most feces collected during dry seasons. The seasonal distribution of ungulates, carnivores, small mammals, and birds in the diet of leopards differed (see table 8.8), suggesting that some species were more difficult for leopards to capture during certain seasons. Leopard response to trap baits and the kill frequency of impala during dry seasons indicated that impala were difficult to capture during dry seasons. Pienaar's (1969) data also suggested that leopards killed fewer impala but more other ungulates during dry seasons. Leopards killed proportionately more reedbuck, duiker, and small mammals from August through October, the late dry season, than at any other period.

Sex and age of prey.   Of forty-one leopard-killed impala in the study areas, 39% were males and 61% females. This sex ratio was not significantly different from the observed sex ratio (42% males, 58% females), so apparently leopards did not selectively kill impala according to sex. Slightly more male impala were killed by leopards in the study areas during dry (44%) than wet (36%) seasons, but the difference was not significant. Data on kills throughout the park (Pienaar 1969) also suggest that more male impala were killed by leopards during the dry seasons (51%) than wet seasons (42%). Male impala probably become vulnerable to leopard predation during dry seasons because of the rut. During this time, territorial males may force subordinate males to use marginal habitat (Anderson 1972), where they may become vulnerable to leopards. This probably occurred on the leopard study areas because territorial male impala frequently selected open brackish flats for their territories. In contrast, subordinate males and bachelor male herds inhabited the dense *Grewia* thickets during the rut. These dense thickets could have provided leopards the necessary stalking cover to capture males. During the rut, the territorial males would be less vulnerable to leopard predation because of the sparse stalking cover they inhabited.

In the Serengeti, where leopards killed proportionately more male than female Thompson's gazelle, Schaller (1972) speculated that males were more vulnerable than females to leopard predation because nonterritorial males used riparian habitats where leopards often hunted. Schaller was not able to explain a similar high proportion of female kills among reedbuck by Serengeti leopards.

Females are apparently killed by leopards more often than males among the larger ungulates in KNP. Females made up 63%, 73%, 67%, and 60%, respectively, of the kills of waterbuck, kudu, zebra, and wildebeest of known sex (Pienaar 1969). Males of these species are generally larger, more aggressive, or have potentially dangerous horns. Most of these large ungulates taken by leopards were young or subadult individuals of both sexes. Regardless of age, the smaller size and perhaps the behavior of females may make them more vulnerable than males to leopard predation.

The ages of impala killed by study area leopards suggested that in this medium-size ungulate younger males were killed more often than younger females. Ten of sixteen (63%) male impala, compared to four of twenty-one (19%) female impala, were less than two years old when killed by leopards. Six of sixteen (37%) male impala killed by leopards were less than one year old, but only three of twenty-one (14%) females. Because the proportion of young male impala killed by leopards was greater than that present in the population (see table 8.9), leopards were apparently selecting young male impala. Jarman and

TABLE 8.9 *Relative Ages of Impala*

| | Impala | | | |
| | Males | | Females | |
Age	Population Sample	Killed by Leopards	Population Sample	Killed by Leopards
Adult	157	6	176	17
Young[1]	55	10	112	4
Total	212	16	288	21

[1] 0–2 years old

Chi-square test: Chi-square = 9.77, *df* = 1, *P* < 0.010 (male impala: population sample versus number killed by leopards).

Chi-square test: Chi-square = 3.28, *df* = 1, *P* < 0.10 (female impala: population sample versus number killed by leopards).

Jarman (1973b) discovered that male impala less than two years old were subject to higher natural mortality rates than females of the same age.

For some reason, impala one year or younger and impala more than four years old may be especially vulnerable to leopard predation. Of 182 impala killed by leopards in the nearby Sabi-Sand Game Reserve, the majority (23.6%) were four to five years old (Graupner and Graupner 1971). More impala one year or younger were probably killed by leopards than their data indicate; remains of young impala are difficult to discover. To determine whether leopards in the study area regions of KNP and nearby Sabi-Sand Game Reserve selected impala by age, I compared ages of Graupner and Graupner's leopard-killed impala to the ages of impala in KNP (Fairall 1969). Assuming the age structure of impala in both areas were similar, this comparison suggested that leopards in the study area regions also selected impala four to five years old and older (see table 8.10).

I could not explain why four- to five-year-old impala were apparently more susceptible to leopard predation. If the majority of four- to five-year-old impala kills were males, behavior could be a contributing factor. Social status of males in bachelor herds is age dependent (Jarman and Jarman 1973b). Perhaps when males four to five years old leave the bachelor herds to establish territories, they become more vulnerable to leopard predation. Older males may maintain better territories (Anderson 1972), so perhaps prime territories provide better visibility to escape predators.

If leopard-killed four- to five-year-old impalas were primarily females, their increased vulnerability could be related to reproductive behavior. Schenkel (1966) noted that some female impala leave the safety of large herds to temporarily join bachelor male herds to elicit courtship behavior before lambing. Females also frequently give birth to young in secluded areas. If courtship and parturition is most prevalent among females four to five years old, then such females may be more subject to leopard predation than other females because

TABLE 8.10 *Comparison of Ages of Impala*

Age Class of Impala	Collected in Kruger National Park (Fairall 1969)		Killed by Leopards in Sabi-Sand Game Reserve (Graupner and Graupner 1971)	
	n	Occurrence	*n*	Occurrence
0–1	267	33	33	18
1–2	180	22	31	17
2–3	133	16	17	9
3–4	113	14	14	8
4–5	73	9	43	24
5–6	25	3	19	10
6–7	8	1	14	8
7–8	5	1	8	4
8–9	3			
9–10	2	1	3	2
Total	809		182	

Chi-square test: Chi-square $= 120.06$, $df = 9$, $P < 0.010$ (ages of impala: numbers in population in Kruger National Park versus numbers killed by leopards in the Sabi-Sand Game Reserve).

they temporarily leave the protection of the herds. The age at which female impala are more reproductively active is unknown.

Sex and age of leopards killing prey.   Remains of prey in feces from known-age leopards revealed significant differences in food habits among leopards. Feces collected from traps containing captured individuals revealed that the greatest difference in food habits occurred between adult males and subadult females and the least between adult females and subadult males (see table 8.11). Ungulate remains, for example, occurred most often in feces of adult leopards, whereas small mammal and bird remains occurred more frequently in feces of subadult leopards, especially females.

These data suggested that younger leopards took smaller prey more frequently than did adult leopards. Perhaps younger leopards are not large enough or lack the experience to kill the larger prey often taken by older leopards. If experience is a prerequisite before younger leopards can successfully capture

TABLE 8.11 *Prey Remains in Feces of Known-age Leopards*

Sex	Age	Number of Prey Occurrences		
		Ungulates	Small Mammals	Birds
Male	Adult	5	0	0
Male	Subadult	2	1	0
Female	Adult	5	1	0
Female	Subadult	0	3	1

TABLE 8.12 *Impala Killed by Known-age Leopards*

Leopard		Number of Impala			
		Male		Female	
Sex	Age	Adult	Subadult	Adult	Subadult
Male	Adult	0	1	9	0
Male	Subadult	2	0	1	0
Female	Adult	1	3	1	1
Female	Subadult	0	1	2	0

large prey, smaller prey such as mammals and birds may be an important component of a leopard's environment. Muckenhirn and Eisenberg (1973) suggested that there may have been separation in effective predation pressure by different sex and age classes of Asian leopards in Wilpattu National Park. Although they speculated that younger leopards may have been more successful in preying on small mammals in trees, my data suggested that size of leopards was not the only influence. Difference among weights of adult and subadult females, both of which had significantly different food habits, was not as great as that between subadult males and females, with similar food habits. My data suggested that age and hunting experience of leopards had an equal or greater influence on food habits of leopards than did the size of the leopard.

Age and sex of impala killed by known-age leopards revealed that adult male leopards killed fewer adult and more subadult male impala than did subadult male leopards (see table 8.12). Adult male leopards also killed more adult female and fewer adult male impala than did adult female leopards. Surprisingly, adult male and subadult female leopards killed impala of comparable sex and age. Why these two classes of leopards killed prey of similar sex and age was unknown. Perhaps these particular leopards discovered adult female impala were the most vulnerable to kill. Perhaps subadult female leopards frequently killed female impala because they were unable to kill the larger male impala, which would not explain why adult males killed more female impala.

Kill Rates

To determine how often leopards killed large prey, I monitored two radio-collared leopards almost daily for forty consecutive days. Kills were confirmed by approaching the leopards on foot until a carcass or remains of prey were observed. Movements of other radio-collared leopards were also used to determine intervals between suspected kills. In analyzing movements I excluded intervals if two or more days elapsed between location fixes, or when leopards fed on baits. Females with cubs were also excluded because they often remained with their cubs for several days in the same location. Because leopards were sometimes found as much as 1 km from their kills, and because of the angular error in the tracking system, I chose a movement of 0 to 400 m between consecutive daily locations of leopards as the criteria for suspecting a kill. Seventy-five to 91% of forty-eight suspected kills were visually verified. Because of the possibility that leopards reduced their movement for other reasons, kill rates based on these criteria may be slightly greater than actual rates.

Leopards averaged about one kill, usually an impala, per week. The average interval between ninety-four suspected kills was 6.9 days and between nine confirmed kills of two leopards, 7.9 days. These two differently determined kill rates were not significantly different. Although the intervals between kills of large prey varied from one to nineteen days, leopards could also have killed smaller prey between these detected kills of large prey. Some leopards also fed on baits near traps between kills of large prey. Male 14 killed six impala between June 11 and July 20, 1975. In addition, he once fed on a trap bait and twice ate bait inside traps after being captured. Male 26 fed on five impala, scavenged the remains of another impala, and fed on bait in traps three times between May 20 and June 28, 1975.

Size of prey influences kill rates of leopards. In the Serengeti where Thompson's gazelle formed the bulk of the leopard's diet, Schaller (1972) estimated that leopards killed forty to forty-eight gazelle per eight-month period. This averaged about five to six days between kills of prey weighing up to 23 kg. In Zambia a leopard killed two reedbuck, a bushpig, and a duiker during a twenty-day period (Mitchell et al. 1965). This averaged five days between kills and probably provided the same amount of meat had the kills been the size of a gazelle. Pienaar (1969) estimated that leopards in KNP killed twenty-five head of large game or fifty head of smaller game annually. Pienaar's upper limit corresponds to a kill rate of 1/7.3 days, a rate similar to that in my leopard study areas. In the Londolozi Game Reserve near KNP, a female leopard was seen with twenty-eight carcasses of prey weighing more than 10 kg in 330 days, or 1 carcass each 11.8 days (Le Roux and Skinner 1989).

Some differences were noted in kill rates among leopards. Leopards averaged a kill once every 7.1 days in the dry seasons, compared to 6.8 days in the wet seasons. This suggested that leopards may have had a more difficult time killing large prey, primarily impala, in the dry seasons. During the study period leopards characteristically killed prey more often during each wet season than during the previous dry season (see table 8.13). Leopards also tended to kill large prey more frequently each consecutive season throughout the study.

Increased density of stalking cover in the study areas could also have contributed to an increased kill rate for leopards. Before my study began, rainfall in the district had been low and vegetation growth slow. After heavy rains at the beginning of the study, vegetation became and remained dense throughout the study period. Seasonal fluctuations in density of vegetation occurred, but there

TABLE 8.13 *Interval Between Suspected and Confirmed Kills of Leopards*

Leopard		Season											
		1973 Dry		1973–74 Wet		1974 Dry		1974–75 Wet		1975 Dry		Total	
Sex	Condition	n	Days	n	Days	n	Days	n	Days	n	Days	n	Days
Male	Old			2	6.5			2	5.7			4	5.9
	With mange					5	6.0					5	6.0
	Prime	8	8.8	22	6.4	6	7.3			9	7.9	45	7.2
Female	With mange					14	4.0			4	3.5	5	3.6
	Prime			14	8.7	24	7.3	3	4.3	4	3.5	40	7.5
Total		8	8.8	38	7.3	35	7.0	9	5.2	13	6.5	103	7.0

was a general increase in vegetative cover throughout much of KNP during this period (Smuts 1976). This increase in stalking cover may have enabled leopards to capture prey more readily.

No significant difference among kill rates of healthy male and female leopards was apparent. Males killed large prey once every 7.2 days; females killed once every 7.5 days (see table 8.13). Leopards with mange spent more time feeding at kills than did healthy leopards (see table 8.13). Old leopards and leopards with mange fed more often than healthy leopards. In the Kalahari Desert male leopards killed prey once every 3 days and females with cubs once every 1.5 days (Bothma and Le Riche 1984). As their hunger increased, leopards in the Kalahari moved increasingly longer distances per day to increase prey contacts (Bothma and Le Riche 1990).

If healthy male leopards killed large prey, of which 87.3% were impala, once every 7.2 days, they would have killed an average of 45 impala/year. Their kills would also include an average of three other ungulates, usually warthog, bushbuck, steenbuck, or duiker, and three smaller mammals. Female leopards would kill an average of forty-three impala, three other ungulates, and three smaller mammals per year. These are merely crude average estimates, because leopards undoubtedly killed more smaller mammals than indicated by kill data. Some kills of leopards were probably lost to other scavengers, and others may have been scavenged from other predators. Kill rates probably vary considerably with seasons and individuals. Younger leopards probably killed more smaller mammals and fewer impala than did older leopards. Leopards may also have killed more newly born impala than my data suggested. My estimates of annual kill rates by leopards suggested that leopards killed considerably more ungulates, not fewer, than the 25 head/year estimated for leopards in the park by Pienaar (1969).

Consumption Rates

The amount of food consumed by leopards depends on size of prey and the proportion of kills and edible proportion of each kill lost to scavengers. Kills cached in trees were relatively safe from scavengers; therefore I assumed that loss to other predators were insignificant. I also assumed that leopards killed most of their own prey.

Because I could not weigh carcasses of prey killed by leopards, I estimated uneaten portions left behind. If 11% of an average impala is digestive tract contents, 5% is hide (Ledger 1968), and 14% is skeleton (Young and Wagener 1963), about 30% of each impala was inedible. The weight of meat left behind on the carcass after feeding was assumed to equal the amount of skin and bone consumed by leopards. For simplicity, I also assumed that 30% of the carcasses of other prey species were also inedible, although smaller mammals and birds were probably entirely eaten.

Two male leopards that I monitored almost daily during the dry season probably killed and consumed prey at a maximum rate. Adult M14 killed two adult male impala, one adult female, and one seven-month-old male; he also ate baits in traps three times in forty days. I found evidence that this male also killed two other impala but I did not locate their remains. If the two unidentified impala were adult females, M14 killed 259 kg of impala or 6.5 kg/day. Assuming that 30% was inedible, and not more than 5 kg of bait was eaten during the period, this leopard consumed 186 kg, or 4.7 kg/day.

During another forty-day period in the dry season, subadult M26 killed one adult male and one adult female impala, fed several days at each of three other kills of impala, scavenged the remains of an impala killed by another leopard, and fed on bait at traps three times. Assuming the three unidentified impala were females and that only M26 fed on them, he killed about 236 kg, or 5.9 kg/day. If one assumes 30% wastage from the five impalas, that 5 kg was scavenged, and 5 kg was consumed from the trap baits, M26 consumed 175 kg, or 4.4 kg/day.

Estimated metabolic requirements of leopards were also used as an alternate method of determining consumption rates. Using Lamprey's (1964) method, I estimated that an average 52.8 kg male leopard required 3,888 cal/day and the average 37.5 kg female leopard required 3,024 cal/day. If the meat of impala supplies 1,020 cal/kg (Ledger 1968), and the caloric content of other prey was equivalent to impala, then male leopards would have to consume 3.8 kg/day and female leopards 3.0 kg/day. Assuming 30% wastage, male and female leopards would then have to kill 5.4 and 4.3 kg/day, respectively.

Consumption rates were also estimated from average kill rates of leopards. Based on the sex and age ratio of impala killed by leopards in the study areas, twenty-two of the forty-five impala killed annually by male leopards were adult females weighing 44 kg; five were juvenile females weighing 23 kg; seven were adult males weighing 60 kg; and eleven were juvenile males weighing 23 kg. If three smaller ungulates totaling 70 kg and three other mammals totaling 41 kg are added to 1,756 kg of impala, a total of 1,867 kg of prey were killed annually by male leopards. This is equivalent to 5.1 kg/day and, assuming 30% wastage, a consumption rate of 3.1 kg/day. Using the same sex and age ratios of prey for the forty-nine estimated annual kills by female leopards, they killed about 4.8 kg/day and consumed 2.9 kg/day.

Estimates of consumption rates of leopards can also be obtained from captives. Eisenberg and Lockhart (1972) reported leopards required 950 kg/year, or 2.6 kg/day, at the National Zoo. But Crandall (1964) reported captives were maintained by feeding them only 1 to 1.2 kg/day. However, it is likely that captive leopards require fewer calories than leopards in the wild; I assumed that adult male leopards in the study areas killed at least an average of 5 kg and consumed at least 3.5 kg/day. Adult female leopards probably killed at least 4 kg and consumed 2.8 kg/day. In the Serengeti, Schaller (1972) estimated leopards killed only 2.7 to 3.3 kg/day. Leopards in southern Africa, however, may have slightly higher consumption rates than those from East Africa because of their larger size and the cooler temperatures of the more southern latitudes.

Impact of Leopard Predation on Prey Populations

The impact of predators on prey populations in KNP is of significance because a management objective is to ensure that a balance exists between the numbers of animals and their habitat (Fairall 1969). Despite the park's large size, traditional migrations and seasonal feeding areas have long been altered by game-proof fences and artificial sources of drinking water (Smuts 1982). Because of these factors, park officials have taken an active role in management, using techniques such as culling, controlled burning, translocation, reintroduction, and disease control. Although predator control has been used as a management technique in the park, it probably had little impact on the park's overall predator population (Smuts 1982). More recently, certain species of ungulates were given the opportunity to increase by temporarily culling lions and spotted hyenas. Because of these concerns about predation, I assessed, in a crude manner, the impact leopards may have had on several ungulate populations in my study areas.

IMPALA

Because impala were taken by leopards more often than any other ungulate in the study areas, one might assume if leopard predation impacted any prey populations, impala were likely candidates. Predation's influence on impala

TABLE 8.14 *Impala Population and Recruitment Taken by Leopards and Other Predators*

Study Area	Year	Estimated Annual Impala Population[1]	Recruitment[2]	Proportion (%) taken by Predators — Annual Population — Leopards	All Predators	Recruitment — Leopards	All Predators
Sabie River[3]	1973	2,533	1,039–1,356	9	21	17–22	39–51
	1974	1,572	678–842	15	33	28–34	62–78
	1975	1,000	431–535	23	53	43–54	98–122
Nwaswitshaka[4]	1973	3,240	1,398–1,735	13	29	24–30	54–68
River	1974	6,496	2,803–3,478	6	15	12–15	27–34
	1975	13,252	5,718–7,096	3	7	6–7	13–17

[1] Derived from table 4.3.

[2] Recruitment = (Population) (Percent Females) (Percent Adults) (Percent adult females parturating). I assumed 58% were females, 61% of which were adults (study area data), and that 90.6% of the adult females parturated (Fairall 1971).

[3] Sabie river study area: Predation by leopards = [1.9 male leopards × (45 impala/year)] + [3.4 female leopards x (43 impala/year)] = 232 impala/year; predation by spotted hyenas = [15 hyenas × (10.6 impala/year)] = 159 impala/year; predation by lions = [3 lions × (16.3 impala/year)] = 49 impala/year; predation by wild dogs = [6.6 wild dogs × (55 kills/year) (0.93 impala) (17.7 km²/338 km²)] = 76 impala/year; predation by other predators (cheetah, jackals, martial eagles, crocodiles, pythons) = 10 impala/year; Total impala removed by predators = 526 impala/year.

[4] Nwaswitshaka River study area: Predation by leopards = [2.4 male leopards × (45 impala/year)] + [7.3 female leopards × (43 impala/year)] = 422 impala/year; predation by spotted hyenas = [28 hyenas × (10.6 impala/year)] = 297 impala/year; predation by lions = [6 lions × (16.3 impala/year)] = 98 impala/year; predation by wild dogs = [12.2 wild dogs × (55 kills/year) (0.93 impala) (81 km²/452 km²)] = 112 impala/year; predation by other predators (cheetah, jackals, martial eagles, crocodiles, pythons) = 15 impala/year; Total impala removed by predators = 944 impala/year.

NOTE: Spotted hyena kill rates = (1.98 kg/day) (365 days) = 72.27 kg/year (0.615 impala) = 444.46 kg/impala/year/41.93 kg = 10.6 impala/year/hyena.

Lion kill rates = (6.00 kg/day) (365 days) = 2190 kg/year (0.313 impala) = 685.47 kg/impala/year/41.93 kg = 16.3 impala/year/lion.

population growth was estimated by calculating the annual kill of impala by leopards and estimating what proportion of the potential annual recruitment this represented. Influence on population composition was estimated by calculating numbers of male and female impala of different ages taken by leopards. Predation on impala by spotted hyenas, lions, wild dogs and other predators was also estimated. I assumed that predator populations remained relatively stable during the study period, an assumption probably valid for leopards, but unknown for other predators. I averaged monthly impala population estimates for an annual estimate and used female impala productivity data from my study area, as well as that of Dasmann and Mossman (1962) and Fairall (1969), to estimate annual recruitment into the impala population.

Predation by leopards on impala in the study areas accounted for less than the estimated annual recruitment into the impala populations. Leopard predation varied from a low of 6% of the estimated annual recruitment into the impala population (NRSA, 1975) to a high of 54% (SRSA, 1975) (see table 8.14). Thus leopard predation alone appeared unable to prevent the impala population from increasing during the study period. Even when predation by other species of predators was considered, predation seemed unable to prevent the impala population from increasing. The only exception during the study period was in 1975 when all predators were estimated to take 98 to 122% of the recruitment in the impala population in the SRSA. This would have stabilized the impala population or perhaps caused a decline.

My information supported earlier conclusions of Pienaar (1969) and Hirst (1969b) on the impact of leopard predation on impala in KNP and the Timbavati Game Reserve, respectively. They also concluded predation alone was insufficient to regulate impala populations. Apparently impala periodically increase to levels determined by their food supply, and then die directly from starvation or become increasingly vulnerable to predation because of their poor condition. Starvation usually occurs at the end of the dry season (Hirst 1969b) and is often triggered by heavy rains. This was so in my study area in September 1973.

Predation by leopards on impala in the study areas appeared selective on young males and old females. Although predation on male impala was proportional to their numerical abundance in the population, most of the overall predation on males (63%) was on individuals less than two years old. Hirst (1969b) also reported predation was great on partly-grown male impala. Because males that age make up about 26% of the male population before lambing, they were more than twice as prevalent in leopard kills as in the population. Female impala less than two years old make up only 19% of the female impala killed by leopards, which also indicated that male impala of that age were much more vulnerable to predation. Selective predation pressure on young male impala could thus have contributed to the observed low proportion of male impala (74:100) in the study areas.

Leopard predation was also selective on older female impala. Eighty-one percent of all female impala killed by leopards were adults, yet adults made up only 61% of the population before lambing. Leopards, therefore, appeared to be selecting older females. If selection was for impala more than four years old, as the age structure data suggested, females this age probably were most prevalent in the kills of female impala by leopards. I was unable to determine if predation on older females altered the age structure of the female segment of the impala populations in the study areas. Females this age probably were more vulnerable to leopard predation because during lambing they often retire to secluded areas in dense brush (Jarman and Jarman 1973b) where leopards are more apt to capture them.

Other Ungulates

Four other species of ungulates regularly preyed upon by leopards in the study areas were steenbuck, bushbuck, gray duiker, and warthog. The kill rates, composition of kills, and population densities of leopards suggested about fifteen small ungulates were killed annually in the SRSA and twenty-nine in the

TABLE 8.15 *Impact of Leopard Predation on Selected Small Ungulates*

Study Area	Prey	Number Killed/Year by Leopards	% of Predation on	
			Population	Recruitment
Sabie River[1]	Steenbuck	5	71	143
	Bushbuck	5	83	167
	Warthog	5	83	56
Nwaswitshaka[2] River	Steenbuck	10	24	48
	Gray duiker	10	29	57
	Warthog	10	31	21

[1] Assume the following populations: steenbuck = 7, bushbuck = 6, and warthog = 6. Assume that 50% of populations are females, each female steenbuck and bushbuck produces 1 young/year, and each female warthog produces 3 young/year.

[2] Assume the following populations: steenbuck = 42, gray duiker = 35, and warthog = 32. Assume 50% of the populations are females, each female steenbuck and gray duiker produces 1 young/year, and each female warthog produces 3 young/year.

NRSA. For simplicity, I assumed that leopard predation was equally distributed among the most common small ungulates in the study areas. Based on this assumption, leopards were estimated to kill about five each of steenbuck, bushbuck, and warthog annually in the SRSA and ten each of steenbuck, gray duiker, and warthog in the NRSA. These kills were then expressed as proportions of the total estimated populations and potential recruitment into the populations.

If my assumptions were valid and population estimates accurate, leopards were taking a substantial proportion (71% to 83%) of the steenbuck, bushbuck, and warthog populations in the SRSA (see table 8.15). When this predation was expressed in terms of potential ungulate recruitment removed, leopards may have been controlling the steenbuck and bushbuck populations in this area because predation was potentially removing more individuals (143% to 167%) than were being added to the populations each year. Because steenbuck and bushbuck populations were present, leopards were obviously not able to eliminate their populations. But they may have prevented their populations from increasing.

Leopard predation had less impact on steenbuck, gray duiker, and warthog populations in the NRSA. There, they took only 24% to 31% of the estimated populations, or 21% to 57% of their potential recruitment (see table 8.15). The difference in impact of predation on small ungulates in the study areas was probably related to leopard and prey densities. The higher leopard densities in the SRSA had a significant impact on steenbuck and bushbuck because they occurred at relatively low densities compared to impala. Although the high leopard population was being maintained by the high impala population, the number of leopards in the area was also having a significant impact on the low population density of steenbucks and bushbucks. This impact would probably be even greater should the impala population suddenly decline. The "buffering" effect of impala would then lessen and other species would be preyed upon by leopards even more intensively.

The buffering effect of impala on the impact of predation on other species is probably significant. Although large numbers of impala can support high leopard densities and not be controlled by leopard predation, other prey species sharing the same habitats with impala may be impacted by leopard predation. Should numbers of impala decline, leopards would be forced to take more alternate prey and might significantly impact low population densities, at least temporarily, until a new and lower density of leopards is attained. Then, leopard densities would decline and their impact on the prey would also decline.

Although leopard predation in the study areas may have affected the sex and age composition of impala populations, it did not appear to control their population growth. But leopard predation may have been sufficient to impact steenbuck and bushbuck populations in the SRSA, perhaps even to control their population growth. Populations of small ungulates in the NRSA were not controlled by leopard predation because leopards in that area occurred at much lower densities.

Summary

1. Leopards hunted mainly at night using sound and vision to locate prey. Leopards were opportunistic hunters, seldom pursued prey a long distance, and captured some impala by ambushing them along game trails. Only about 16% of the observed daytime hunting attempts of leopards were successful.

2. Nineteen of fifty kills of leopards had conspicuous wounds about their necks. Although thirteen of these had been bitten in the throat, I was unable to determine the actual killing method used by leopards.

3. Leopards dragged nearly all the carcasses of large kills to cover before feeding. Eighty-four percent of all discovered kills of leopards were found cached in trees.

4. When feeding on impala, leopards usually fed first at the groin and hindquarters, then progressively ate towards the head. Impala were not disemboweled first, but all viscera were eventually eaten. Leopards stayed an average of 2.4 days near each impala carcass, feeding mainly in darkness.

5. Leopards readily scavenged impala carcasses when they were available. Scavenged impala were sometimes cached in trees for several days before leopards returned to feed on them. Leopards wasted more flesh than usual from scavenged impala, and scavenged carcasses were sometimes left on the ground instead of being cached in trees. Leopards also scavenged from kills of other leopards.

6. Ninety-four percent of fifty-five kills of leopards and 64% of leopard feces were ungulates or contained ungulate remains. Sixty percent of the leopard feces contained remains of impala, and 97% of the kills of leopards were impala.

7. Small mammals ranked second in importance to ungulates in the leopard's diet. Although only 4% of the kills were small mammals, 29% of the feces of leopards contained remains of small mammals. Vervet monkeys, cane rats, and porcupines were important small prey of leopards. Small mammals were eaten more often by leopards during dry than wet seasons.

8. Leopards killed male and female impala according to their occurrence in the impala populations. More male impala were killed by leopards during dry than wet seasons, and more young male impala were killed by leopards than expected from the numbers estimated in the populations. For unknown but speculated behavior reasons, impala in the four to five year age class and older appeared more vulnerable to predation by leopards based on their occurrence in the impala population.

9. Small mammal remains were found most often in the feces of subadult leopards, especially females. Feces of adult leopards contained a larger proportion of ungulate remains. Adult male leopards killed more adult female impala and fewer adult male impala than did adult female leopards.

10. Radio-tracking information suggested that leopards killed one large prey, usually impala, once about every seven days.

11. Adult male leopards killed an average of at least 5.0 kg/day and consumed an average of 3.5 kg/day. Adult female leopards were estimated to kill and consume an average of at least 4.0 and 2.8 kg/day, respectively.

12. Leopard predation alone was not great enough to stabilize or decrease the study area impala populations. But impala were a buffer species, reducing predation on other, less numerous, ungulate prey. Leopard predation may have affected the sex and age composition on the study area impala populations.

13. Leopard predation may have significantly influenced the steenbuck and bushbuck populations in the Sabie River study area because of the high population density of leopards in the area.

9

Social Organization of Leopards

THE social organization of leopards describes how leopards interact and space themselves relative to each other in time and space. Certain classic spatial concepts, such as territoriality, which were first developed to describe avian social systems, are difficult to apply to solitary felids. For example, territoriality has usually been associated with aggression and fighting to actively defend an area that has distinct, patrolled boundaries. In large, long-lived, and potentially dangerous species, such as the leopard, fighting can be fatal and home ranges are often too large to have distinct, frequently patrolled boundaries. Because of differences in defining classical territorial behavior (Nice 1941; Emlen 1957; Pitelka 1959; Etkin 1967; Wilson 1971), I first describe the interactions among leopards I observed and the leopard's apparent land tenure system. Using home range overlap, scent marking, and agonistic and avoidance behavior criteria, I examine whether leopards in my study areas exhibited a form of territoriality.

Degree of Sociality

Many attempts have been made to establish criteria for, or define features of, animal sociality (Crook 1970; Brereton 1971; Wilson 1975; Leuthold 1977). Most of these have failed because, according to Wilson (1975), no two authors have agreed upon what qualities of sociality are essential. One criterion often used is the type or size of group, or social unit, animals form that usually includes

TABLE 9.1 *Radio-collared Leopards in and near the Sabie River Study Area*

Leopard	Total Days Located	Days Located with Leopard							Total Days	% of Total
		M1	M2	M3	F4	M5	M7	F12		
M1	268	—	0	4	8	0	0	1	13	5
M2	268	0	—	0	0	0	0	0	0	0
M3	256	4	0	—	21	0	0	3	28	11
F4	260	8	0	21	—	0	0	2	31	12
M5	35	0	0	0	0	—	0	0	0	—
M7	148	0	0	0	0	0	—	0	0	—
F12	142	1	0	3	2	0	0	—	6	4

single individuals. Although a completely unsocial lifestyle is impossible (Leyhausen 1965), many animals, including most felids, live solitary lives. Because interactions must occur even among solitary animals, if only during mating and the rearing of young, the degree to which solitary individuals interact is a measure of sociality. The frequency and description of social interactions among leopards is thus important in understanding their social organization.

FREQUENCY OF INTERACTIONS

Leopards seldom interact socially with other leopards. Of 135 visual observations of 142 leopards during my study, 95% were of solitary individuals. Seven radio-collared leopards with overlapping home ranges in the SRSA monitored for 1,377 "leopard-days" were located together only 6% of the days (see table 9.1), and eight leopards monitored for 1,035 leopard-days in the NRSA were located together only 4% of the time (see table 9.2). No individual radio-collared leopard was located near or with its radio-collared neighbor on more than 12% of the days it was monitored. However, measures of frequencies of interactions between leopards may be biased if based solely on infrequent radio monitoring because brief interactions, lasting only several hours, may not be detected.

Observations of leopards in other areas confirm their solitary nature. Schaller (1972) reported 98% of 155 visual encounters with adult leopards in the Serengeti National Park were of solitary individuals, as were 87% of the 150 sightings

TABLE 9.2 *Radio-collared Leopards in the Nwaswitshaka River Study Area*

Leopard	Total Days Located	Days Located with Leopard								Total Days	% of Total
		F6	M10	F11	F13	M14	F17	M18	F21		
F6	197	—	0	4	0	1	0	0	0	5	3
M10	117	0	—	0	0	0	0	0	0	0	—
F11	204	4	0	—	0	11	0	0	0	15	7
F13	130	0	0	0	—	5	0	0	0	5	4
M14	205	1	0	11	5	—	0	0	1	18	9
F17	44	0	0	0	0	0	—	0	0	0	—
M18	66	0	0	0	0	0	0	—	0	0	—
F21	72	0	0	0	0	1	0	0	—	1	1

of leopards in Tsavo National Park by Hamilton (1976). Both areas also supported moderate to high populations of lions and spotted hyenas. In Wilpattu National Park, Sri Lanka, where the Asian leopard is the dominant predator and often seen during the day, 83% of 169 visual encounters of leopards were of solitary individuals, and the remainder were of two leopards together (Muckenhirn and Eisenberg 1973). In Ruhuna National Park, also in Sri Lanka, 84% of thirty-two visual observations of leopards were of lone individuals (Santiapillai et al. 1982).

Leopards appear more solitary than other larger felids. The sociable lion usually forms prides of four to thirty-seven individuals (Wright 1960; Mitchell et al. 1965; Pienaar 1969; Schaller 1972; Elliot et al. 1977; Bertram 1978; Hanby and Bygott 1982; Smuts et al. 1982). Only 6% of 3,123 lions observed in the Serengeti National Park were solitary (Schaller 1972). Cheetah, which are much less sociable than lions (Frame and Frame 1981), were solitary only 27% (Graham 1966) to 52% (Schaller 1972) of the time, but many of the observed social groups were females and their offspring. Female cheetahs are usually solitary, but males may form groups of one to four individuals with which they associate more than half the time (Caro and Collins 1987). Tigers are usually solitary but seem to be more social than leopards. Eight radio-collared tigers located 1,422 times were found near (within 0.4 km) each other 11% of the time with individual variations between 6% to 18% (Sunquist 1981).

Cougar interactions revealed a degree of sociality comparable to tigers, but they still interacted slightly more often, on the average, than did leopards in the study areas. Of 924 location days of radio-collared cougars, they were together on 88 days (10%) with individuals varying between 5% to 13% (Seidensticker et al. 1973). Such comparative data reveal that leopards, cougars, and tigers are highly solitary, but of the three, leopards, at least in my study areas, were the most solitary.

Socialization among solitary felids sometimes occurs at carcasses of large prey. Although I recorded only three, possibly four, interactions among adults at kills during 135 visual observations, Hamilton (1976) observed at least two leopards together at 12 of 122 (10%) baits and speculated they were usually an adult male and female. An adult male and two female tigers were observed at kills on six different occasions in Kanha National Park. The interactions between the male and a tigress with four cubs were friendly, but those between the two females were agonistic. From these and other observations Schaller (1967) speculated that a tiger had priority rights to its own kill, even in the presence of a larger or stronger individual. Sunquist (1981) reported that although tigresses and young subadults frequently socialized at baited kills (19%), few associations (5%) among adults occurred at natural kills. Of seventeen interactions between cougars, at least ten (59%) possibly twelve (71%), were associated with a kill (Seidensticker et al. 1973).

TABLE 9.3 *Radio-collared Leopards with Overlapping Home Ranges and Unmarked Leopards*

Leopard Association	Number of Days Leopards Together			
	Radio-collared Leopards	Unmarked Leopards	Total Days	% of Total
Adult male–adult female	39	3	42	62
Subadult male–adult female	8	0	8	12
Adult male–subadult male	4	1	5	7
Adult female–adult female	4	0	4	6
Adult male–subadult female	3	0	3	4
Adult female–subadult female	2	1	3	4
Subadult male–subadult female	1	0	1	1
Subadult male–unidentified	0	1	1	1
Unidentified–unidentified	0	1	1	1
Adult male–adult male	0	0	0	—
Subadult male–subadult male	0	0	0	—
Total	61	7	68	—

INTERACTIONS BETWEEN LEOPARDS

Observations of more than one leopard are usually of adult males and females courting or of females with offspring. Arranging interactions among leopards in order of decreasing frequency revealed that the adult male-adult female interaction was the most common, followed by a subadult male-adult female interaction (see table 9.3). Radio-collared adult male leopards were never located or seen together, nor were subadult males. Of only seven observations of more than a solitary leopard, three were of adult males and females traveling together or courting. This is comparable to sightings reported by others. Most of the nineteen interactions reported by Hamilton (1976) were adult males and females together (74%) or a female and her offspring (5%). Of twenty-eight contacts with two leopards together in Wilpattu National Park, 54% were of adult males with females and 25% of adult females with offspring (Muckenhirn and Eisenberg 1973).

Adult male and female interactions.   Most leopard interactions I recorded (62%) were between adult male and female leopards. Although most associations lasted only one day, resident M3 and resident F4 once remained together five days (see fig. 9.1). Male and female leopards stayed together seven days in Tsavo National Park (Hamilton 1976), up to seventeen days in Wilpattu National Park, Sri Lanka (Eisenberg and Lockhart 1972), and less than one day in the Kalahari Desert (Bothma and Le Riche 1984). Cougar pairs have been known to stay together up to sixteen days (Seidensticker et al. 1973). The peak period of interactions between adult male and female leopards in my study areas occurred in the late dry seasons, specifically July, August, and October (see fig.

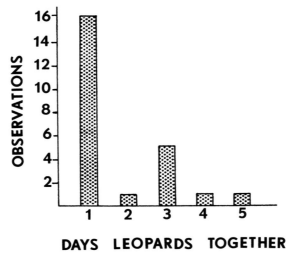

FIGURE 9.1 Number of observations of leopards together for one- to five-day intervals.

9.2). If these interactions were related to reproduction, and if mating occurred successfully, a peak in leopard births would have occurred in September, October, and December.

Only resident male and female leopards frequently interacted. In the SRSA nine interactions occurred between resident M3 and resident F4. Thirteen interactions occurred between resident M14 and four females, resident F6, F11, F13, and F21, in the NRSA (see fig. 9.3).

Male 3 was located with F4 on nine different occasions (twenty-one days) between February and October 1974. Changes in F4's movement and recapture

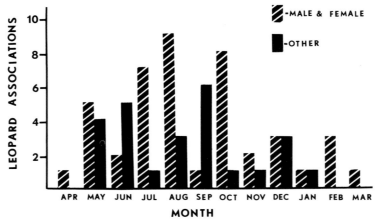

FIGURE 9.2 Cumulative monthly distribution of numbers of associations observed between leopards.

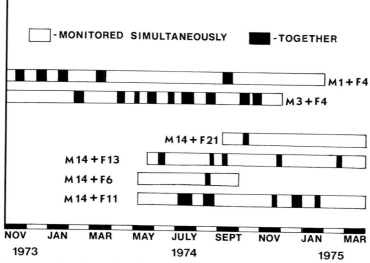

FIGURE 9.3 Periods when simultaneously monitored male and female leopards were together.

pattern after this strongly suggested she successfully mated and had cubs in December 1974. Male 14 was located with four different female leopards thirteen times during eighteen days between June 1974 and February 1975. On August 9, 1974, M14, F6, and F11 were all located together, but the following day F6 was alone. Male 14 and F11, however, remained together for two more days before separating. Ten days later, M14 was located with F13. These observations indicated that adult male leopards probably knew of the whereabouts of the resident female leopards within their home ranges and could easily associate with them if they desired. The observations also suggested that once a female had cubs, the male rarely associated with them, especially if the cubs were less than a year old. Whether the male or female determined direction of movements after they associated was often unknown. However, of three visual observations of male-female interactions, the female took the lead and the male followed. On at least one occasion, the female (F11) responded to the male's (M14) vocalizations, joined him, and they moved off together. Although aggression was not recorded between male and female leopards during my study, aggression may occur. In the nearby Londolozi Game Reserve, a chance encounter between a resident female leopard and an unknown male resulted in a fight, with both leopards sustaining superficial injuries (Le Roux and Skinner 1989).

Subadult male and adult female interactions. Although this was the second-most frequent interaction among leopards, it involved only two individuals, subadult M1 and adult F4. These individuals were located together eight times between November 1973 and September 1974, but never for more than one day. This contrasted with the interactions between resident M3 and F4, which sometimes lasted as long as five days. Although the relationship between M1 and F4 was unknown, I did not believe they interacted to mate. Male 1 may have been an offspring of one of the females in the area, perhaps F4. Female leopards may periodically interact with their offspring even after the offspring become independent (Le Roux and Skinner 1989). Schaller (1972) observed a female leopard associating with her independent male offspring when it was twenty-two months old. That subadult M1 made exploratory trips outside his home range and later dispersed from the study area supported the belief that he was probably reared in the study area. But it is also possible that adult female leopards occasionally associate with subadult males, whether or not they are capable of breeding. Hamilton (1976) reported that the adult female leopard in his study area interacted with at least three males on separate occasions. Eisenberg and Lockhart (1972) reported that a tracker once saw two male leopards with one female leopard. Additional observations are needed to determine if a resident female will breed with nonresident males during estrus.

Adult male and subadult male interactions.   Adult males seldom interacted with subadult male leopards. Of four potential situations where different radio-collared adult and subadult males could have interacted because of their overlapping home ranges, only four interactions between two males (M1 and M3) were documented. It should be emphasized that in three of four situations, the subadult male's home range lay almost entirely within the adult male's home range, but despite this overlap, the males seldom interacted.

All known interactions occurred in September 1973, when M3 interacted at least four times on four different days with subadult M1. No unusual behavior resulted from these interactions that I could determine, and M1 utilized the same areas used by adult M3 before and after the encounters. This pattern of home range use was similar to that among other adult and subadult males: the subadult male avoided the adult male, and the adult male apparently tolerated the younger male within his home range.

The only other documented interaction between an adult male and subadult male occurred when radio-collared subadult M26 scavenged from an impala killed by an unmarked adult male leopard (probably M30 who was later captured). The adult male appeared to dominate M26, who was concealed in the bushes and did not approach the adult nor attempt to feed until the adult abandoned the kill. The adult was aware of the subadult because he snarled at least once in M26's direction. When the adult male left, he traveled away from the subadult male.

I did not observe any aggressive interaction between adult and subadult male leopards, although on one occasion a leopard of unidentified sex attempted to appropriate a kill from another unidentified leopard. A brief struggle occurred with the trespasser leaving the kill to its original claimer. Eisenberg and Lockhart (1972) reported they once observed an adult male chasing a subadult male from a kill and recorded two instances when subadult males were temporarily displaced from their home ranges. One displacement was by an adult male, the other by an adult male-female pair. Hamilton (1976) reported eight probable conflicts between what he considered to be adult male leopards. However, one of his adult male leopards (M2) was smaller and may have been a subadult. This particular leopard was apparently attempting to establish himself in an area already occupied by adult male leopards. The rather frequent aggressive encounters and fights this leopard had with a neighboring adult male (M3), the fact that he left his home range on an exploratory trip after an encounter with a larger male leopard using the same area, the number of other exploratory trips he took during the monitoring period, and the nearly complete overlap of his home range with another larger leopard (M8) all suggest that he was not an established resident but a subadult trying to become established in Hamilton's study area.

The aggressive encounters between male leopards reported by Hamilton and my observations of encounters between adult and subadult male leopard differ. However, I simultaneously monitored only two adult resident male leopards. Most adult males I captured had old scars or recent wounds, so fighting between adult male leopards may have occurred.

Adult female interactions.   Adult female leopards in my study areas seldom interacted. Of five adult female leopards simultaneously monitored in the NRSA, only two (F6 and F11) interacted on four separate occasions in May, June, July, and August 1974. These interactions occurred after F6 had cubs and before F11 gave birth. Most of the encounters between F6 and F11 occurred near the Msimuku tributary and the Nwaswitshaka River near a small koppie that both females used at different times to conceal their small cubs. Because I did not visually observe these interacting females, I was not able to record their behavior. The only visual observation of female leopards near each other involved F4, subadult F12, and an unmarked female (later F29) at an impala carcass cached in an isolated tree (see photo 9.1). Female 4 passed within 5 m of F12 and her mother who were on the ground nearby. Female 12 and her mother climbed the tree and fed after F4 passed by. Although I never observed any aggression between adult females, four captured female leopards bore punctures and laceration scars from fighting. One had bite marks on her neck, another at the base of her tail, and three females had old bite marks on their forelegs. However, the wounds could have also been inflicted by other predators or dangerous prey such as ratels, warthogs, or baboons. My monitoring and visual observations suggested adult females were intolerant of or avoided each other.

Only 4 of 946 locations of eight adult female leopards indicated any interactions, which emphasized that interactions among adult females occurred infre-

9.1. The tree in which a female leopard (F29) and her offspring (F12) were feeding on an impala carcass when another female leopard (F4) walked near during the night.

quently. Neither Eisenberg and Lockhart (1972), Schaller (1972), Hamilton (1976), or Bertram (1978) reported adult female leopard interactions. But Le Roux and Skinner (1989) documented a fight between two female leopards and the killing of a nearly three-year-old female by another female leopard.

Adult male and subadult female interactions.   I recorded only three instances of an adult male leopard (M3) interacting with a subadult female (F12). Two of these encounters occurred in June and one in October 1974. Each encounter lasted only one day, and the female was probably sexually immature. These interactions were interesting, however, because they suggested that as a female leopard matures she encounters the resident male leopard who inhabits her area. If she remains in or near her natal area, she may eventually breed with the male that was also her father. This assumes that a resident male leopard remains in the same area for at least two to three years.

Adult female and subadult female interactions.   Leopard interactions in this category were probably between females and their offspring. Because females were so secretive in concealing their cubs, however, I gained little knowledge of early mother-young relationships among leopards. Information from Adamson (1980), Seidensticker (1977), Bertram (1978), and Schaller (1972) suggest that leopard cubs are born in a secluded place, often a rocky outcrop, koppie, or among cliffs and caves when they are available. Even in my study areas, where rock outcrops were scarce, the leopard's preference for rocky areas was evident. At least two of the three places where female leopards had hidden their cubs were in koppies. Another female (F6) may initially have used the burrow of an aardvark before moving her cubs to the koppie.

If disturbed, female leopards may relocate their cubs even if they are only four days old. Movements of cubs from one secluded place to another may occur frequently, as often as every two to four days, depending on the level of disturbance. Adamson's leopard, Penny, carried her two cubs at least 2 km on different trips when they were only thirteen days old. Apparently the female leaves her cubs behind in protected places and hunts alone during the first two to three months (Seidensticker 1977), perhaps as long as six months, after the cubs are born.

I did not observe the cubs of my radio-collared female leopard until they were about six months old. At that age the cubs apparently will follow their mother to a kill. A kill was present when I first saw F6's cub climb a tree for protection. According to Eisenberg and Lockhart (1972), six-month-old cubs stay with the female constantly, even when she crosses wide openings in the vegetation.

Cubs remain with their mother at least twelve months (Eisenberg and Lockhart 1972), perhaps as many as twenty months (Schaller 1972) before they become independent. Although cubs eleven to eighteen months old will seek out and stalk prey, mainly small prey such as hares, birds, hyraxes, (Bertram 1978), they appear to be unsuccessful hunters most of the time. But some cubs are capable of killing large prey at the age of twenty months, when most are independent but still confined to their mother's home range. Female cubs may mature more rapidly than male cubs. Schaller (1972) observed that a female cub thirteen months old began to travel independently, whereas a male cub remained with its mother until it was twenty months old. Even after cubs separate from their mother they may occasionally join her for brief periods until they are 2.5 to 3 years old (Bertram 1978).

My data suggested that female leopards with cubs may associate with other leopards after their cubs are three to six months old. Such interactions, however, seldom occur near the cubs' hiding places. My data also suggested that at least some female cubs may remain within their mother's home range even after they become independent. Male cubs almost always disperse out of their mother's home ranges. When this occurs, after the males are twenty-four to thirty-six months old, the female has probably bred again and is rearing another litter.

Adult male interactions.   The adult male leopards I monitored were never located together. If encounters between adult male leopards occurred, they were brief. Although adult males rarely interacted, they each knew of the others' presence because of their distinct home range boundaries, which were recognized and identified by behavioral means. Perhaps frequent interactions between neighboring adult males occur when a male is first establishing residence in an area. Later, after they recognize each other, frequent encounters are unnecessary to maintain spacing. That subadult males were able to remain in an adult male's home range without apparent conflict suggested that they did not become, or were not capable of becoming, established residents in occupied areas. Infrequent interactions with the resident females suggested subadult males were not competing with the adult resident males for breeding with females.

Although fighting between adult male leopards was not documented during the study, scars and wounds indicated captured males had previously been injured. When initially captured, M23 had twelve puncture wounds (teeth marks?) on his neck and a laceration on a hind leg. However, hyenas could have inflicted these wounds because M23 periodically attacked, killed, and ate hyenas. Some injuries might occur if leopards tried to kill hyenas, especially if a ratel was able to wound a leopard. Fresh wounds including 15 cm of missing tail, indicated M5 had been in a fight during the six-day period between his initial capture and his recapture. A subadult male (M7) captured nearby only six days later was in excellent condition, suggesting he had not fought with M5. Adult M10 and subadult M18 also had old scars when initially captured.

Violent fights between male leopards were reported by Hamilton (1976). On at least eight occasions when different male leopards met, conflicts occurred perhaps six times. These conflicts resulted in broken or removed radio collars, deep canine puncture wounds on the head and face, and claw marks on the head and neck. Several encounters occurred at game trails and the edge of an escarpment, where one male may have surprised the other. Another fight occurred at a kill. All conflicts involved a young male leopard that may have been attempting to occupy an area used by a resident male. Another encounter between two male leopards involved no conflict, indicating that they tolerated each other.

I suggest that violent conflicts between adult male leopards occur primarily when a new male is attempting to become established in an area. If successful, he will have encountered neighboring resident males and will know their approximate home range boundaries. Thereafter, conflicts between males should be rare unless a new male arrives and challenges the established male or a resident male dies. In stable leopard populations where adults live many years and are familiar with each other, little fighting probably occurs. In unstable populations, with a rapid turnover of resident males, fighting might be more common if there is a sufficient influx of males to occupy vacancies left by the deaths of residents. Knowledge of each others' home range boundaries and core areas of intensive use probably contributes to a stable, peaceful society of solitary leopards. The spatial and temporal relationships among leopards can thus provide valuable insights into their social organization.

Land Tenure System

As shown in the previous section, leopards spend most of their time alone. Observed associations between leopards are usually of courting or breeding adults or females with their offspring. Interactions between nonbreeding adult leopards are apparently uncommon and of brief duration. This section on leopard social organization examines the land tenure patterns that result from these apparently infrequent, direct interactions and other, more subtle, indirect interactions between leopards.

ADULT LEOPARDS

In this section I examine the spatial relationships between four resident males and two female leopards in and adjacent to the SRSA and two resident males and six female leopards in the NRSA. I also compare these relationships to those of other radio-collared leopards as well as unmarked leopards.

Resident male leopard home range overlap.   Of four male leopards captured and monitored between the Sabie and Sand Rivers, only M3 and M23 were considered adult residents. Leopard M3, captured early in the study (August 30, 1973), occupied most of the study area but seldom ventured into the southwestern region. After monitoring M3 for several months, I suspected that another large periodically observed unmarked male leopard used the southwestern region avoided by M3. Although I saw this male's distinctive tracks on many occasions, I did not capture him until December 5, 1974. By that time I had lost radio contact with resident M3. But reports of others as well as my own observations indicated this second adult, M23, was a resident in the study area during the time I monitored his neighbor, M3.

Resident M3 and M23 mutually used an area totaling 4.9 km² in the southwestern and southeastern regions of their home ranges, respectively. This encompassed 13% of M3's and 30% of M23's home ranges (see fig. 9.4). Use of this overlap varied seasonally. During the 1973 dry season M3 was never located in the area of overlap and was only twice located there during the 1973–74 wet season. During the 1974 dry season, however, he traveled more frequently into M23's home range and was located within the overlap area on twenty-four (25%) of ninety-seven days. His use of the overlap area remained constant (24% of locations) during the 1974–75 wet season after which radio contact with him was lost.

The number of 1-km² quadrats jointly used by resident M3 and M23 also

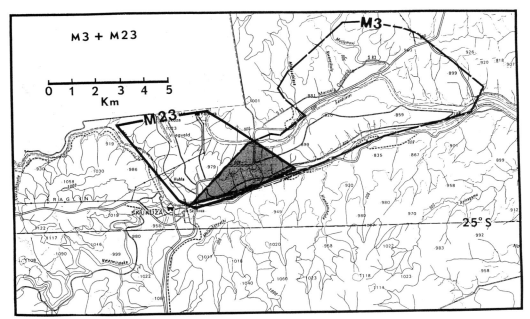

FIGURE 9.4 Overlap (shaded) of areas used by resident male leopards M3 and M23 in the Sabie River study area.

indicated the degree of spatial overlap of home ranges of resident male leopards. Of forty-seven quadrats used by M3, nine (19%) were also jointly used by M23, and of twenty-three quadrats used by M23, nine (39%) were also used by M3. Thus both methods of estimating home range overlap indicated residents M3 and M23 did not maintain totally exclusive areas. But the greater part of each male's home range, or core area, was used to the exclusion of the neighboring resident male.

Two other older male leopards, presumably residents, were also monitored south of the SRSA. Male 9, who was monitored from February to June 1974, utilized an area of at least 29.3 km^2, and M22, monitored during September and October of 1974, used an area of at least 3.7 km^2 northeast of M9's home range. Although these two adjacent males were not monitored simultaneously, overlap between their home ranges was evident. About 3% of M9's home range and 22% of M22's home range were jointly used. Using the 1-km^2 quadrat method of overlap analysis, one of nineteen quadrats (5%) used by M9, and one of six quadrats (17%) used by M22 were jointly used. The overlap area shared by these two older male leopards was proportionately similar to that jointly used by males M3 and M23 within the SRSA.

Two resident male leopards with adjacent home ranges in the NRSA also jointly used substantial portions of their home ranges. These two males, M10 and M14, were monitored simultaneously for one year. During this period M10

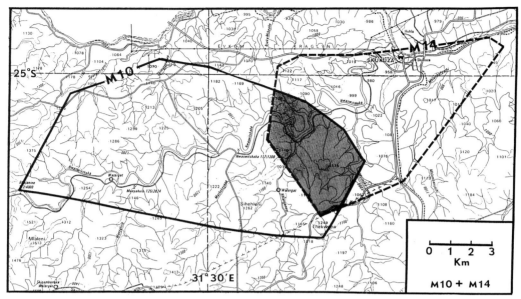

FIGURE 9.5 Overlap (shaded) of areas used by resident male leopards M10 and M14 in the Nwaswitshaka River study area.

used an area of 96.1 km² and M14 an area of 56.4 km². An area of home range overlap totaling 15.8 km² represented 16% and 28% of M10 and M14's home ranges, respectively (see fig. 9.5). The number of 1-km² quadrats used jointly by both males was ten of fifty quadrats (20%) for M10 and ten of fifty-three quadrats (19%) for M14.

The spatial relationships documented among resident male leopards had at least two common features. First, the home ranges of radio-collared resident males were not maintained to the total exclusion of neighboring resident males. The percentage of home range overlap between neighboring resident males based on the polygon method of determining home ranges averaged 19% with a range of 3% to 30%. Second, older resident male leopards shared a greater proportion of their home ranges with younger resident male leopards. Although some information may have been biased, data from M3 and M23 and from M9 and M22 indicated that the older males (M22 and M23) had comparatively smaller home ranges than their younger male counterparts and that a larger proportion of their home range was shared with the younger males than vice versa. Male 10 and M14 were probably closer in age than the other four males; M10, apparently the older of the two, had the larger home range.

The seasonal distribution of impala, the leopards' principal prey, may have contributed to the home range overlap between resident M10 and M14. When impala were more widely but evenly distributed during the 1974–75 wet season, both resident male leopards jointly used only one 1-km² quadrat. This represented only 3% and 4% of all the quadrats used by M14 and M10, respectively. During the 1974 dry season, when impala concentrated along the Nwaswit-shaka River and a few waterholes, M10 and M14 jointly used four 1-km^2 quadrats, which represented 11% and 14%, respectively, of the total quadrats they used during that season. The proportion of each male's home range in the overlap area decreased three-fold from the dry to wet season as impala dispersed and became more evenly distributed throughout the surrounding area.

A more refined analysis of the 1-km^2 quadrats jointly used by resident M10 and M14 revealed less spatial overlap than was initially suggested. By subdividing the 1-km^2 quadrats into four quarters prior to analysis, even less spatial overlap of home ranges was apparent. During the 1974 dry season both resident males jointly used only two (13%) of sixteen quarter sections. During the 1974–75 wet season they mutually used only one (25%) of four quarter sections. The data thus suggested that these particular males maintained a greater degree of exclusive use than was initially apparent and emphasized the need to examine location data in greater detail.

Resident female leopard home range overlap. Home ranges of monitored resident female leopards also overlapped considerably with those of neighboring resident females. Two older resident leopards (F4 and F16) in the SRSA had completely overlapping home ranges. The capture of three more older female leopards in the SRSA late in the study indicated the SRSA was jointly used by at least four to five resident female leopards. The relatively small size of the SRSA and the home range sizes of just two resident females (F4 and F16) monitored there suggested that a high degree of home range overlap existed among resident female leopards in the SRSA. This overlap was greater than that exhibited by the resident female leopards in the NRSA.

Of six resident female leopards monitored in the NRSA, five shared home ranges with three to four other resident female leopards (see fig. 9.6). The degree of home range overlap among resident female leopards in the NRSA varied with the individual. Female 6, who had cubs during the study, shared parts of her home range with F11, F13, F17, and F21 (see table 9.4). One of these females (F11), who also had cubs, shared a majority (68.5%) of her home range with F6. Female 6 shared only 2.2% of her home range with F17, whose movements were confined primarily to the Nhlanganeni tributary of the Nwaswitshaka River. Home range overlap was greatest among female leopards that intensively used the lower reaches of the Nwaswitshaka River and its tributary, the Msimuku. The home ranges of F17 and F25, which were associated with the Nhlanganeni tributary and the Nwaswitshaka waterhole areas, respectively, did not overlap. Neither did F25 overlap with F21's primary home range nor F17 with F11's. Of twenty-two overlap area combinations among six radio-collared resident female leopards in the NRSA, the average proportion of home ranges jointly used with other resident female leopards was 18% compared to

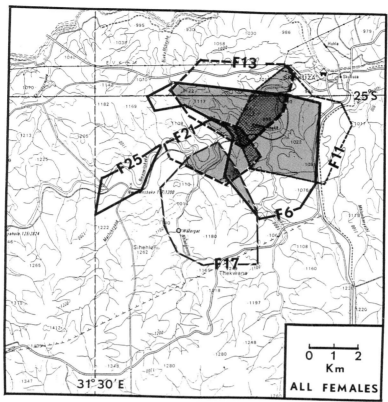

FIGURE 9.6 Overlap (shaded) of areas used by resident female leopards F6, F11, F13, F17, F21, and F25 (all females) in the Nwaswitshaka River study area.

TABLE 9.4 *Resident Female Leopard Home Ranges in the Nwaswitshaka River Study Area (1973–75)*

Resident Leopard	% of Female Home Range in Overlap Area Shared with the Female on the Left					
	Resident Leopard					
	F6	F11	F13	F17	F21	F25
F6	—	68.5	57.7	3.2	23.2	NA
F11	45.8	—	29.5	0	7.5	NA
F13	28.2	22.1	—	NA	16.4	NA
F17	2.2	0	NA	—	54.6	0
F21	6.2	3.0	8.8	21.5	—	0
F25	NA	NA	NA	0	0	—

NA = not adjacent.

19% for all monitored resident male leopards in both study areas. Although the average size of the overlap areas among resident female leopards ($n = 10$, mean $= 2.2$ km^2) was less than that among all monitored resident male leopards ($n = 3$, mean $= 7.2$ km^2), the proportions of home ranges jointly used were similar.

I used the quadrat occupancy method to examine seasonal home range overlap among four resident female leopards that were simultaneously monitored during the 1974 dry season in the NRSA. The average proportion of home range that overlapped for twelve possible combinations was 19.2% per female. A similar comparison for three resident females simultaneously monitored during the 1974–75 wet season revealed an average 23.8% overlap among six home range combinations.

Some resident female leopards spent a high proportion of their time in overlap areas. During the 1974 dry season F6 was located 87% of the time in the overlap area shared with F11. She was also located thirteen days in overlap areas shared with F13 and six days with F17. At the same time, F11 was located 76% of the time in F6's overlap area, 33% of the time in F13's overlap area, but only 1% of the time in F17's overlap area. Female 6 and F11 often jointly used an area that included dense riparian vegetation along the lower Msimuku tributary and the Nwaswitshaka River, and a koppie used as a den east of the Msimuku.

One female without cubs (F13) frequently visited areas jointly used by F11 and F21. Of fifty-two monitoring days during the 1974 dry season, thirty-two (62%) were in F11's overlap zone. And during the 1974–75 wet season, she was located thirty-seven of sixty-five days (57%) in the area used by F21. Female 17, however, seldom visited the areas jointly used with the other resident female leopards. During the 1974 dry season she was located only four of thirty-five days (11%) in F6's overlap zone and one of thirty-five days (3%) in F11's. Female 21, who had cubs during the 1974–75 wet season, often visited the overlap zone used by F13 (73% of days) and F11 (22% of days).

The mutual use of overlap areas among resident female leopards suggested that resident female leopards in my study areas did not maintain completely exclusive areas. The high degree of overlap and frequency of joint use of these overlap areas also suggested that essential resources were not evenly distributed throughout the female leopards' habitat. These essential resources were probably available prey, dense stalking cover, and rocky outcrops, or koppies, needed to conceal cubs. Impala distribution probably influenced the degree of home range overlap among resident female leopards. Between the 1974 dry season, when impala concentrated along the Nwaswitshaka River, and the 1974–75 wet season, when impala dispersed to upland areas, use of the overlap areas by the resident female leopards decreased. During the dry season F11 visited the area used by F13 on 33% of the 109 days she was monitored. During the subsequent wet season her use of the same area declined to 26%. Female 13, who was monitored simultaneously, exhibited similar behavior. During the

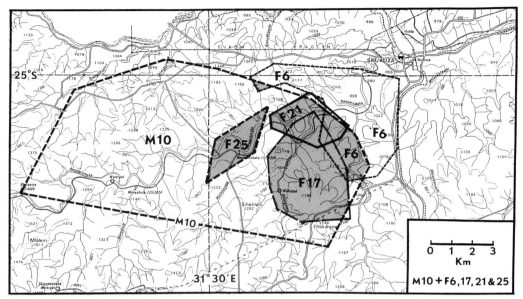

FIGURE 9.7 Overlap (shaded) of areas used by resident male leopard M10 and female leopards F6, F17, F21, and F25 in the Nwaswitshaka River study area.

1974 dry season, she visited the area used by F11 on 62% of the fifty-two days she was monitored, but her use of the area decreased to 40% during the 1974–75 wet season.

Resident male-female leopard home range overlap.   The home ranges of resident male leopards overlapped, often completely, the home ranges of several resident female leopards. In the NRSA the home ranges of resident female leopards F6, F11, F13, F17, and F21 were completely overlapped by the home range of resident M14. And the home ranges of these resident female leopards, as well as resident F25, were overlapped by the home range of resident M10 (see fig. 9.7). Male 14's home range almost completely overlapped the home ranges of four radio-collared females (F6, F11, F13, and F21) and the major portion of F17's home range (see table 9.5). Additional unmarked female leopards were probably within the northeastern region of M14's home range; I once heard courtship vocalizations there when I approached M14. A conservative estimate is that the home ranges of at least six resident female leopards lay totally or partially within resident M14's home range. At least three resident female leopards, probably more, were encompassed by the home range of resident M10.

Although fewer resident female leopards were monitored in the SRSA, data indicated that the home ranges of resident male leopards also encompassed those of several resident female leopards. Female 4's home range lay almost entirely (87%) within the home range of resident M3, as did that of old F16. The

TABLE 9.5 *Resident Female Leopard Home Ranges in the Nwaswitshaka River Study Area*

Resident Leopard	% of Female's Home Range Within Resident Male Leopard's Home Range	
	Leopard M10	Leopard M14
F6[1]	22 (4.4 km^2)	95 (18.7 km^2)
F11[1]	4 (0.6 km^2)	100 (15.0 km^2)
F13[1]	4 (0.4 km^2)	97 (10.6 km^2)
F17[1]	99 (14.0 km^2)	70 (9.8 km^2)
F21[1]	87 (4.8 km^2)	97 (5.3 km^2)
F25	100 (2.8 km^2)	— (none)

Size of overlap area is in parenthesis.
[1] Home range overlapped by both males.

home ranges of female leopards F27, F28, and F29, captured late in the study, almost certainly lay within M3's home range. Thus the home ranges of at least five female leopards were overlapped by M3's home range. A portion of F4's and F16's home ranges also lay within the home range of resident M23. Assuming that part of F28's home range also lay within M23's home range and that at least one additional unmarked female leopard lived within the northwestern portion of M23's home range (but outside the SRSA), at least four resident female leopards probably lived totally or partially within resident M23's home range (see fig. 9.8).

SUBADULT LEOPARDS

Subadult male leopards. Superimposed over the mosaic of home ranges of resident male and female leopards were those of subadult male leopards. Depending on the age of the subadults, these home ranges were either smaller than those of resident males (and confined wholly or partially within the home ranges of resident males), or they were larger than those of resident male leopards. Subadult male leopards used the same areas used by resident males and females and subadult females. The younger subadult males had smaller home ranges than those of resident females, but older subadult males roamed over vast areas during their exploratory movements.

The home ranges of some subadult male leopards were completely overlapped by those of resident males. The home range of subadult M18 was completely inside that of resident M10 (see fig. 9.9). This 1.5- to 2-year-old male, probably the offspring of resident F25, was monitored from June 1974 to May 1975. He neither left the study area nor traveled outside the home range of resident M10 during that period. Part of M18's home range was also overlapped by that of resident M14, and another part by that of subadult M19.

The home ranges of other subadult male leopards extended beyond the

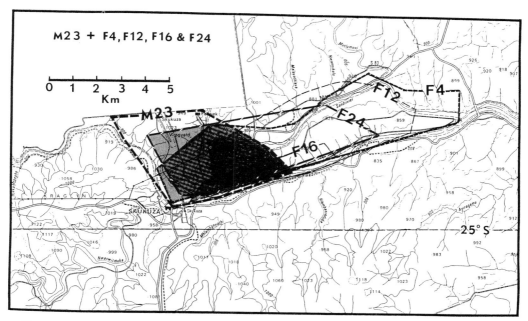

FIGURE 9.8 Overlap (shaded) of areas used by resident male leopard M23 and female leopards F4, F12, F16, and F24 in the Sabie River study area.

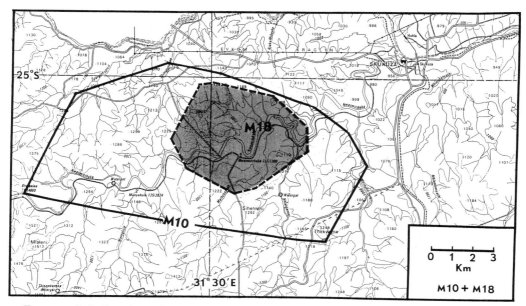

FIGURE 9.9 Overlap (shaded) of the areas used by adult male leopard M10 and subadult male leopard M18 in the Nwaswitshaka River study area.

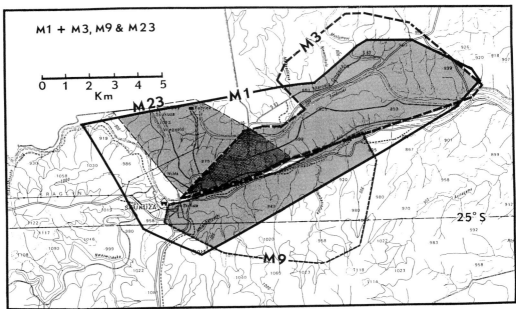

FIGURE 9.10 Overlap (shaded) of areas used by subadult male leopard M1 and male leopards M3, M9, and M23 in and adjacent to the Sabie River study area. Darker shaded area is where M1, M3, and M23 all overlapped.

boundaries of those used by resident males inhabiting the same areas. The home range of subadult M19 extended well beyond the home ranges of M10 and M14. Although the majority of M19's home range was overlapped by those of M10 and M14, approximately 17% of M19's known home range lay beyond those boundaries. Similarly, in the SRSA approximately 44% of the home range of subadult M2 was overlapped by resident M23. Of about 16 km² not overlapped by M23, about 23% lay within resident M9's home range. The home range of subadult M26, also in the SRSA, was completely overlapped by that of residents M23 and M3. As subadult males matured their movements and home range size increased. Initially, their movements merely took them beyond the boundaries of their natal home range. Later, their exploratory movements took them beyond the larger home ranges of the resident male leopards inhabiting the area.

As the exploratory movements of subadult male leopards increased, the degree of home range overlap with those of resident males and females decreased. However, overlap probably increased with unmarked leopards adjacent to the study areas. Subadults M1 and M7 exhibited this movement pattern. At first, M1's home range corresponded closely with, and was almost completely overlapped by, that of resident M3. But as he matured he moved outside of M3's home range into areas occupied by M9, M23, and unmarked males adjacent to the study area (see fig. 9.10). Similarly, M7's home range initially

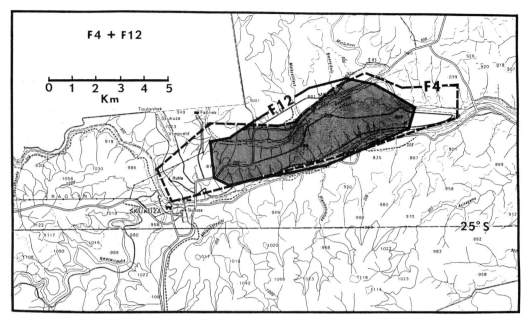

FIGURE 9.11 Overlap (shaded) of areas used by resident female leopard F4 and subadult female leopard F12 in the Sabie River study area.

corresponded with, and was largely overlapped by, that of resident M5. As he matured, however, his home range extended well beyond M5's into the home ranges of other resident male leopards including M3, M9, M14, and M23.

Subadult male leopards did not have stable home ranges comparable to those of resident leopards. Instead, their home ranges changed with the age of the leopard, the density of leopards, abundance and distribution of prey, and habitat quality. The home ranges of subadult males were always overlapped by those of resident leopards in and near the study areas. Subadult male leopards apparently did not settle in an area already occupied by resident leopards unless they were able to successfully compete for that space or a resident male that already occupied the area died. These subadult males probably settle in an area if suitable unoccupied habitat is discovered. Whether they breed with resident female leopards is unknown. Behavioral information (scent marking and vocalizations) suggested these subadult males behaved differently from sexually active resident males. They may not successfully breed until they gain resident status. This process may include acceptance by the resident female or domination of the resident male leopard.

Subadult female leopards. The home ranges of subadult female leopards were completely overlapped by those of resident males and females and subadult males. Most of subadult F12's home range was overlapped by that of resident F4 (see fig. 9.11). Only her exploratory movements took her outside the area used by resident F4. Subadult F12's home range was also overlapped by residents M3, M9, M23, and F16, and subadults M1, M2, M26, and F24. Subadult F24 and subadult F12 jointly used an area of about 14 km² during the brief period F24 was monitored. Female F24's home range was also significantly overlapped by residents F4, F16, M3, and M23; and subadults M1 and M2.

A major difference between the home range overlap pattern of subadult males and subadult females appeared to be that females wandered less than males. Because subadult females appeared to remain in their natal ranges, their home ranges were more stable than those of subadult males. Subadult F12 used the same area used by her mother (F29) and resident F4. Except for several exploratory movements, F12 seldom left the study area during the period she was monitored. Adjustments in home range overlap patterns among resident and subadult females appeared to be related more to home range vacancies following the deaths of residents than to changes initiated by the subadult females themselves.

Spatiotemporal Relationships Among Leopards

Although home range overlap analysis indicated neighboring leopards shared variable portions of their home ranges, it revealed little about the spacing of leopards in time. To obtain temporal information, I analyzed minimum distance to nearest neighbor data (Clark and Evans 1954) and intraspecific distances between radio-collared leopards located on the same day. Most of the data were collected during the day; thus they represent distances between resting leopards. Comparable data from active leopards during the night was not available.

SPATIOTEMPORAL DISTRIBUTION OF LEOPARDS

Data from three radio-collared resident male leopards in the SRSA indicated they were usually spatially distributed in a nonrandom pattern over a 79 km² area (see table 9.6). An average R value of 1.6 indicated that M1, M2, and M3 were spaced nonrandomly despite their home range overlap. Spacing was greatest ($R = 1.7$) during wet seasons (November through February), and least ($R = 1.2$) in September, at the end of the dry season.

When data from May through July for F4 and F12 were compared with that of the three males, their spacing approximated a random distribution ($R = 1.2$). This suggested male leopards were avoiding each other but not the females, or the females were not avoiding the males.

Comparable data for radio-collared female leopards in the NRSA were obtained over a 106-km² area from June 1974 through August 1974. During this period the three female leopards were spatially distributed in an aggregated pattern ($R = 0.6$). The clumped R-value and high degree of home range overlap

TABLE 9.6 *Dispersion Values (R) of Leopards in the Sabie and Nwaswitshaka River Study Areas*

Year	Month	Leopards	Number of Comparisons	Average "R"-value
1973	September	M1-M2-M3	20	1.24
	October	"	20	1.60
	November	"	15	1.68
	December	"	10	1.71
1974	January	"	11	1.65
	February	"	12	1.73
	March	"	15	1.49
	June	"	11	1.43
	July	"	6	1.80
	Average	"	—	1.59
1974	May	M1-M2-M3-F4-F12	2	1.03
	June	"	6	0.97
	July	"	3	1.45
	Average	"	—	1.15
1974	June	F6-F11-F13	16	0.47
	August	"	6	0.59
	September	"	7	0.67
	Average	"	—	0.58

($R = 1$ in a random distribution, $R = 0$ in a maximum association, and $R = 2.15$ in maximum spacing).
[1]Leopards M1 and M2 = subadults, M3 = adult, F4 = adult, F12 = subadult, F6, F11, and F13 = adults.

among these females were probably caused by their preference for the narrow riparian zone adjacent to the Nwaswitshaka River and the Msimuku tributary. Their *R*-value also suggested that resident female leopards did not avoid each other to the same degree as male leopards.

INTRASPECIFIC DISTANCES BETWEEN LEOPARDS

Leopards with adjacent overlapping home ranges, regardless of sex, social status, or season of year, were seldom located closer than 1 km during the day. The average distance between adjacent leopards varied between 2.9 km to 8.0 km in the SRSA and 1.7 km to 8.1 km in the NRSA (see tables 9.7 and 9.8). Comparable-aged males were separated by the greatest distances, and different-aged females were separated by the least distances. In the NRSA the average distance between two resident males (M10 and M14) with overlapping home ranges was 8.0 km. Both males were old, but M14 was probably slightly younger than M10. In the SRSA the average distance between comparable-aged M1 and M2 was 8.0 km.

These data and the home range overlap data indicated that male leopards with the least home range overlap were also separated by the greatest distances. The lowest average distance between leopards was usually between older and younger female leopards that shared the same area. In the SRSA older F4 and

TABLE 9.7 *Average Distances Between Leopards Having Overlapping Home Ranges in the Sabie River Study Area*

Sex Comparison	Leopards	Season									
		Dry 1973		Wet 1973–74		Dry 1974		Wet 1974–75		Average	
		n	km	n	km	n	km	n	km	n	km
Male vs female	M1 & F4	—	—	101	2.9	53	4.1	23	2.8	177	3.2
	M1 & F12	—	—	—	—	45	3.9	23	3.8	68	3.9
	M2 & F4	—	—	111	7.0	67	6.1	—	—	178	6.7
	M2 & F12	—	—	—	—	52	6.4	—	—	52	6.4
	M3 & F4	—	—	96	3.1	80	2.9	13	1.4	189	2.9
	M3 & F12	—	—	—	—	68	3.0	10	2.9	78	3.0
Male vs male	M1 & M2	25	6.8	105	8.7	38	6.6	—	—	168	8.0
	M1 & M3	23	2.5	95	3.4	52	4.3	12	2.4	182	3.5
	M2 & M3	24	6.7	98	6.8	58	6.0	—	—	68	3.9
Female vs female	F4 & F12	—	—	—	—	69	2.7	21	3.2	90	2.9

Leopards were located on the same day.
  n = number of comparisons.

young F12 stayed closer together (2.9 km), on the average, than any other pair of radio-collared leopards, except F4 and M3, who were also separated by 2.9 km. In the NRSA older F6 and younger F11, with greatly overlapping home ranges, were separated, on the average, by 1.7 km.

Distances between leopards with overlapping home ranges varied with the season. Male 1 and M2, and M2 and M3 in the SRSA were spaced farther apart during the wet seasons of 1973–74 than the dry seasons of 1973 and 1974. The

TABLE 9.8 *Average Distances Between Leopards Having Overlapping Home Ranges in the Nwaswitshaka River Study Area*

Sex Comparison	Leopards	Season				Average	
		Dry 1974		Wet 1974–75			
		n	km	n	km	n	km
Male vs female	M10 & F17	17	4.6	3	7.4	20	5.0
	M10 & F21	1	5.5	28	5.4	29	5.4
	M14 & F6	47	2.6	2	2.6	49	2.6
	M14 & F11	66	2.0	49	2.4	115	2.2
	M14 & F13	33	2.8	33	2.5	66	2.7
	M14 & F17	24	4.6	9	6.0	33	5.0
	M14 & F21	6	2.4	38	3.3	44	3.1
Female vs female	F6 & F11	73	1.7	1	1.8	74	1.7
	F6 & F13	37	2.9	1	3.7	38	2.9
	F11 & F13	48	2.7	40	2.3	88	2.5
	F17 & F21	2	2.8	6	3.3	8	3.2
Male vs male	M10 & M14	32	8.0	23	8.2	55	8.1
	M10 & M18	15	3.8	12	3.4	27	3.6

Leopards located on the same day.
  n = number of comparisons.

average distance between the two males with greatly overlapping home ranges (M1 and M3) increased during the 1973–74 wet season, increased more during the 1974 dry season, and then decreased during the 1974–75 wet season. In the NRSA, M10 and M14, who had slightly overlapping home ranges, were located only slightly farther apart during wet than dry seasons. The average distance between M10 and M18, who had greatly overlapping home ranges, decreased between the 1974 dry season and the 1974–75 wet season.

These seasonal distance-between-leopard patterns among male leopards had several common features. Older resident males, or males of comparable age with overlapping home ranges, expanded their home ranges during wet seasons when impala, their principal prey, dispersed from the rivers. This movement pattern usually increased the distance between male leopards. Because male leopards with greatly overlapping home ranges usually expanded their home ranges simultaneously, the distance between such individuals increased more rapidly than that between males with only slightly overlapping or adjacent home ranges. The distance relationships between two males in the SRSA (M1 and M3) departed from this pattern only after M1 began exploratory movements.

The spacing between female leopards was also greater during wet than dry seasons. As with male leopards, the distance between female leopards with greatly overlapping home ranges (>50% shared) was usually greater during wet than dry seasons, compared to females with slightly overlapping home ranges (<50% shared). Between the 1974 dry and 1974–75 wet seasons, the average distance between F4 and F12 in the SRSA and between F17 and F21 in the NRSA increased 19% and 18%, respectively. These females had greatly overlapping home ranges. In contrast, the distance between two females (F11, F13) with slightly overlapping home ranges actually decreased 15%.

These distance-to-nearest-neighbor measurements suggested that leopards spaced themselves apart even when they had greatly overlapping home ranges. Data from leopards in both study areas indicated this distance was generally more than 1 km, with slight differences dependent on sex and age. Male leopards generally were spaced farther apart (5.4 km) than female leopards (2.6 km), and males and females were spaced farther apart (4.0 km) than were females.

These data were similar to those reported for a female and two male leopards and for male leopards in Tsavo National Park (Hamilton 1976). Seven males were spaced, on the average, 5.4 km apart, the same distance as males in my study areas. The Tsavo National Park female and two males were spaced only 3.0 km apart, compared to 4.0 km for males and female in my study areas.

The observed ranges and home ranges of three radio-collared adult male leopards in the mountains of the southwestern Cape Province in South Africa (Norton and Henley 1987) overlapped in a manner similar to those of male leopards in my study areas. Overlap of observed ranges varied from 6.5% to

54.2%. The greatest overlap (54.2%) was between a younger male (WW) who was apparently displacing an older male (OS). Overlap of these two males with a well-established resident male (DD) was less, between 6.5% and 39%. The old male also had the smallest (53.5 km^2) of the three observed ranges (73.2 and 127.7 km^2).

The distances between cougars in Idaho were much greater than between leopards, probably because the cougars had much larger home ranges (Seidensticker et al. 1973). Cougars were located close together during winters when their home ranges coincided with deer and elk wintering areas, and they were spaced farther apart during summers when deer and elk dispersed over their summer ranges. In one sense leopards exhibited a similar pattern, but the differences were not as striking as with cougars. The principal prey of leopards, did not seasonally migrate as far as deer and elk.

As I monitored resident leopards inhabiting the study areas, I sensed that they knew where their neighbors were and could avoid them or rapidly associate with them if they desired. In stable leopard populations, individuals probably become familiar with each other and soon become aware if a neighboring resident dies or a new individual arrives in the area. Direct encounters between neighbors or between residents and newcomers, although they may occur infrequently, probably have long-lasting effects in maintaining an area's stable social organization.

VACANCIES IN THE LAND TENURE MOSAIC

The death of a resident leopard either caused other residents to adjust their home ranges to fill in the void or allowed other leopards to settle in the area. During the study period ten leopards died, three were suspected of dying, and one disappeared from the study area. Of these fourteen leopards, four were resident males, three resident females, three subadult females, three subadult males, and one a large cub. The vacancies left in the mosaic of home ranges in the study areas were quickly filled by other leopards.

During the study period an ample number of subadult male leopards in the study areas were apparently able to fill any vacancies resulting from the deaths of resident males. After a resident male died, a new, previously uncaptured male was usually observed or captured in the former male's home range. After radio contact was lost with resident M3 in November 1974, a large unmarked male and subadult M26 were observed well within M3's former home range. A large male (M30), probably the same one observed earlier, was captured nearby on July 11, 1975. He was well within M3's former home range and could have filled the vacancy. No large male was captured or observed within M3's home range before his disappearance. The other male, subadult M26, captured on May 5, 1975, confined his movements to the western portion of M3's former home range. The capture of the two males (M26 and M30) within M3's former home range and the movements of subadult M26 suggested that either of these two males could have occupied the area formerly used by resident M3. Subadult M26 could also have occupied resident M23's home range after M23 was killed by a crocodile in late May 1975. By then, M26 had already used nearly half (46%) of the area formerly occupied by resident M23, and he was still in the area when I removed his radio collar on June 28, 1975, nearly a month after M23's death.

Additional observations also indicated that a sufficient number of subadult male leopards were available to replace resident male leopards dying in the SRSA. After M5 died just southwest of the study area, a large unmarked male was observed scent marking just south of M5's home range. Similarly, after M22 disappeared, a large unmarked male leopard was observed scent marking along the Lower Sabie Road within M22's former home range.

The home ranges of resident and subadult female leopards that died were also taken over by other female leopards. After old F16 died in July 1974, three other female leopards (F27, F28, and F29) were captured in the same region. At least two of these (F28 and F29) were observed in the areas used by F16 before her death. After subadult F24 died, subadult F12 continued to use the overlap area previously shared with F24.

Land Tenure Rights

Land tenure among leopards insured certain rights. Resident male leopards appeared to have exclusive breeding rights; no known subadult or transient males associated with resident females long enough to breed. Only resident males, usually older individuals, associated with females often and long enough to breed. The breeding rights of males were associated with permanent and stable land occupancy and behavior (scent marking and vocalizations) characteristic of socially dominant individuals. But when a resident male died, another, often younger, male quickly assumed his area and presumably bred with the resident females. Successful occupancy of an area and the social status apparently needed to maintain that occupancy ensured the owner's right to breed.

Land tenure among resident males appeared to be based on prior use. I did not document any instance where an established resident male was driven out of his occupied area by another male. If an older established male becomes physically disabled, however, and is unable to advertise his presence by scent marking or vocalizations, another male may assume that role. The former resident would then have to assume a socially subordinate role or leave the area. Such a situation was apparently later documented in the southwestern Cape Province of South Africa (Norton and Henley 1987). There, a younger male leopard apparently displaced an older, probably resident male in poor condition. The older male eventually abandoned his area, became a stock-killer, and was later killed by a farmer well outside his previous range.

The primary right associated with land tenure among male leopards appeared to be access to, and successful breeding with, resident female leopards. The area occupied by resident males was much larger than that needed to obtain sufficient prey and encompassed the home ranges of up to six females. These observations indicated access to prey was not the principal factor responsible for the large areas occupied by resident males.

The primary advantage, or right, associated with land tenure among resident female leopards appeared to be access to high-quality habitat. High-quality habitat was characterized by abundant and stable prey populations, adequate escape and resting cover, and secluded places to rear cubs. All high-quality habitat in both study areas was occupied by resident females. Marginal, upland habitats in the NRSA, where prey was scarce, were avoided or infrequently used by resident females. The size of the areas required by females to successfully rear young varied with habitat quality. Females in poorer habitats generally had larger home ranges than those in high-quality habitat.

Resident female leopards seemed much more likely than males to share their areas with, or adjust use of their areas to accommodate, other females, at least in the high-quality habitats of the study areas. The other females included their own independent offspring, neighboring resident females, and perhaps the female offspring of neighboring resident females. This sharing of resources exhibited itself in the use of prey from the same area and in the use of the same sites (but at different times) to rear offspring.

Mutual avoidance, as generally described in cougars (Hornocker 1969), rather than aggressive encounters and fighting, appeared to be the principal mechanism by which leopards spaced themselves. Avoidance between resident males and between resident females was facilitated by olfactory, auditory, and visual signals. Nonbreeding individuals seldom, if ever, advertised their presence. They apparently avoided the residents, and residents avoided each other.

The relatively peaceful associations among, and stable land tenure system of, study area leopards may have been characteristic of leopard populations composed of older, well-established residents. Because the residents were already familiar with each other and their surroundings, unexpected and potentially dangerous encounters with unfamiliar leopards were probably infrequent. Furthermore, their continued presence was made known to neighbors and to any newcomers through various modes of communications, especially olfactory and auditory signals.

Communication Among Leopards

Intraspecific communication among leopards probably serves several functions. First, it may allow them to avoid each other and thus separate themselves in space and time. Second, it may also allow them to attract each other, an important function during courtship and breeding. Third, leopards may use subtle modes of communication to distinguish each other by individual, sex, age, and social status.

Scent Marking

Scent marking was the most frequently used mode of communication among leopards. Scent marking has several obvious advantages over other forms of communication. First, scent marks probably convey information for extended periods—days, weeks, or perhaps months under certain conditions. Second, a scent mark's radius of effectiveness extends well beyond the immediate area. I could sometimes detect the odor of leopard scent somewhere within an area of about 30 m^2, but could not locate its source. Leopards were quite likely to detect scent at even greater distances. Third, a scent mark probably conveyed specific information about its sender. Subtle differences in odor were often detectable. Sometimes leopard scent was extremely powerful and musky smelling; at other times, even within minutes after it was expelled, the scent had little odor. Several types of scent marking by leopards were observed.

Urine spraying.   Both male and female leopards sprayed urine to scent mark. Of ten leopards I observed spraying urine, six were females and four were males. These included two radio-collared adult males (M3 and M23) and one adult resident female (F6). Leopards sprayed urine during all seasons, at all times of day, and in a variety of contexts.

Leopards usually sprayed urine while leisurely walking along trails and roads. Leopards periodically stopped by a bush, clump of grass, or a tree, swiveled their rump close to the bush and with tail raised high and tip wiggling, ejected a stream of fluid. Sometimes, leopards rubbed their cheek against the object before scent marking (see fig. 9.12).

> July 27, 1974, Msimuku tributary, 11:20 A.M. to 11:35 A.M.: Radio-collared F6 lies in a small dry stream watching some impala about 100 m away. Several minutes later she gets up and walks along the stream parallel to the Msimuku. I follow behind in my vehicle. After reaching the top of the bank, she pauses by a bush, sprays urine, and continues walking to a large granite rock, where she lies down to watch me. After several minutes she gets up, walks about 30 m, rubs her cheek against another bush, swivels around and sprays urine on it. She then goes to the base of a tree 30 m away, rubs her cheek against the side of the tree without spraying urine, walks 10 m to another tree, rubs her cheek again and squirts urine on the tree before walking into dense cover.

Female 6 did not scrape while spraying urine, but another female did.

> March 3, 1974, Doispan Road, 5:53 P.M. to 6:00 P.M.: An unmarked female leopard stands beside the Doispan Road where a small dry stream

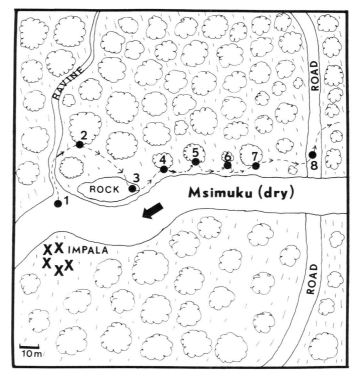

FIGURE 9.12 Route (dashed line) taken by adult female leopard F6 and the locations where she scent marked (numbered black dots) along the Msimuku tributary on July 27, 1974 (11:20 A.M. to 1:35 P.M.). She first lay watching impala (1), then walked up a small ravine, climbed an embankment and sprayed urine on a bush (2), lay down on rocks (3), rubbed her cheeks on and then sprayed urine on another bush (4), rubbed cheeks on a tree (5), rubbed cheeks on another tree and then sprayed urine (6), walked under another tree (7), and crossed a road (8) before I lost sight of her in dense vegetation.

passes through a culvert. With her tail raised high and the tip vibrating, she swivels her rump to a leafy bush and sprays urine. She then crosses the road with her tail still held high, scrapes several times in front of another bush with her hind feet, and again sprays urine. Recrossing to the opposite side of the road, she returns to within 5 m of the first scent-marked bush, rubs her cheek against another bush, and sprays urine for the third time. She then leaves the road, crosses the dry streambed and leisurely moves through the brush, rubbing her cheek and body against the vegetation as she walks. As she passes a raisin bush (*Grewia hexamite*), she raises her tail, vibrates it, and sprays urine. She climbs on a fallen tree, pauses briefly to look around, steps down, and disappears into the brush.

Broad-leaved deciduous *Grewia hexamite* were frequently scent marked by leopards. These 2- to 4-m high shrubs often grew beside roads and dry streambeds that leopards used as travel routes. They were scent marked by leopards more often than other vegetation, perhaps because of their dense foliage close to the ground level and their retention of leaves well into the dry season. Leopards scent marked on *Grewia hexamite* every month except August and October, including December when the shrub's leaves were absent or developing.

Leopards responded to the odor of their own scent as well as to that of other leopards. On several occasions leopards carefully investigated scent on vegetation along their travel route. Sometimes this investigation was so subtle that it almost escaped notice. I observed radio-collared resident leopards of both sexes carefully investigating vegetation.

February 15, 1974, Sand River Road, 3:26 P.M. to 3:47 P.M.: Attracted by the alarm calls of a group of vervet monkeys, I stop to watch a vervet cautiously climb down a tree while peering downriver into some dense vegetation. Driving forward 10 m, I discover radio-collared resident M3 concealed in the vegetation. The male gets up and moves slowly upstream, but reverses his travel and bolts across the road when I back up my vehicle. He is moving toward another road; I leave and intercept him about seven minutes later. He is standing near the junction of the road and a dry streambed. He carefully smells the vegetation for several seconds, raises his tail, and sprays urine. He then reverses his direction of travel and moves down the dry streambed out of sight.

Male 3's tracks in the sand revealed that he walked up the streambed to the road, investigated the vegetation, scent marked, and returned in the direction he had come.

I also observed other leopards investigating scent on vegetation. On April 20, 1975 I watched radio-collared adult F25 slowly walk along the edge of a dirt road and carefully investigate the vegetation, although she did not spray urine. On May 13, 1975 I watched an unmarked female leopard for nearly two hours. During this time she stopped at least twice and carefully smelled leafy green vegetation with her tail held high, but did not spray urine. On two other occasions leopards carefully investigated scent at the base of a tree, but departed without scent marking.

After spraying urine, leopards usually just walked away without smelling the vegetation. Once, however, a large male leopard carefully investigated the vegetation after he sprayed it with urine. This occurred during twenty-three minutes of scent marking when he scraped, apparently deposited scent from his anal glands while assuming the defecation posture, and sprayed urine. Once he sprayed urine on the same bush at least three times and carefully investigated the scent after each event before he finally departed. Although I was not usually close enough to see a leopard's face as it investigated scent marks, on one occasion a leopard grimaced (flehmen) immediately after smelling scent at the base of a tree.

June 15, 1974, Nwaswitshaka River Road, 5:15 P.M.: I surprise a female leopard sitting in the middle of a firebreak road; she runs off, circles, and comes back on the road before traveling along the road. Soon after, she steps off the road, walks to a tree 3 m from the road and investigates the vegetation at the base of the tree. As she smells she grimaces but does not scent mark. Later, she returns and walks down the road.

Several minutes later the leopard ran down the road toward me and climbed a tree to escape from a group of lions pursuing her. My observations of leopards spraying urine were similar to those reported for several other large felids. Both male and female leopards spray scent. Although my sample size was small (n = 10), I observed female leopards more often spraying urine (60%) than males (40%). Females were also observed investigating scent (57%) slightly more often than males (43%) (n = 7). Schaller (1972) noted that female leopards appeared to spray urine more often than female lions. During Hamilton's (1976) study he observed a leopard, an adult male, spray urine only once. In Wilpattu National Park, Eisenberg and Lockhart (1972) reported that both male and female leopards sprayed urine. They were unable to determine which sex sprayed more often.

Among some of the larger felids, males appear to spray urine more frequently than females. Schaller (1972) observed that male lions sprayed urine often, but lionesses sprayed only seven times during a three-year study. He also reported that male cheetahs sprayed urine more often than females. Eaton (1974) reported that female cheetahs did not scent mark. Male cheetahs scent marked by spraying urine backwards on trees or other prominent objects or squatted and urinated on the ground and scraped repeatedly with their hind feet over the urine (Caro and Collins 1987).

Urine spraying by both males and females was the most frequently used mode of scent marking among free-ranging tigers (Smith et al. 1989). Most urine spraying (89%) occurred on the same trees—trees that were large and leaning at an angle. Tigers often sprayed urine on the underside of these trees, perhaps to protect the scent from rain. Two female tigers marked 25.8 trees/month on 5.1 visits/month; males marked 13.0 trees/month on 4.4 visits/month.

Scraping.   Scrapes often accompanied the scent marks of leopards. When a leopard scraped it lowered its rump, slightly arched its back, and clawed the ground with its hind feet, alternating between left foot and right foot. If the soil was soft, the grass was usually removed to bare soil and a rectangular area up to 50 cm wide appeared (see photo 9.2). If the scrape was fresh, clear imprints of the leopard's hind feet were visible at the rear of the scrape. If the soil was hard, only the claw marks were visible on the soil's surface.

Scent-marking leopards often urinated during or after scraping. At many recent scrapes, urine could often be seen in the soft dirt, at or on the small

9.2. Close-up of a recently made leopard scrape in short grass near a firebreak road used as a leopard travel route.

mound of dirt at the rear of the scrape. The strength of the odor of leopard urine at scrapes varied, probably declining with age. Occasionally I initially detected scrapes by the odor of leopard urine. At other times I could not detect the odor of urine at a scrape even though it appeared recent.

Leopards may scrape frequently within a short distance. One evening I had the opportunity to watch a leopard scrape on three occasions.

March 3, 1974, Kruger Road, 7:00 P.M. to 7:25 P.M.: While driving after dark, I observe a male leopard lying in the grass beside the road. I stop, and several minutes later the leopard gets up and slowly walks away. I follow 20 m behind. After trailing only a short distance, the leopard steps 1 m off the north edge of the road into some dry grass, scrapes several times, and then slightly lowers his rump over the scrape and deposits scent (urine?). He then steps back on the road, crosses it, and proceeds along the road's edge. About 150 m down the road he stops, steps 1 m off the south side of the road, scrapes, squats, and deposits scent. After walking another 150 m, he steps 1 m off the north side of the road, scrapes seventeen times with his hind feet, squats, and deposits scent on the scrape. After walking a short distance on the road, the leopard steps off and lies down in the grass to watch me. Several minutes later he gets up, walks another 200 m along the roads, turns, and disappears into the brush.

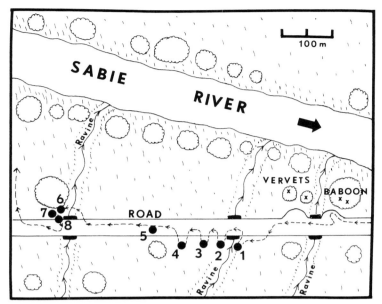

FIGURE 9.13 Route (dashed line) taken by an uncollared male leopard and locations where he scent marked along the Sabie River on July 21, 1975 (5:47 P.M. to 6:00 P.M.). Baboons and vervet monkeys alarm called when they saw the leopard. Male scent marked and scraped at locations (1), (4), (7), and (8); probably sprayed urine at locations (2) and (3); smelled elephant dung at location (5); and sprayed urine at location (6).

Each scrape was 30 cm to 35 cm long and 15 cm to 20 cm wide. Each had a strong urine odor, and all were within a 300-m section of the road. No older scrapes were near the recent ones. Scrapes were sometimes associated with spraying urine, but usually scraping and spraying occurred separately.

July 21, 1975, Lower Sabie Road, 5:47 P.M. to 6:10 P.M.: While returning to Skukuza after dark, I observe a large male leopard leisurely walking on the road ahead of me. The leopard crosses the road by a large fig tree where baboons are sleeping. After the baboons alarm call, the leopard moves back onto the road. After the leopard walks a short distance, vervet monkeys also alarm call, but the leopard ignores them. He crosses the road, walks 1 m off into the grass, scrapes with his hind feet several times, squats three seconds after scraping, and deposits scent. He then comes back on the road, walks several meters, goes off the same side of the road to spray urine and then resumes walking. About 10 m down the road the leopard sprays urine again, resumes travel another 7 m, walks off the road 1 m, scrapes, and again squats to deposit scent. After he resumes travel, he stops to briefly smell some elephant dung. He then crosses the road, walks up to a leafy green bush, sprays urine two or three times. He walks

1 m, scrapes five or six times, but does not squat or urinate at the scrape. He walks several meters, stops, scrapes several times, and squats and urinates at the scrape. He continues on but does not scent mark again before he leaves the road and disappears into the brush. (See fig. 9.13.)

Male leopards appeared to scrape more often than females. On sixteen occasions when I was able to identify the sex of the leopard making scrapes, thirteen (81%) were made by males. At least four radio-collared resident male leopards were responsible for making scrapes on nine of these occasions. The status of the other male leopards was unknown, but they were also large adults. All three females that scraped were radio-collared resident females.

Thirty-six of fifty-six scrapes (64%) were made by leopards during the wet seasons and the remainder during dry seasons. During dry seasons, more scrapes (eight) were observed in July than any other month. Possibly scrapes were more visible and conspicuous during wet seasons because the soil is often moist and the scratch marks prominent, an observation also noted by Eisenberg and Lockhart (1972) in Wilpattu National Park.

Other felids also scrape to scent mark. Both male and female snow leopards scraped along common travel corridors and frequently revisited scrape sites. Scrape sites were found at prominent topographic features such as large rock outcrops, sharp ridges, and promontories (Ahlborn and Jackson 1988). The intensity of scraping peaked during the mating season; one 750-m transect had more than one hundred scrapes, and some tufts of bunchgrass were surrounded by more than twenty-four scrapes (Hillard 1989). Male jaguars frequently scraped along roads serving as common boundaries between territories, often (68%) defecated on their scrapes, but apparently did not urinate at scrapes (Rabinowitz and Nottingham 1986).

Tigers scraped less often (21%) than they sprayed urine (78%) to scent mark (Smith et al. 1989). Urine, feces, and no odors were present at 54%, 17%, and 19% of tiger scrapes, respectively. Scraping and spraying urine were used interchangeably; scrapes were more evident in grassy areas where scent posts were scarce.

Fecal marking.   Throughout my study areas uncovered leopard feces were commonly observed along leopard travel routes, especially dirt firebreak roads. These feces were usually deposited in very conspicuous places. If the strip in the middle of the road was covered by grass, leopards seldom defecated there. But if the strip was grass-free, leopards often defecated there. Otherwise, leopards defecated in one of the ruts made by vehicles' wheels or on the bare soil near the edge of the road. Another place I observed leopard feces was near kills, and on one occasion many different-aged feces of leopards were found along a game trail in a small clearing. Latrines with numerous feces of leopards were not observed, nor did I find feces in or near scrapes of leopards as reported by Schaller (1977) for leopards inhabiting the Karchat Hills in Pakistan. Perhaps leopards only use feces to mark scrapes in extremely dry environments, a behavior also exhibited by some bobcats (Bailey 1972). The high evaporation rates in dry environments would quickly evaporate urine and possibly shorten the effective life span of scrapes. The odor from feces at scrapes in dry environments may last a long time, perhaps several months.

Feces of leopards were encountered more often during dry than wet seasons. Of forty feces estimated to be less than twenty-four hours old when first discovered, thirty-seven (93%) were encountered during the dry seasons and the remaining three during the wet seasons. The majority of feces (72%) were collected in June ($n = 16$) and July ($n = 13$). Dung beetles, which are more active during wet than dry seasons, may have removed some leopard feces during wet seasons.

That leopards respond to feces they encounter was exemplified by one observation.

> August 8, 1974, Sand River Road, 6:07 P.M. to 6:08 P.M.: While I observe radio-collared resident M3, he walks parallel to the road, later comes onto the road, crosses it, and stops to investigate something on the ground 30 cm from the road's edge. After smelling, he scrapes with his hind feet but does not squat or scent mark with urine. After scraping, he leaves the road, crosses a dry ravine, and disappears into the darkness. Vervet monkeys suddenly alarm call near the leopard.

When I examined the scraped area later, I discovered the feces of a leopard, feces apparently not deposited by M3. These observations suggested that leopards used feces to scent mark common travel routes, especially during dry seasons. Other felids that use feces to scent mark include female bobcats (Bailey 1972), male jaguars (Rabinowitz and Nottingham 1986), Spanish lynx (Robinson and Delibes 1988), and male and female tigers (Smith et al. 1989).

Tree scratching.   I did not observe leopards scratching trees as Eisenberg and Lockhart (1972) reported for Asian leopards in Wilpattu National Park. They reported that leopards there repeatedly scratched several trees in their study area and described this behavior in detail. Scratch trees were usually leaning, or had large limbs 1.8 m to 2.4 m above the ground. The scratches of leopards rendered the tree conspicuous by the visual appearance of the scratch marks and sap that exuded from the wounds made in the bark of the tree.

According to Eisenberg and Lockhart leopards that scent marked trees usually smelled at the base of the tree, climbed rapidly upward, and scratched with either the front or the hind feet. Leopards paused to smell at the scratch marks from time to time and often sprayed urine at the base of the tree. Eisenberg and Lockhart postulated that exudates from the paws may have been left on the scratch marks and that odor from the leopard's body may have been left behind when the leopard stretched or reclined on the tree.

I knew of only one tree in the study areas that may have been a leopard scratch tree. This tree was located in the sandy area bordering the Sabie River and showed signs of repeated use. The scratch marks were too small and low for a lion and too large to be made by a smaller cat. I periodically visited this tree for one year to see if leopards often used it; I never observed leopard tracks nearby or any fresh scratch marks in the bark. None of the other trees that I observed leopards smell had any visible scratch marks on their trunks.

Both male and female tigers occasionally clawed trees in the Chitwan (or Chitawan) National Park, Nepal (Smith et al. 1989). Claw marks were found on eight trees, 2 to 3 m above ground, and a leaning tree was repeatedly clawed by two territorial male tigers during a one-year period.

Dung rolling.   Rolling in smelly substances is considered to be a form of scent marking (Ewer 1968). Male tigers rolled in and flattened vegetation on at least three occasions during Smith et al.'s (1989) study of scent marking in tigers. The only feces that leopards appeared attracted to, besides their own, was elephant dung. A large male leopard once paused briefly to smell fresh elephant dung but did not roll in it. An adult female smelled something not visible to me on the road and rolled on her back over the spot several times before departing. Once, when a male and female leopard were courting, the female rolled in fresh elephant dung and elephant urine on a road. She rolled repeatedly in the dung, saturating the fur on her back and her sides. Five minutes later the male leopard briefly copulated with the female.

Locations of scent marks.   Most scrapes (79%) were along travel routes frequently used by leopards such as roads and dry streambeds. The remainder (21%) were near livetraps (16%), kills (3%), or at sites where leopards encountered each other or other predators (2%). Scrapes were probably present along other travel routes, such as game trails, but I did not regularly monitor these travel routes. All observations of leopards spraying urine also occurred along leopard travel routes.

Leopards frequently scent marked at conspicuous places along travel routes, such as intersections of roads and dry streambeds, bridges, and culverts. Because dry streambeds were also used by leopards as travel routes, higher concentrations of scent marks occurred at intersections than along the travel routes. A well-used scent mark location along the Naphe Road was at the Msimuku tributary where the road crossed a small bridge (see fig. 9.14). Many leopard scrapes were present at the east end of the bridge, as well as 0.5 km east of the bridge near the intersection of a small dry stream. Scrapes of leopards were also found at the Sabie River bridge 10 km east of Skukuza and where several dry streams passed under the Sand River road.

Leopards scent marked along tarmac tourist roads more often than along dirt roads, firebreak roads (see photo 9.3), or abandoned roads. Although some scent marking was evident along the hard surface roads paralleling the Sabie

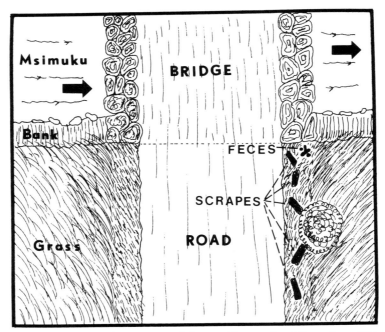

FIGURE 9.14 Locations of scrapes and feces of leopards at the edge of Naphe road at the Msimuku tributary bridge in the Nwaswitshaka River study area. Leopards, including M14, periodically scraped and scent marked at this location throughout the leopard study.

and Sand Rivers, scent marks were much more evident along the Naphe and Kruger roads. The Sabie and Sand River roads, which were adjacent to those rivers, often formed common boundaries between nonoverlapping or slightly overlapping leopard home ranges. In contrast, the Naphe and Kruger Roads were not near natural barriers. This allowed for extensive home range overlap and may have increased scent marking among leopards because the roads served as common travel routes for neighboring residents.

It was also evident that the most frequently scent marked travel routes were at the peripheries of leopard home ranges rather than deep within them. Exceptions included roads adjacent to natural barriers such as the Sand and Sabie Rivers. Other frequently scent-marked roads did not lie adjacent to natural barriers. These observations suggested that fewer scent marks were made along roads adjacent to natural barriers because of a reduced likelihood for encounters to occur among leopards using such routes. Routes shared by neighboring leopards had more scent marks because interactions were more likely to occur.

Other studies of free-ranging felids have revealed that scent marking is most common along roads and trails that serve as common boundaries between territories (jaguar [Rabinowitz and Nottingham 1986], tiger [Smith et al. 1989])

9.3. The scrape of a leopard in the center grassy strip of a firebreak road used by leopards as a travel route.

or that are used by several individuals sharing an area (snow leopards [Hillard 1989]). Scent marks were also used to mark special areas such as natal and auxiliary den sites (bobcats [Bailey 1972, 1979]) and are more likely to be found at trail intersections than along trails (Spanish lynx [Robinson and Delibes 1988]).

Function of scent marks. Scent marking by leopards appeared to serve several functions: (1) to delineate their home ranges; (2) to communicate specific information about their social and reproductive status; (3) to prevent or enhance encounters among leopards, especially along common travel routes; and (4) to identify places of special interest such as kill sites or encounter sites between leopards or other predators. The actual response of leopards to various scent marks was seldom observed, but these functions were derived from the following evidence.

1. Leopards used scent to delineate boundaries of their home ranges. Certain places, often at known home range boundaries, were periodically marked with scent. Scrapes, scent, and conspicuous feces were used to mark home ranges. Leopards periodically visited some of these places to renew or freshen up their scent marks. Scrapes were monitored at two sites for M14 and at one site for F6 in the NRSA, and at one site for M3 in the SRSA. I regularly checked these four sites for leopard scrapes. Each site was checked weekly for 111 to 168 days from August 1974 through March 1975. Each new scrape was identified and marked with a small L-shaped wire. If a leopard scraped over the previous scrape, the small hidden wire pivoted. By checking the position of the inconspicuous wire, I could determine if a scrape had been renewed.

   New leopard scrapes were discovered at the scrape sites every seven to one hundred days. The average interval between new scrapes was 25.9 days. Six scraping intervals at the Msimuku Bridge varied from 7 to 40 days with an average of 21.7 days between new scrapes. About 0.5 km away, near the culvert and ravine, new scrapes were renewed by leopards every seven to twenty-six days. Once, after resident F6 scraped at a game trail and road intersection, another new scrape was not observed there until twenty days later. No new scrapes were found at the site for at least ninety-one days. Resident M3 scraped at one site on August 27, 1974 and again fourteen days later. These data suggested that scent and scrapes remained effective for relatively long periods of time and that some scrapes were periodically renewed by leopards; others were not.

   Leopard feces apparently were not used to mark specific sites, but were merely deposited anywhere along selected travel routes within the leopard's home range. Feces and scrape marks may have signaled other leopards that an area was already occupied. This could repel, attract, or otherwise alter the movements of other leopards depending on their social status and other factors. Presumably, feces and scrape marks lasted longer as signals than urine spraying and were thus more suitable for delimiting areas for longer periods.

   Among tigers, the frequency of scent marking was highest in narrow contact zones at territorial boundaries (Smith et al. 1989). These marking zones shifted as territorial boundaries shifted and marking was associated with territory establishment. Scent marking was five times higher at territorial boundaries than in the middle of territories. The authors concluded that scent marking appeared to be the primary means of shaping and maintaining tiger territories and that tigers must use an area frequently to establish a strong claim to their territories.

2. Leopards' scent marks probably communicated specific information about their social and reproductive status. I never observed young or subadult leopards scent marking—only established resident leopards. Although a leopard raised by Joy Adamson (1980) sprayed scent when it was only eight

months old, its social behavior was influenced by man. Scent marking among leopards appeared similar to that of lions and tigers in that only older, sexually mature individuals or individuals establishing territories scent marked. Lionesses did not scent mark until they were about 2 years old and males did not scent mark until they were 3.3 to 3.5 years old (Schaller 1972). A captive tiger did not begin obviously to spray urine until it was eleven months old (Brahmachary and Dutta 1991), and young female tigers began to scent mark extensively when they began establishing breeding ranges (Smith et al. 1989).

Like other felids, younger leopards may be able to ascertain whether older, sexually mature, and perhaps socially dominant leopards are already occupying an area they want to settle in. Scent can also signal when a female leopard is in estrus. My observations suggested that female leopards scent marked more frequently during the peak mating period. Leyhausen (1979) observed that most female cats in estrus generally sprayed urine more often than during other periods. Scent marking by female tigers increased just before their estrus, but declined during estrus, and male tigers increased their scent marking while the female in their area was in estrus (Smith et al. 1989).

3. Scent marking may also facilitate encounters among leopards. Male and female leopards of all ages, including the resident male and female and the female's offspring, often utilized the same travel routes. To prevent encounters, scent could function like railroad signals (Leyhausen and Wolff 1959): an old mark might signify to a leopard that another has not recently passed by and it is safe to proceed along the trail. A recent scent mark could mean that a leopard was in the vicinity and that the intruder should proceed with caution if it desires to use the same travel route. A fresh scent mark could signal that another leopard is close by and that by proceeding along the same route an aggressive encounter might occur. The same scent mark could also enhance encounters if a male and female leopard were searching for each other, or if a female or her offspring were searching for each other.

The effective life span of scent marks probably varies with weather conditions and the chemical components comprising the scent. Scent did not persist long during the wet seasons because of frequent thunderstorms, but during the dry seasons I could detect leopard scent for at least one month. Schaller (1967) detected tiger scent on a tuft of grass after one week, and on tree trunks for three to four weeks; in one instance on a tree trunk, the odor was detectable for three months. Smith et al. (1989) detected tiger scent for at least three weeks up to 3 m away. Eaton (1974) reported that cheetah urine marks more than twenty-four hours old did not appear to prevent individuals from using the same travel routes.

4. Scent marking was also apparently used by leopards to claim temporary ownership of or to identify places of special interest, such as kills, sites of intraspecific or interspecific encounters, waterholes, and rest trees. Scent marks at kills could signify ownership should another leopard arrive at the

scene, especially if the kill's owner was away at the time. It was common for leopards to temporarily leave their kills to attend to cubs and for other reasons. Scent marks at sites of previous aggressive encounters could identify the site as a place for caution should the leopard pass through the area again in the future. For example, I observed leopard scrapes where leopards encountered each other and where a leopard encountered several spotted hyenas at a warthog carcass.

VOCALIZATIONS

Although leopards do vocalize, they seldom did so in my study areas. Their vocal repertoire was limited to the long-range coughing call commonly referred to as rasping and to a variety of grunts, growls, hisses, and snarls. In the field I heard only four types of sounds: the rasping call, a "woofing" when leopards were surprised, snarling between leopards or other predators, and grunting during copulation. Trapped leopards also snarled, hissed, and growled.

Rasping call.   The most common vocalization of the leopard was a harsh rasping call that is frequently described as a sound resembling uneven sawing with a crosscut saw. Reportedly, considerable effort is required to give the call, with the animal dropping its head and showing visible physical effort while calling (Turnbull-Kemp 1967). On three occasions I observed leopards at close range rasping. On one occasion a large male had just finished scraping and scent marking. He walked about 3 m, paused, and then stood, with his head held low, and called before disappearing into the darkness. Identical behavior was observed after another male leopard scent marked. Both observations occurred during the night. Another leopard, a female, was observed calling during the daytime.

> January 23, 1974, Lower Sabie River Road, 5:15 P.M. to 5:45 P.M.: While I am parked in my vehicle along the Lower Sabie Road, a leopard calls about 100 m away. Driving ahead, I see the leopard come out of the brush along the road. Upon arriving at the road, the leopard lowers its head and while slowly walking, begins to rasp several times. The leopard appears to be hot as its head is held low and saliva drips from its partially open mouth. After rasping the leopard continues slowly, almost sluggishly, ahead. As I approach more closely to ascertain the leopard's sex, the leopard doubles back and disappears into the undergrowth beside the road. As it walks away, I discover it is a female. Several minutes later the leopard rasps six to seven times about 30 m off the road. During the next twenty minutes it remains silent, so I leave the area.

The leopard did not stop walking while it called and it ignored my vehicle parked only 15 m away.

The range of the leopards' rasping call varied with cover and terrain. During

TABLE 9.9 *Vocalizations of Leopards*

Characteristic of Rasping Vocalizations of Leopards	Males			Females			Total		
	n	*x*	*SD*	*n*	*x*	*SD*	*n*	*x*	*SD*
Number of strokes per call	19	8.8	2.0	9	12.9	2.6	39	9.9	3.1
Number of calls per calling period	8	3.3	2.6	4	5.5	4.4	25	3.6	2.7
Interval in minutes between calls	14	5.6	6.4	12	7.0	8.2	39	6.7	6.8
Total length of calling period (in minutes)	4	19.5	23.7	2	26.5	19.1	13	17.7	14.8

Males and females are radio-collared leopards. Totals include unmarked leopards of unidentified sex.

wet seasons, on flat, densely vegetated terrain, calls could be heard only about 1 km away. During the dry season, on a slight, sparsely vegetated ridge above the Sand River, I once heard a leopard call about 3 km away. The low-volume moaning call of tigers carries from only about 60 m to 400 m (Schaller 1967), but the louder roar can be heard 3.2 km away according to Powell (1958). Lions can be heard roaring 3 to 4 km away on the open plains (Schaller 1972).

Determining the location of calling leopards was often difficult because of the call's ventriloqual effect. Its undulating amplitude, low intensity, and abruptness contributed to this feature. The call, however, was highly conspicuous. One night while sitting on our veranda, I heard a leopard call 300 m across the river. In the silence of the night, the sudden call of a rasping leopard immediately gained one's attention.

Male and female leopards I monitored had individually distinct calls. Known female leopards had more strokes per call, more calls per period, longer intervals between calls, and longer total call periods than did known male leopards (see table 9.9). I was also able to recognize certain individual leopards in my study areas by their calls. In the NRSA resident M14's call was usually 8 to 10 strokes/call ($n = 17$, $x = 9.25$, $SD = 2.7$); adjacent M10's call was 5 strokes/call. In the SRSA F4's call was about 14 strokes/call ($n = 8$, $x = 13.6$, $SD = 1.4$);, an old male (M23) in the area called about 7 strokes/call. On several occasions I first predicted which individual leopard was calling and then verified its identity with the radio receiver. Possible recognition of leopards by their calls was suggested by Eisenberg and Lockhart (1972) after they noticed the difference in calls between two captive leopards.

Radio-collared subadult leopards did not call (rasp) in my study areas. On numerous evenings during peak calling periods, I waited within hearing distance of leopards to determine whether subadult leopards were calling; I never heard them do so. I periodically heard most of the resident leopards call. I estimated that in my study areas female leopards did not call until they were about twenty-four to thirty months old. Lions do not begin roaring until they are at least 2.5 years old (Schaller 1972).

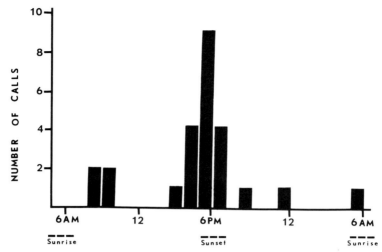

FIGURE 9.15 Cumulative hourly distribution of the number of recorded rasping calls of leopards.

The peak calling period of leopards was in the early evening just after sunset between 4:00 P.M. and 7:00 P.M. Another less-pronounced peak occurred in the early morning hours between 8:00 A.M. and 10:00 A.M. (see fig. 9.15). I did not hear leopards calling during midday nor during the middle of the night. The peak calling period of leopards coincided with the onset of their evening activity period, and leopards seemed to be signaling their presence just before they began moving about their home ranges. Turnbull-Kemp (1967) noted that leopard calls were not restricted to night, and could be heard on overcast or misty days or during midday in brilliant sunlight. Similarly, leopards in Tsavo National Park called at any time of day, but most calls were at dusk and dawn (Hamilton 1976).

Leopards called more often during dry than wet seasons. Leopards called most frequently in May and June and least often between August and January (see fig. 9.16). Observations also suggested that calling and scent marking dominated at different times of the year. Leopards called frequently when few scent marks were observed and scent marked frequently when calling was minimal. Different environments might select for seasonal variations in the use of signals. Calls should carry longer distances during dry seasons when leafy vegetation is minimal, and scent marks should be most effective during the wet seasons when high humidity reduced evaporation of scent.

Leopards appeared to call throughout their home ranges. But when leopards called, they were usually traveling well-used routes such as firebreak roads. Female leopards did not call near koppies or other places near their cubs, nor did any leopards call near their kills.

Leopards also did not call while hunting although on one occasion a hunting

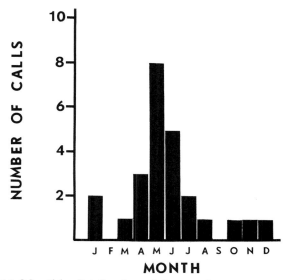

FIGURE 9.16 Monthly distribution of the numbers of rasping calls of leopards.

leopard did call, but only after it was detected by a herd of impala. On the evening of March 14, 1974, resident F4 was 0.5 km from a group of impala that had bedded for the night in an open area. The signal from her radio collar indicated that she was stationary until thirty minutes after darkness. At that time the leopard began moving toward the bedded impala. An hour later, the impala suddenly burst into activity and began to alarm call. The leopard then left the vicinity, calling several times. She then traveled along a firebreak road until she encountered and was captured in one of my traps.

The primary function of calls appears to be advertisement. Leopards may respond to calls of other leopards by joining, avoiding, ignoring, or answering the caller. One observation revealed how leopards call to locate each other in dense cover.

> December 28, 1974, Nwaswitshaka Study Area, 5:49 A.M. to 7:30 A.M.: Resident F11 is about 100 m away when another leopard suddenly calls 100 m to the south. A quick check on the radio receiver reveals the caller is resident M14 who uses the same area. Upon hearing the male's call, F11 moves rapidly toward M14. Within eleven minutes their radio signals are together. When I stalk the leopards on foot, I hear them growling at each other nearby but cannot observe them because of dense vegetation. Finally, the leopards leave together and I return to my vehicle. On the way back I discover a fresh leopard scrape beside a game trail used by the leopards. (See fig. 9.17.)

Individual recognition was suggested because the female responded to the call of the male and went directly and rapidly to him. These leopards were

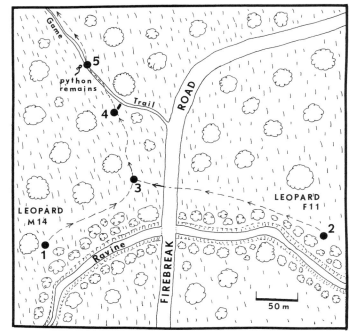

FIGURE 9.17 Routes (dashed lines) taken by and locations of (numbered black dots) adult male leopard M14 and adult female F11 on December 28, 1974 (5:45 A.M. to 7:30 A.M.). The male called at 5:49 (1) when F11 was nearby (2). By 6:00 they were together (3), and there was much snarling and growling. After the leopards departed together, a fresh scrape was found near an aardvark burrow beside the game trail (4). Remains of a freshly killed python were found at (5), but the python may not have been killed by the leopards.

located together on previous occasions, and M14 probably sired F11's offspring that she lost earlier in the year. Female 11 may have been in estrus at the time of this observation. A female that called on January 23, 1974 along the Lower Sabie Road may also have been in estrus. Female leopards, like tigresses and lionesses (Schaller 1967, 1972), probably call more frequently during estrus than at other times.

Calls of leopards may also promote avoidance. This was suggested during observations of three leopards. Two of the leopards were sharing an impala kill cached in a tree beside a firebreak road.

21 May 1974, Sabie River Study Area, 9:00 A.M. to 10: 15 P.M.: I observe two spotted hyenas fleeing from the base of a tree in the SRSA and find a recently killed six-month-old impala with only a small portion of one hindquarter eaten. I return at 3:15 P.M. but no leopards or hyenas are in sight. At 4:55 P.M. a leopard calls about 1 km to the east. At 5:25 P.M., after impala alarm call nearby, I see a radio-collared leopard approaching

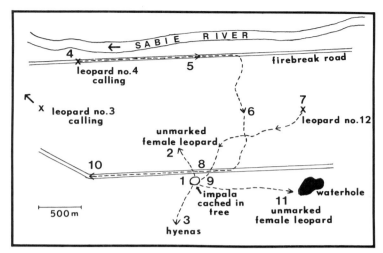

FIGURE 9.18 Movements (dashed lines) of an unmarked female leopard (mother of F12) and female leopards F4 and F12 on May 21, 1974 (9:00 A.M. to 11:00 A.M. and 5:15 P.M. to 10:15 P.M.). The unmarked female fled (2) after she was accidentally flushed from an impala carcass she was feeding on in a tamboti tree (1). Two hyenas also fled (3). Six hours later, leopard M3 and F4 began to call. Female F4 traveled along the firebreak road (4, 5) and crossed over to another firebreak road (6). Meanwhile F12 (7) came to the kill and was joined by the unmarked female (9). Female F4 passed by them, calling, as she moved along the road (8, 10). After the unmarked female and F12 fed, the unmarked female left (11), but F12 remained to feed on the impala while one hyena waited at the base of the tree.

the kill. It is subadult F12 who lives in the area. However, F12 does not approach the kill, perhaps because another leopard, now verified as resident F4, begins calling at 5:40 P.M. Female 4 calls fourteen times as she approaches and passes under the tree (at 6:47) with the kill. At 7:00 P.M. a hyena comes to the tree but departs when two leopards, F12 and an unmarked female, approach the tree. The unmarked female and F12 alternate as one feeds in the tree and the other waits on the ground. A hyena appears after both leopards leave at 8:38 P.M. but departs at 10:00 when a leopard growls nearby. I leave at 10:15 P.M. to release a captured leopard.

Neither the unmarked female nor F12 called during the episode. But F4 called continuously as she approached and passed within 5 m of the kill (see fig. 9.18). It is difficult to believe that F4 did not notice the kill or the other leopards, but apparently no encounter occurred between F4 and the other two leopards. The unmarked female and F12 did not return to the kill until F4 had passed by. Both leopards appeared to avoid F4, whose location could be determined from her frequent calling. The unmarked female was undoubtedly F12's mother.

By vocalizing, resident leopards shared common areas and used common travel routes, while avoiding each other. Calling reduced or prevented sudden or unexpected encounters. The following observations illustrate how one leopard called before using a common travel route.

June 18, 1975, Naphe Road, 4:50 P.M. to 6:20 P.M.: Radio-collared resident M14 is about 200 m off the road and moving toward me. He calls nine times seven minutes after sunset, nine times two minutes later, and twelve times three minutes after that. Several minutes later he appears on the road 50 m away, crosses it, and lies down. Seconds later a spotted hyena appears, obviously following the leopard. The hyena circles the resting leopard but runs off when the leopard snarls. The leopard then gets up, moves 30 m, and lies down again. The hyena sniffs where the leopard was first lying. The leopard is still lying down when I leave ten minutes later.

The leopard called only before it appeared on the road (a common travel route). It did not call earlier, nor later after it crossed the road. The hyena was apparently attracted to the calling leopard. This observation occurred near the periphery of M14's home range, at the same location where I once observed another large, unmarked male leopard. The unmarked male traveled along the same section of road as M14.

Dueting is a term used by Eisenberg and Lockhart (1972) to describe the calling and answering between two male leopards. They believed that on at least two occasions, one of the leopards deviated from its travel route to avoid the other during the duet calling. I did not record duet calling by male leopards in my study areas, but neither did I record two resident males traveling near each other.

Although resident and subadult male leopards shared common areas in both study areas, I never heard a subadult male call. Resident males called, however, and any subadult males inhabiting the area would know of the resident male's whereabouts. But the resident male would have difficulty determining if subadults were present, because subadults did not scent mark or vocalize. These data suggested that subadult males avoided resident males.

Growling and snarling.   Leopards I observed growled and snarled at me, at hyenas, and at other leopards. All captured leopards in traps snarled when I approached them, and their snarls and growls usually grew in intensity the closer I approached. A few remained silent until eye contact was made apparently because they felt hidden in the well-camouflaged traps.

On two occasions a free-roaming, radio-collared female leopard growled and reacted aggressively toward me and my vehicle. On the first occasion, I stepped from my vehicle when F25, who had been concealed, charged 10 m toward me, growling. She stopped with her front feet widespread and her tail held high and twitching. Seconds later she calmly turned away, walked 10 m, and lay down. Nine days later when I stopped near her, she growled aggressively at my vehicle and twitched her tail but did not get up or leave.

On three other occasions while I was observing kills of leopards, I heard leopards growl at spotted hyenas nearby. Two of the leopards were resident males; the third was an adult female. Usually, hyenas fled when a leopard approached, but on these occasions, the hyenas fled only after the leopards growled. One leopard, M23, a hyena killer, charged two hyenas near a warthog carcass, but the hyenas fled unharmed. Leopards also growled at hyenas that were following them. Male 14 growled after a hyena circled him, and I once heard M3 repeatedly growl in some tall grass, but I could not determine the cause.

Leopards also growled and snarled at each other. I once observed a leopard feeding on a bushbuck when another rushed up the tree and attempted to appropriate the kill. As the two leopards struggled in the tree, they growled and snarled at each other. After the encounter, the intruding leopard jumped from the tree and fled.

Leopards also uttered a low gruntlike growl during courtship. On two occasions when adult male and female leopards were traveling together, one uttered a low, deep-throated growl that sounded like a loud purr. Because the leopards were close together, I could not determine which one growled. The growl appeared to be associated with impatience or annoyance rather than aggression or defense. In both instances, the pair of leopards continued walking along the trail after one growled.

On another occasion adult F11 and M14 growled in the darkness during courtship behavior. The sounds were identical to those I heard when a male copulated with a female on the Lower Sabie Road. They were low-rolling, deep-throated growls, which, according to Leyhausen (1979), are the pantherine equivalent of purring intensified to maximum volume. I heard the same sounds on another occasion when M14 and F11 were together, and also when M14 was with an untagged, presumably female leopard near the northeastern periphery of his home range.

BODY POSTURES AND FACIAL EXPRESSIONS

Because leopards did not associate socially as frequently as lions or cheetahs, the relative importance of leopard body postures and facial expressions in intraspecific communication is uncertain. Because of their solitary nature, the leopard's most frequently used modes of communication should be those that help keep leopards aware of each others' presence, but separated in time and space. Thus body postures and facial expressions (both of which are visual signals that require close encounter) may not be as important to leopards as

9.4. The low, crouching body posture often assumed by leopards when they cross open areas.

vocal or olfactory signals. Because leopard body postures and facial expressions appeared similar to those of lions, I used Schaller's (1972) descriptive terminology that he derived from cats (Leyhausen 1979) and primates (Van Hooff 1967).

Body postures. Many of the leopards' nonsocial body postures were postures that made leopards appear inconspicuous. The response to prey included remaining motionless, the stalking walk, the crouching walk, and the crouch. Even when leopards crossed open areas or were disturbed by me, they usually assumed a low profile (see photo 9.4).

Leopards usually walk with their tails curved down, with the end turned slightly upward (see photo 9.5). However, when leopards are investigating scent marks or walking through tall grass the tail is held much higher, and when leopards are spraying urine, the tail is almost vertical. The tip of the tail is often vibrated during urine spraying, although not vigorously. The only time leopards lashed their tails was when I approached them in livetraps. I was unable to determine if the dazzling white tip of the tail was used by females to provide a "flag" for their cubs to follow as suggested for other felids by Schaller (1967) and Leyhausen (1979). However, the upturned tip of the leopard's tail is highly visible and conspicuous when seen from behind in dense cover or in darkness.

9.5. The normal body posture of a traveling leopard showing the downward pointing tail with upturned tip.

Two body postures associated with sexual behavior were observed during leopard courtship. These included "presenting" by the female and the "flirting chase" (Leyhausen 1979). Once at night on the Lower Sabie Road, a female leopard presented herself to a male. Included in her repertoire were head rubbing against the male and the ground, rolling a short distance in front of the male, flank rubbing against the male with trembling and twitching of the tail, and lying down in front of the male with her vulva pressed against his nose. He immediately mounted the female and copulated but did not bite her nape. On another occasion, a captured female leopard purred and rolled over when I approached the trap, but later became extremely defensive. On two occasions I observed a male cautiously pursuing a female in what appeared to be a "flirting chase" (Leyhausen 1979). This usually occurs before a female is receptive. Once, the male uttered murmuring sounds deep within his throat as he trotted past me (hidden in the bushes) in pursuit of the female.

I did not observe the strutting or head-low and head-twist postures Schaller (1972) described for lions. Whether leopards assume this position in encounters with other leopards is unknown. If they do, observing such behavior in the wild would be difficult.

9.6. A defensive leopard in a livetrap baring its teeth.

Facial expressions.   The defensive, bared-teeth threat with lips pulled back showing the teeth (see photo 9.6), or pupils of the eyes enlarged and ears flattened against the head was seen only on captured leopards. Most free-roaming leopards had a relaxed face, with ears erect, eyes opened, and mouth either closed or slightly opened.

The next most frequently observed facial expression was the alert face. This occurred when leopards watched nearby prey or responded to sounds. Two leopards had an alert face when they first saw impala and a steenbuck, and another displayed an alert face when it heard a ratel digging in the darkness. When leopards showed the alert face, the erect ears pointed forward, the eyes stared straight ahead, and there was a sense of tension in the way the leopard held its head and positioned its body.

The facial expression of a vocalizing (rasping) male leopard included the head held level or slightly below normal, the chin slightly stretched out, the ears pulled back, and the mouth partially open. He did not lift his head slightly upward as do male lions when roaring (Schaller 1972). A rasping female leopard walked with little strain evident on her face, but saliva dripped from her open mouth perhaps because of the intense heat.

The tense open-mouth face with the backs of the ears turned forward was observed only once, when F25 threatened me. She rushed forward, emitting a coughing growl; lashed her tail; then turned back into the brush. Lions usually exhibit an open-mouth face when another lion approaches too closely, usually at kills (Schaller 1972). The grimace face was observed only once, when a female leopard smelled scent on bushes beside a road. In cats, this gesture, also known as flehmen, usually occurs in a sexual behavior context when males smell the urine of receptive females (Leyhausen 1979). However, Schaller (1972) observed that lions grimaced after smelling their own scent as well as other things such as wounds and dead animals, including a lion cub.

Territoriality

Several criteria have been proposed to define territoriality. Etkin (1967) defined territoriality as any behavior on the part of an animal that tends to confine the movements of the animal to a particular locality. Hornocker (1969) documented and emphasized the importance of avoidance behavior rather than overt aggression in the maintenance of territories among cougars. Brown and Orians (1970) proposed three criteria for territoriality: (1) little overlap between home ranges, (2) scent marking behavior, and (3) agonistic interactions. Any site-dependent behavior resulting in conspicuousness and in avoidance by other similar behaving individuals was proposed as evidence of territoriality by Fretwell (1972). Based on most of these definitions and criteria, the data I collected on leopard spacing behavior indicated certain leopards in my study areas were territorial.

Resident male leopards in my study areas maintained intrasexual territories within which several (one to six) female leopards resided and from which other males were precluded from breeding. Evidence for this breeding-type of territoriality among resident male leopards included (1) little overlap between adjacent home ranges of neighboring resident males, (2) site-dependent scent marking and vocalizing behavior that resulted in the conspicuousness of resident males, (3) avoidance behavior between males, and (4) lack of known breeding by nonconspicuously behaving males.

Overlap between adjacent home ranges of all monitored resident male leopards (M3, M9, M10, M14, M22, and M23) averaged 19%. However, the man-made boundaries I determined for leopard home ranges may not have depicted actual boundaries that were behaviorally significant to the leopards. If only the proportions of the home ranges of the younger resident male leopards (M3, M14, M9) that overlapped those of the older males (M10, M22, and M23) were considered, the overlap averaged just 15%. If only the proportions of the home ranges of the resident males with the larger home ranges that were overlapped by the males with the smaller home ranges were considered, the average overlap was reduced to just 11%. In two of the three situations of adjacent resident males, the younger males (M3 and M9) also had the larger home ranges.

Of the male leopards using my study areas, only resident males were known to scent mark and vocalize within their territories. If subadult, or floater, males scent marked or vocalized while within the territories of the resident males, they did so infrequently and were not conspicuous. These apparently non-breeding males appeared to be inconspicuous and avoided the resident males.

Male leopards monitored in the study areas avoided each other. The distance between all male leopards monitored on the same day averaged 5.1 km in the SRSA and 5.8 km in the NRSA. The greatest average distance (8.1 km) between male leopards located simultaneously on the same day was between two adjacent resident males (M10 and M14) in the NRSA. Spatial analysis of radio-locations of male leopards also indicated they were spacing themselves apart ($R = 1.6$). In the spacing of males, avoidance behavior appeared more significant than aggressive behavior. If fighting, as a form of aggression, occurred among male leopards in my study areas, it occurred infrequently and was very brief because it was never observed. However, others (Corbett 1947; Hamilton 1976) have documented occasional fighting among male leopards.

Associations between male and female leopards indicated that only the resident males spent significant periods of time with the resident females, courting and mating. Nonresident males either associated only briefly with females or avoided them. The only documented association between an older resident female (F4) and a subadult male leopard (M1) during the study may have been between a mother and her independent offspring.

Evidence for territoriality among resident female leopards in my study areas was unclear. One difficulty was determining the relationships among females—relationships that could be established only by following known individuals over several years. My study was too brief to ascertain such relationships. Female tigers, for example, often share their territories with their female offspring before the offspring become breeders; they sometimes fight with their female offspring; they may adjust or sometimes even leave their territories, which are then either partially or entirely taken over by their female offspring (Smith et. al. 1987). Thus some of the female leopards I monitored may, without my knowing it, have been previous offspring of other females I monitored in the same area.

One criterion for territoriality—little home range overlap—was confusing to determine for female leopards. Some females (F4 and F16) had completely overlapping home ranges; other adjacent females (F11 and F17, F17 and F25, F21 and F25) had no home range overlap. Complete home range overlap appeared mostly among female leopards in the high-leopard-density, prey-rich SRSA. But several females (F6 and F11, F6 and F13) in the lower reaches of the Nwaswitshaka River in the NRSA also shared considerable portions of their home ranges. Such observations of home range overlap could have been between older female leopards and their previous female offspring. My brief observations may have sampled only segments of longer periods that are required for mother and daughter leopards to sort out land tenure rights.

Female leopards in the NRSA exhibited less home range overlap than resident male leopards. Average home range overlap (18%) for six female leopards

monitored in the NRSA (see table 9.4) was slightly less then that for all monitored resident male leopards (19%) in both study areas. Furthermore, if F6 and
F11 (who may have been mother and daughter) were excluded from analysis,
average home range overlap for NRSA females was reduced to 14%. Even
where home range overlap occurred, adjacent female leopards avoided each
other and kept a considerable average distance (1.7 km to 3.2 km) between
themselves on a daily basis. Because fewer females were simultaneously monitored in the SRSA and their relationships were uncertain, land tenure among
female leopards in the SRSA and its relationship to female leopard territoriality
remain unclear.

Another criterion of territoriality—site-dependent scent marking and vocalizations—was observed only among older female leopards in both study areas.
The only monitored younger female leopard (F12) did not display such conspicuous behavior. Because the boundaries of most female leopard home ranges
did not coincide with accessible and obvious trails or roads, I was not able to
discern if the marking behavior of females occurred more frequently along
boundaries than in the centers of home ranges.

A third criterion for territoriality—agonistic behavior—was not observed among
female leopards in my study areas. After my studies concluded, however,
others (Le Roux and Skinner 1989; Ilany 1990) reported agonistic behavior,
fighting, and even deaths from fighting among female leopards. Since my
studies, female leopards have also been documented fighting, chasing, and
eventually expelling their daughters from their natal home ranges (see appendix C).

Available evidence suggests that a form of territoriality may exist among
female leopards under certain conditions. Its functions appear to ensure a form
of restricted rights to prey, protective cover, and denning sites needed by the
female and her offspring. Where prey is plentiful and evenly distributed, cover
abundant, and sites to rear young adequate, as in the SRSA, competition for
resources and female territoriality appear to be less significant. Where prey is
scarce, cover sparse, and sites to rear young limited, as in the upper reaches of
the Nwaswitshaka River in the NRSA and perhaps in desert areas (Ilany 1990),
competition for resources and female territoriality appear more significant.

Summary

1. Leopards spent most of their time alone. Interactions occurred most frequently among radio-collared resident male and female leopards (12% of
   the monitored days) and the least (none) among resident males. Most
   male-female interactions were associated with courtship, lasted up to 5
   days/interaction, and occurred in the late dry seasons.
2. Fighting among leopards in the study areas was never observed and
   probably rarely occurred. Although adult male leopards were never lo-

cated together, old scars and wounds on several captured males indicated fighting had occurred with other predators, defensive prey, or possibly with other leopards.

3. Spatial relationships among adult male leopards indicated little home range overlap (average 19%). Home range overlap generally decreased from dry to wet seasons in the Nwaswitshaka River study area and was believed to be related to the availability of impala.

4. The home ranges of resident female leopards overlapped more in the Sabie River study area than the Nwaswitshaka River study area. Adjacent radio-collared female leopards in the Nwaswitshaka River study area shared an average of 18% of their home range with another resident female.

5. Resident male leopard home ranges overlapped, sometimes completely, the home ranges of as many as six resident female leopards.

6. Home ranges of subadult male leopards were dynamic and often completely overlapped those of resident males. As young males matured, their home ranges expanded beyond those of the resident males and resident females (probably their mothers).

7. Home ranges of some subadult female leopards were highly overlapped, sometimes completely, by the home ranges of adult males, adult females, and subadult males.

8. Analysis of temporal location data revealed a nonrandom spacing of three male leopards in the Sabie River study area and an aggregated distribution of six resident female leopards in the Nwaswitshaka River study area. Female leopards concentrated their activities in the high-density-prey riparian zones of that study area.

9. All monitored leopards, regardless of age, sex, or degree of home range overlap, were usually at least 1 km from their nearest radio-collared neighbor during the day.

10. Vacancies in the home range mosaic of resident leopards were quickly filled either by subadults already in the study areas or by leopards immigrating into the study areas. Apparently, more males than female leopards immigrated into the study areas.

11. The primary rights or advantages associated with land tenure among resident male leopards was access to and successful breeding with females. Male land tenure appeared to be based on prior use. Land tenure among female leopards appeared related to access to prey and high-quality habitats needed to successfully rear young.

12. Scent marking and vocalizations were believed to be important modes of communication in maintaining spacing and reducing the chances of aggressive encounters between leopards.

13. Scent marking was an important and frequently used mode of communication among leopards. Female leopards were observed more often spraying urine (60%) than male leopards, but males often scraped. Although scrapes were usually associated with urine spraying, some were not. Fecal marking varied seasonally, and tree scratching was not observed.

14. Most scent marks of leopards were along frequently used trails, by con-

spicuous places such as trail intersections, edges of bridges or culverts used by leopards, and at intersections of game trails and dry streambeds used as travel routes.

15. Neighboring leopards repeatedly visited scent-marked sites. The average interval between new scrapes was 25.9 days, and leopard scent could be detected by its odor for up to a month. The possible functions of scent marks are discussed.

16. Study area leopards did not frequently vocalize (rasp); when they did, they called primarily in the early evening, just after sunset, and more often during dry than wet seasons.

17. Male and female leopards had different calls. Female calls had more strokes per call, more calls per calling period, longer intervals between calls, and longer total calling periods than males. Some males and females had distinctive calls that permitted individual identification.

18. Although body postures and facial expressions of leopards were not frequently observed, they seemed to be similar to those of lions and other felids in general.

19. Resident male leopards exhibited a form of breeding territoriality that appeared to give them exclusive rights to breed with from one to six resident females.

20. Resident female leopards, especially in the upper reaches of the Nwaswitshaka River, exhibited a form of territoriality that seemed to give them restricted rights to limited prey, cover, and denning sites. In the Sabie River study area, territoriality among resident females seemed less significant and may have been nonexistent.

10

Other Predators and Their Relationships to Leopards

LEOPARDS were only one of several species of large predator inhabiting the KNP study areas. Other large predators included the spotted hyena, lion, wild dog, and cheetah. The extent to which these other predators interacted and competed with leopards for prey and other resources could have influenced leopard distribution and abundance in the park. Schaller (1972) suggested that a riverine forest may not be the preferred habitat of leopards, but a refuge from lions. The high incidence of spotted hyenas at kills of leopards in the KNP study areas suggested that hyenas may have significantly influenced the prey caching strategy of leopards, their preferences for certain habitats, and their utilization of prey.

Other lesser-known predators in the park were also important in the leopard's environment. Crocodiles and pythons were interesting to observe; they also may have had a direct impact on the survival rates of leopards. Smaller predators, such as genets, civets, and ratels, sometimes became leopard prey.

Spotted Hyenas

Spotted hyenas are probably the most abundant predator in KNP (Pienaar 1969). At one time, however, intensive hunting, trapping, and poisoning reduced spotted hyenas almost to the point of extinction. Hyena populations in the park have also exhibited periodic fluctuations. When spotted hyena populations were extremely low, an epidemic may have further reduced them. The population gradually increased after 1925, despite continuing control measures. After control measures were terminated in 1960, some populations remained low; other populations in different locations in KNP (Kingfisherspruit, Tshokwane, and Malelane) increased (Pienaar 1969).

ECOLOGY

Spotted hyenas are most abundant in the central and southern districts of the park and least numerous in the northern district (Pienaar 1969). Smuts (1975) reported a conspicuous absence of spotted hyenas in the savanna country east of the Crocodile bridge in the Lower Sabie region and speculated that hyena distribution was dependent on impala distribution.

The number of spotted hyenas in KNP is unknown. Pienaar (1963) estimated there were several thousand in the park in the early 1960s and Smuts (1975), working in the central district of the park ten years later, estimated 1,932 hyenas occupied an area of 5,560 km^2, or 1 hyena/2.9 km^2, a density three times that for lions. This density was less than that of hyenas in the Ngorongoro Crater (1 hyena/0.6 km^2), but greater than that reported for the Serengeti Plain (1 hyena/8.3/km^2) (Kruuk 1972).

The sex ratio of hyenas in KNP suggests there are more males than females. Of 209 hyenas killed between 1954 and 1960, the ratio was 1 male/0.5 females (Pienaar 1969). However, in the Ngorongoro Crater and Serengeti Plains, Kruuk (1972) reported 1 male/1.1 female and 1 male/1 female, respectively. He speculated a sex ratio in favor of males may result if hyenas are sampled around carcasses, because females are less likely to scavenge.

Most female hyenas in the park produce 2 young/litter. Lactating female hyenas cropped in the central district had two dependent young and others had two embryos or two placental scars. Of twenty-two females, nineteen had twins, two had singles, and one had triplets (Smuts 1975). In September 1974 I observed four adult hyenas with four young at a culvert along the Lower Sabie Road. One with black pelage was probably less than two months old (Kruuk 1972). A month later along the Naphe Road, about 12 km from the other observation, I observed two adult female hyenas with three cubs at least three months old.

Because of the nocturnal habits of hyenas, I seldom observed them during the day. Of fifty-one free-roaming hyenas observed in the SRSA, thirty-one (61%) were seen after dark. Of twenty-four observed in the NRSA, only seven (2%) were seen after dark. I also observed thirty-seven hyenas at twenty-seven kills of leopards and captured hyenas thirteen times in leopard traps (see table 10.1). Spotted hyenas seldom associated in large groups in the leopard study areas. The largest group observed in the SRSA was eight. Seven hyenas were once seen feeding on a lion-killed giraffe, and seven were once observed together along the Lower Sabie Road. The largest group observed in the NRSA

TABLE 10.1 *Other Species Captured in Livetraps*

Species	Study Area Sabie River	Study Area Nwaswitshaka River	Total
Genet	37	19	56
Spotted hyena	9	4	13
Lion	6	5	11
Civet	8	2	10
Banded mongoose	0	5	5
Crested partridge	0	5	5
Slender mongoose	4	0	4
Leguan	3	0	3
Tawny eagle	0	2	2
Ratel	0	1	1

Total trap effort in the Sabie River and Nwaswitshaka River study areas was 1,530 and 882 trap days, respectively.

was four hyenas. Average group size in the SRSA (3.6 hyenas/group) was slightly higher than that in the NRSA (3.0 hyenas/group). Similarly, hyenas in the central district were seldom seen in groups larger than three, but observations at carcasses sometimes revealed one to thirty hyenas (Smuts 1975).

Hyenas were more abundant than either lions or leopards in the leopard study areas. They frequently visited livetraps set for leopards, and between August and September 1973, before the traps were placed in trees, hyenas were captured more frequently than leopards and lions. In the SRSA, an average of 26.7, 31.2, and 46.7 trap nights were required to capture a hyena, leopard, and lion, respectively. Only 3.8 trap nights were required to capture a hyena in the NRSA. Independent evidence also suggested that hyenas were abundant in the leopard study areas. During a survey of lions, Smuts (1975) reported that the area of which the SRSA was a part supported one of the highest densities of hyenas in the central district of KNP.

If hyena densities in the SRSA were comparable to the Nwanetzi-Gudzani area, which had 1 hyena/2.8 km^2 (Smuts 1975), at least six hyenas should have been in the 17 km^2 study area. Because I saw eight hyenas on one occasion, the population included at least that many for an average density of 1 hyena/2.1 km^2. But if my hyena trapping success was an indicator of hyena abundance, probably twelve to fifteen hyenas were present, for a density of 1 hyena/1.1 to 1.4 km^2. This density was lower than the 1 hyena/0.6 km^2 in the Ngorongoro Crater (Kruuk 1972). Hyena densities were probably lower in the NRSA because of its lower prey densities. If the NRSA supported hyena densities comparable to the average in the central district (1 hyena/2.9 km^2) (Smuts 1975), the 81 km^2 area supported about twenty-eight hyenas.

In KNP the spotted hyena is recognized as a formidable predator, as well as a scavenger. Hyenas are regularly observed hunting and killing their own prey in addition to scavenging from kills of other predators (Pienaar 1969). Impala

TABLE 10.2 *Spotted Hyena Kills in Kruger National Park Between 1936–46 and 1954–66*

Prey Species	Number	Relative Frequency
Impala	110	60.4
Waterbuck	25	13.7
Wildebeest	21	11.5
Kudu	19	10.4
Buffalo	2	1.1
Pangolin	2	1.1
Zebra	1	0.6
Reedbuck	1	0.6
Sharpe's grysbuck	1	0.6

SOURCE: Pienaar 1969.

are taken most frequently by hyenas, followed by waterbuck, kudu, and others (see table 10.2). Adult kudu, zebra, wildebeest, and waterbuck are sometimes killed by packs of up to forty individuals (Pienaar 1969). Smuts (1975) reported that the three main prey of hyenas cropped in the central district were, in decreasing order, impala, zebra, and wildebeest. Sixty-nine percent of all the hyenas had fed on impala and up to thirty hooves of impala were found in the stomach of a single hyena.

In a later study Henschel and Skinner (1990) confirmed that spotted hyenas in KNP frequently scavenged from carcasses of large ungulates, particularly buffalo. More than half the mass of food consumed was from their own kills, primarily of medium-size ungulates—impala, kudu, and warthog—which became vulnerable because of drought. Adult hyenas consumed about 3.8 kg per day, and female hyenas appeared to have a better net energy economy than males.

Impala lambs constituted a major proportion of the diet of hyenas in KNP during the lambing period (Smuts 1975). In one region, twenty-eight of thirty-two stomachs of hyenas contained 326 hooves of impala, 300 of which were from lambs. Hyenas also preyed on impala in the leopard study areas. I once observed a hyena eating a freshly killed impala lamb and on another occasion watched a hyena chase a small lamb. In December 1973 I saw a hyena eating an impala lamb beside the Lower Sabie Road. When I stopped, the hyena picked up the impala and ran off with it in its jaws as a herd of impala watched from a distance of about 100 m.

On January 27, 1975, I observed a hyena chasing an impala lamb across a road in the SRSA. The lamb, separated from the herd 70 m away, was only 20 m ahead of the hyena. When the lamb crossed the road, the hyena paused and the lamb gained an advantage. However, the hyena eventually crossed the road and pursued the lamb into dense vegetation. Other indicators suggested that hyenas primarily hunted impala or even smaller prey. One was that hyenas seldom associated in large groups except when scavenging prey killed by lions.

Another was that only one to two hyenas were usually present when they scavenged kills of leopards. Kruuk (1972) believed that the size of a hyena group, when hunting, was dependent on the swiftness and aggressiveness of the prey. He reported that the average pack of hyenas when chasing zebra, wildebeest calves, adult wildebeest, and Thompson gazelles was 10.8. 1.6, 1.4, and 1.2, respectively. He concluded that the number of hyenas participating in a hunt was dependent on the behavior of the prey rather than prey size, that is, the greater the prey aggressiveness, usually the larger the hyena group.

If Kruuk's hypothesis is true, the nonaggressive behavior of impala would not require that hyenas hunting them form large groups. Because relatively small impala are the hyena's principal prey in the park, large group sizes seem to be unnecessary to successfully capture them. Large group sizes may, however, be to the hyenas' advantage when scavenging kills of lions because a large pack would be more successful than one or two hyenas in defending a carcass from lions.

Hyenas were killed by other predators, including leopards. On February 27, 1975, I found a dead adult hyena in a patch of torn up grass in the SRSA. Large tooth marks on its neck and lion tracks nearby indicated that the hyena was killed by one or more lions. The carcass was not eaten. Others have reported that lions frequently kill hyenas. Of twenty-eight hyena deaths, including eight adults, reported by Kruuk (1972), 55% were killed by lions. Observations of a leopard that killed hyenas are discussed in the following section.

RELATIONSHIP TO LEOPARDS

Spotted hyenas discovered about half of the carcasses of prey killed by leopards. But despite their large size, abundance, and ability to find kills, hyenas were unable to take kills leopards had cached in trees. Occasionally, hyenas were able to take a leopard's kill before the leopard was able to cache it in a tree; I was unable to determine if this occurred frequently. I once observed a hyena with a young female leopard's kill. The hyena had cached the carcass in a small pool of water and was sleeping 3 m away. The leopard was only 30 m from the sleeping hyena. When I inadvertently frightened the hyena, the leopard reclaimed its kill and cached it in the nearest tree. Tracks indicated the leopard was initially feeding on the carcass under a nearby sicklebush. The leopard, perhaps in response to the hyena, had started to drag the carcass to a nearby tree, but was involved in a struggle before reaching it. The hyena claimed the carcass, which it dragged in the opposite direction to a small ravine (see fig. 10.1).

I seldom observed aggressive behavior between leopards and hyenas at kills. Usually the hyenas stood patiently nearby or under a tree waiting for remains to fall to the ground. Whenever leopards arrived and departed, hyenas avoided

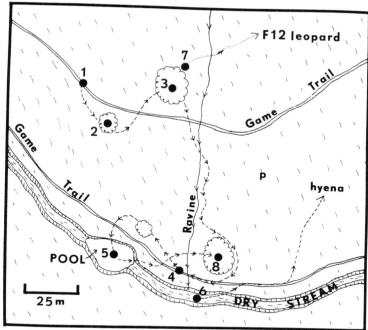

FIGURE 10.1 Movements (dashed lines) of subadult female leopard F12 on November 23, 1974 (9:30 A.M. to 11:00 A.M. and 5:00 P.M. to 5:45 P.M.). She ambushed an impala along a game trail (1), dragged it under a tree (2) where it was apparently disemboweled, dragged it to another tree (3) that was too thorny to climb, dragged it to another game trail where she confronted a spotted hyena (4). The hyena took the impala from the leopard and dragged it into a ravine where it fed and then cached the remains in a pool of water (5). When I first approached the scene, I disturbed the sleeping hyena at (6), and it fled, as did the leopard at (7). When I returned six hours later, the leopard had reclaimed the impala carcass, dragged it to a large knobthorn tree (8), and cached its remains 4 m above ground.

the leopard. Sometimes I heard animals running in the darkness, but could not determine if hyenas were fleeing from or pursuing the leopard.

At least two hyenas were killed and eaten by leopard M23. Both deaths were similar. The hyenas were killed after they were apparently attracted to carcasses of prey, probably impala, killed by M23. On the first occasion, I discovered the remains of a hyena cached in a tree where two days earlier I had seen two hyenas circling M23's kill. Eighteen days later and 4 km away, I again located M23 at a kill. After he left, I examined the area and found the remains of a subadult hyena. Most of the hyena had been eaten. On another occasion, M23 attacked two hyenas that were attempting to scavenge from a warthog he had killed, but both hyenas escaped unharmed.

Single hyenas were reluctant to confront an adult leopard, but two or more hyenas appeared more than a match for a single leopard. A single hyena that once took the kill from a leopard probably weighed 45 kg compared to 29 kg for the leopard. In KNP, Smith (1962) once saw one of nine hyenas chase a leopard into a tree after the leopard approached the carcass of a greater kudu. Kruuk (1972) reported that two hyenas chased a leopard from a carcass of a gazelle that they then carried away. During another incident, he saw a leopard take a kill from a hyena that had taken it from a cheetah. The hyena pursued the leopard, but the leopard climbed a tree where it fed on the carcass. After feeding, the leopard jumped down and charged the hyena. Hyenas, however, can apparently kill leopards. A hyena was seen in the northern district of KNP carrying the carcass of a fully grown young leopard, which it did not eat (Pienaar 1969).

Only a single hyena was present at most kills of leopards I observed. Of twenty-eight kills with hyenas present, only one hyena (68%) was at nineteen kills; two (29%) were at eight; and at least three (4%) were at one . Vultures and hyenas were at kills of leopards four times; tawny eagles and hyenas were at kills twice; and once a lion and a hyena were at a kill. When two hyenas were at a kill, one was usually much larger than the other. This suggested either a female and a male or perhaps a female and her offspring.

Hyenas often rested directly under or near kills cached by leopards when leopards were not feeding. I often startled resting hyenas when I approached such kills on foot. The approach of a leopard to its kill was often indicated by the sudden alertness or disappearance of attendant hyenas. As far I could determine, hyenas never prevented leopards from feeding on kills they had cached in trees.

Resting hyenas usually returned to the cache tree and waited for remains to fall once the leopard began feeding. Small pieces of skin, flesh, and bone dropped by leopards were immediately eaten by the waiting hyenas below. Sometimes when leopards dropped large pieces, such as leg bones, hyenas either ate them underneath the leopard or a short distance from the tree. When more than one hyena was present, they often growled at each other and competed for the remains. Once, when a leopard dropped the entire head of an impala, one of two hyenas immediately ran off with it, with the other hyena in pursuit. Another time, two hyenas fought briefly for some prey remains under a cache tree.

When leopards fed, they usually ignored the hyenas under their cached kills. Once, as a leopard calmly fed above, a hyena patiently waited 2 m below. Other times, leopards stopped feeding if hyenas approached the cache tree, but resumed feeding when they realized the intruders were hyenas. Twice leopards snarled at hyenas waiting below, but the hyenas paid little notice. After feeding, leopards sometimes jumped to the ground within several meters of hyenas, but the hyenas retreated only a short distance. One leopard charged two hyenas feeding on a warthog carcass, but the hyenas escaped unharmed.

The remains of kills apparently provided little food for hyenas because leopards consumed most carcasses. At most kills of impala, hyenas probably ate only 9 kg to 11 kg of skin, bones, and flesh, excluding vegetable material inside the rumina. After hyenas fed, only the teeth, small splinters of bone, and the mandibles remained. If two hyenas shared the remains equally, only 4 kg to 5 kg was available per hyena. Because skin and bones sometimes became lodged in trees instead of falling to the ground, hyenas could not expect to get all of the remains from each kill. It did seem to be advantageous for hyenas to wait until leopards abandoned their kills because the head and neck of impala often fell to the ground when the leopard finishing eating. Leopards provided only 3 (2%) of 148 ungulate carcasses that were eaten by spotted hyenas in a study in KNP between June 1982 and September 1984 (Henschel and Skinner 1990).

Lions

Lions are common throughout KNP, especially in the central district. In 1963 the park's lion population was estimated to be 1,035 with 347, 488, and 200 lions in the northern, central, and southern districts, respectively (Pienaar 1963). Later, Smuts (1975) estimated 708 lions inhabited the central district. He believed the lion population had more than doubled in fifty years because of an increase in numbers of sedentary prey brought about by an increased number of artificial water holes. By 1973 256 artificial waterholes and 46 dams were within the park. The southern district of the park, where I studied leopards, probably supported fewer lions than the northern or central districts. Much of the southern district was dense thornbush thicket, a habitat that favored impala and leopards rather than lions and their prey.

ECOLOGY

During the leopard study I saw lions 301 times on 101 occasions in and near the study areas. I also captured lions eleven times in leopard traps (see table 10.1) and lions often visited the traps or attempted to take the bait. I recorded all lions seen from roads used by park visitors and roads closed to visitor use. Fifty-four percent of the lions I observed were along tourist roads and the rest (46%) along firebreak roads closed to visitor use. These data suggested that lions were just as likely to be seen along tourist roads as along firebreak roads in the more remote areas of the leopard study areas.

Sabie River study area. Of 184 lion sightings on fifty occasions in and near the SRSA, 62 were along the Sand River tourist road, 58 along the Sand and Sabie River firebreak road closed to tourists, 42 along the tourist-used Lower Sabie Road, and 22 along roads bordering the study area. Despite poorer visibility during the wet seasons, lions were more likely to be seen then (103

TABLE 10.3 *Lions per Observation in the Leopard Study Areas.*

| | Season | | | | | | | | | | | | | | |
| | 1973 Dry | | | 1973–74 Wet | | | 1974 Dry | | | 1974–75 Wet | | | 1975 Dry | | |
Study Area	n	x	SD	n	x	SD	n	x	SD	n	x	SD	n	x	SD
Sabie River	5	4.8	2.0	16	3.4	2.8	13	3.1	2.9	12	4.1	2.5	4	4.2	2.1
Nwaswitshaka River	1	2.0	—	10	2.0	1.6	15	2.3	1.8	18	2.4	1.8	7	2.3	1.8

lions) than during dry seasons (81 lions). Included were six captures of lions in leopard traps (see table 10.1). The lion capture rate averaged 170 trap days/ capture when traps were initially set on the ground.

The most common lion observation in the SRSA was of single lions (12 of 50 groups). Of these twelve observations (24%) of single lions, five were adult males, four adult females, and three unidentified. The average number of lions/ observation varied seasonally from 3.1 during the 1974 dry season to 4.8 during the 1973 dry season (see table 10.3). Group size varied from one to eleven lions, with equal-size groups averaging 3.7 lions seen during combined wet and dry seasons. Excluding single lions, most groups consisted of two, three, four, and six lions, in order of decreasing frequency (see table 10.4). Of eighteen observations of groups where all individuals were sexed and aged, the average group consisted of 6.5 lions including 2.4 adult males, 2.1 females, 1.9 cubs under one year old, and 0.1 subadult males.

Lionesses were usually present with groups of lions. At least one lioness was observed with a group on sixteen of eighteen (89%) occasions. In contrast, two males were observed with groups on only seven of eighteen (39%) occasions. Cubs less than a year old were seen with groups on eleven of eighteen (61%) occasions and were always accompanied by at least one female.

Only one pride of lions regularly used the SRSA during the 1973–75 period.

TABLE 10.4 *Lions Observed in Leopard Study Areas*

| Number of Lions Observed per Group | Study Area | | | |
| | Sabie River | | Nwaswitshaka River | |
	n	%	n	%
1	12	24	22	43
2	10	20	14	27
3	8	16	7	14
4	5	10	1	2
5	2	4	2	4
6	5	10	3	6
7	3	6	2	4
8	2	4	0	0
9	0	0	0	0
10	2	4	0	0
11	1	2	0	0

10.1. A lioness with suckling cubs in the Sabie River study area.

The largest number observed together was ten (three adult males, four females, and three small cubs) on July 25, 1975, south of the Sand River. However, while flying on February 19, 1974, I saw eleven lions near the eastern boundary of the study area, but could not determine pride composition.

During my study the Sabie River pride contained at least three cubs. When I first observed the young cubs in July 1974, they were nursing (see photo 10.1). From September to December 1974 I periodically encountered them with the pride. After my studies were completed, two lions were captured and tagged in the SRSA (Smuts 1975). At that time the pride consisted of two adult males, four adult females, and three cubs, matching my observations except for the number of males. Smuts reported that this pride inhabited the SRSA and the adjacent area to the north, east, and west. They did not, in his opinion, use the area to the south because they would have to regularly cross the Sabie River.

Several other prides were observed in the vicinity of the SRSA. One, of at least nine individuals, used the area north of the study area and another of undetermined size used the area south of the Sabie River. This latter pride was observed along the Lower Sabie road on a number of occasions. I observed eight lions (two males, three females, three cubs) on March 24, 1975, 4 km from two other lions (adult and subadult males). The density of lions in the SRSA would have been 1 lion/1.5 km^2 assuming the lions used only the study area. However they used a much larger area; therefore the density estimate was probably 1 lion/3 to 6 km^2, including cubs, or 1 lion/4 to 8 km^2, excluding cubs.

The average density of adult lions throughout this region of the park was about 1 adult/12.2 km² (Smuts 1975); the higher density in the study area was probably attributed to the higher density of prey.

Although little was known about the food habits of the Sabie River lion pride, lions in the central district fed on impala more than any other prey (Smuts 1975). Impala was the most frequently occurring prey in 252 stomachs of cropped lions (15.9%), followed by wildebeest (12.3%), giraffe (7.9%), zebra (5.6%), warthog (4.0%), waterbuck (2.8%), and others (3.6%). I seldom encountered lions on carcasses of large prey in the SRSA, so impala may have been their main prey. Impala remains from lion kills would have been difficult to find because they would have been quickly consumed and scavenged by hyenas, jackals, and vultures.

Giraffe may have been important prey of lions in the SRSA. On September 4, 1974, I observed four lionesses and three cubs from the pride near the study area's western boundary. The following morning they and two males were feeding on a recently killed giraffe. Five days later, and 9 km to the east, they killed another giraffe. These giraffe were the only kills I attributed to the Sabie River pride. Once, however, a hippopotamus was killed, probably by lions, south of the SRSA.

Nwaswitshaka River study area. I recorded 117 sightings of lions on fifty-one occasions in the NRSA. Most (48) were seen along the firebreak road adjacent to the Nwaswitshaka River. Lions were also seen along the Naphe Road (28), Kruger Road (3), Doispan Road (1), roads to the Nwaswitshaka River artificial waterhole (5), and another firebreak road west of the study area (26). Five lions were captured in traps set for leopards (see table 10.1). Like those in the SRSA, more than half (55%) were seen during wet seasons.

The most common observation of lions in the NRSA were of lone lions, and of these, most (59%) were adult males. Adult females were seen alone on only five occasions (23%). Lions were observed in pairs on fourteen occasions followed, in decreasing frequency, by groups of three, six, five, and seven lions. The average number of lions/observation ranged seasonally from 2.0 during the 1973–74 wet season to 2.4 during the 1974–75 wet season (see table 10.3). There was no significant difference between the average number of lions/observation (2.3) during combined wet and dry seasons.

Males were usually present when groups of lions were observed. Males were present on 86% of the observations; 58% of them were adults. At least one female was present 68% of the time, but cubs were seen only 21% of the time. The average group of 3.4 lions consisted of 1.3 females, 1.0 adult males, 0.4 yearlings, and 0.4 cubs less than a year old.

At least two prides of lions used the 81 km² NRSA. One pride intensively used the riparian zone of the Nwaswitshaka River and ranged at least 4 km west of the Nwaswitshaka River waterhole to Skukuza, a distance of 12 km. At least two adult males, three adult females, and two subadult males were in the pride until late 1974. In January 1975, two small cubs less than three months old were seen in the pride. Three months later, when I saw the pride for the last time, both cubs were still alive and in good condition.

The Nwaswitshaka River pride was easy to identify because it had several distinctively marked individuals. One lioness I called Tailless had the tip of her tail missing; she was periodically observed along the Nwaswitshaka River up to 12 km west of Skukuza where, on June 15, 1974, she and her pride chased a female leopard up a knobthorn tree. I first observed Tailless on December 6, 1973 about 0.3 km west of Skukuza. She was destroyed by park officials on June 10, 1975, after she entered the tourist campground and was considered a threat to park visitors.

One of the two adult males in the Nwaswitshaka River pride had a splendid black mane. He was once observed 11 km west of Skukuza and used the area along the Nwaswitshaka River during 1974 and 1975. He was observed courting a lioness of the pride (not Tailless) on November 27, 1974.

Another old male, initially observed in the pride in 1973, apparently left or was forced out of the pride sometime before March 1974. He was with two lionesses on November 5, 1973, when I accidentally captured one of the lionesses in a leopard trap. A month later, I saw him again with two lionesses, including Tailless. He was last seen with the lionesses on February 27, 1974. Before February I captured him three times in leopard traps, but after February, even though he took the baits, he was not captured again. After February I observed him on four occasions. He was always alone and appeared to be declining in physical condition. His occasional appearance near the tourist campground and his bold nature prompted park officials to destroy him on October 31, 1974, for public safety reasons.

After he was destroyed, an examination revealed that his upper and lower left canines were worn to yellow, smooth, round stubs and the tip of his upper left and the entire lower canines were missing. His premolars were heavily worn; a recent wound and blood clot appeared on his right foreleg; old cuts and scratches occurred over his entire body; several claws were missing; his rib cage and pelvis protruded prominently; and his scrotum was abnormally small in size. Two large concentrations of ticks (*Ambylomma spp.*) were found under his right rear and right foreleg. Despite his condition, his stomach contained about 5 kg of flesh and hide of a kudu, which was probably scavenged. He was an extremely old individual, and his social status in the pride had probably been usurped by the splendid black-maned male.

Another pride, the Grand Koppie pride, used the southwestern corner of the NRSA. Neither Tailless nor the old male nor the black-maned male were ever seen in this region or with this pride, which frequented the vicinity of Grand Koppie and two nearby koppies. Because this pride was never observed along the Nwaswitshaka River, I assumed that they focused their activities around an artificial waterhole near the Renoster Koppies and the Vervoerdom 3 km to 8 km southwest of the study area.

The Grand Koppie pride consisted of at least two adult males and three lionesses. On one occasion (November 19, 1974) I observed two subadult males with three lionesses near Grand Koppie. The pride had cubs in late 1973 or early 1974; I saw three small cubs on April 20, 1974. Nearly a year later, on March 31, 1975, I again observed the cubs, now much larger, with the pride 2 km southwest of Grand Koppie.

The density of lions in the NRSA was estimated at 1 lion/13.6 km^2, excluding cubs, and 1 lion/10.2 km^2, including cubs. I assumed the Nwaswitshaka River pride averaged seven lions, or nine including cubs, and that the Grand Koppie pride averaged five lions, or eight including cubs. The Nwaswitshaka River pride was known to use at least 13 km^2 west of, plus the 81 km^2 leopard study area, and the Grand Koppie pride was assumed to use an area of equal size (94 km^2). About 10 km^2 (11%) of the Grand Koppie pride's area was in the study area. I assumed that 86% and 11% of the Nwaswitshaka River and Grand Koppie lion pride areas, respectively, were in the NRSA. Although little was known about the food habits of lions in the NRSA, several observations suggested a varied diet. On three occasions I watched lions unsuccessfully stalk impala and once found a freshly killed impala partially eaten by lions along the Nwaswitshaka River road. On another occasion I watched a lioness unsuccessfully stalk a female kudu and the remains of a kudu were found in the stomach of the old male lion. The Nwaswitshaka River pride was also seen feeding on a freshly killed zebra on February 5, 1975 and on a freshly killed wildebeest on April 3, 1975. I did not observe any lion-killed giraffe in the NRSA, although giraffe were common.

Comparison of lion populations.   Lion populations in the SRSA and NRSA exhibited different characteristics, probably reflecting differences in prey. First, average group and pride sizes were greater in the SRSA (see table 10.5), and prides in the Sabie River region contained more adult males, females, and cubs than prides in the Nwaswitshaka River region. If my estimates of densities of lions were accurate, the SRSA supported more lions than the NRSA. These higher densities probably resulted from the higher prey densities in the SRSA and the fact that the Sabie River pride preyed on giraffe. Because of their large size, giraffe as prey would require lions with a highly developed cooperative hunting technique, or a large pride. Larger males would also be to a pride's advantage for killing giraffe and protecting the carcass from hyenas. Solitary lions were observed three times more often in the Nwaswitshaka River than the SRSA, a difference that was even more striking when seasonally compared. During the 1973–74 wet season of extremely high rainfall, 60% of the lions in the NRSA were alone compared to only 25% in the SRSA. These differences

TABLE 10.5 *Lions in the Leopard Study Areas*

Characteristic of Lion Population	Study Area	
	Sabie River	Nwaswitshaka River
	(*n* = 18 groups)	(*n* = 28 groups)
Average number per group		
Total lions	6.5	3.3
Adult males	2.4	1.0
Subadult males	0.1	0.3
Females	2.1	1.2
Total yearlings	0	0.4
Cubs less than one year old	1.9	0.4
Maximum observed size of pride	11.0	7.0
Average km² per adult lion	4.8	13.6

were probably also a response to prey availability. Single lions would have little difficulty capturing impala, especially during wet seasons in the NRSA, when impala immigrated into the area and were widely dispersed. The greater biomass of prey available to lions in the SRSA was undoubtedly one of the reasons for the observed differences in group size and lion density between the two areas.

The biomass of lion prey was twice as great in the SRSA as in the NRSA. If elephant and hippopotamus were excluded as prey in the SRSA and elephant and rhinoceros in the NRSA (see table 10.6), an average prey biomass of 7,250 kg/km² and 3,675 kg/km² was available in each area, respectively.

TABLE 10.6 *Estimated Biomass (in Kilograms) of Prey in the Leopard Study Areas*

Prey Available to Lions	Study Area	
	Sable River	Nwaswitshaka River
	(17 km²)	(81 km²)
Ungulates		
Large[1]	12,045.5	19,318.2
Medium[2]	6,491.0	47,496.5
Small[3]	99,049.1	221,295.2
Other mammals[4]	5,193.6	9,021.4
Birds[5]	463.9	558.3
Total	123,243.1	297,689.6
Prey biomass per km²	7,250.0	3,675.0

Derived from tables 4.7 and 4.8.

[1] Giraffe, buffalo

[2] Greater kudu, zebra, wildebeest, waterbuck

[3] Impala, gray duiker, warthog, bushbuck, steenbuck, klipspringer

[4] Porcupine, baboon, vervet monkey, hare, cane rat, civet, genet, ratel, mongoose

[5] Guinea fowl, francolin

Lions in the SRSA also appeared to take larger prey, such as giraffe and perhaps hippopotamus, than did lions in the NRSA. Schaller (1972) reported that large prey was often killed by male lions and medium-size prey by lionesses. Although lions in the SRSA had the least disparate adult sex ratio, their larger group size (despite smaller home areas) suggested they may have been taking large prey. Small groups of lions in the NRSA also suggested small- to medium-size prey may have been their major prey.

My observations suggested that impala were the most important prey of lions in the leopard study areas and determined lion group size. These observations included lions stalking impala, the biomass of impala in the study areas, the frequency of lions being seen alone, and the fact that remains of kills of lions were seldom observed. If lions annually kill or scavenge 2,500 kg/year (Schaller 1972) and an average impala weighs 37.5 kg, a diet solely of impala would require each lion to kill at least 67 impala/year. The average lion pride in the SRSA and NRSA would require at least 623 and 570 impala/year, respectively, if impala were their only prey. However, if impala made up only 27.9% of the lions' kills, as reported by Smuts (1975) for lions in the central district, only 173 and 158 impala/year would be required to support lions in the SRSA and NRSA, respectively.

RELATIONSHIP TO LEOPARDS

Given the opportunity, lions will take kills from leopards (Turnbull-Kemp 1967). In Serengeti National Park, Schaller (1972) reported that leopards provided lions with about 5% of their meals. Of fifty-five kills of leopards I examined, lions took only one (2%). In this instance, a female leopard cached an impala carcass only 4 m above ground in a tamboti tree and left most of it uneaten. The rumen and intestine were hanging from limbs nearby. The following morning the carcass was gone and lion tracks, claw marks in the tree, and long hairs from a male lion's mane were found at the site. I suspected that an old, lone male lion I frequently saw in the area had taken the carcass. On another occasion I could not determine if a lioness took a leopard's kill or vice versa.

> August 4, 1974, Lower Sabie Road 4:40 P.M. to 6:54 P.M.: An adult male leopard feeds on a bushbuck carcass in a tree. Most of the carcass has already been consumed. At 4:50 another leopard suddenly rushes up the tree and attempts to take the carcass. In the brief struggle both leopards nearly fall from the tree, but the adult male retains the kill and moves it up higher into the branches (see photo 10.2). The other leopard pauses on a limb as the adult male resumes feeding. At 5:30 a lioness comes out of some brush at the base of the tree, her chest, neck, and forelegs covered with blood. The lioness ambles slowly away, her stomach bulging, to join several other lions 100 m away. After she leaves, the second leopard

10.2. A leopard moves its kill of a bushbuck higher into a tree after a brief dispute with another leopard along the Sabie River.

jumps to the ground and disappears into the brush. As darkness falls, the adult male leopard is still feeding.

Apparently the lioness had just fed nearby, but I could not determine whether the lion or leopard had killed the bushbuck. Although lions are attracted to kills of leopards, they may be unable to find carcasses leopards cache in trees. A tourist in KNP once reported to me that he saw a lion following the trail of a leopard that was dragging an impala. Schaller (1972) reported that lions were not able to recognize motionless carcasses leopards had placed in trees. Kills of leopards cached in trees were usually safe from lions because they were not as adept at tree climbing as leopards. During Schaller's study of lions (1972), he never saw lions climb higher than 7 m, although 80% of the lions climbed 2 m to 5 m. Lionesses were more adept climbers than males and climbed more frequently. Schaller saw lions climb trees to take kills from leopards 3 times; I never saw lions climbing trees. However, I did capture lions in livetraps set 1 m to 2 m above ground for leopards. These lions entered the traps by climbing branches 15 cm to 20 cm in diameter that I had attached to the traps at an angle of 60 degrees to the ground. I never captured lions in traps set more than 2 m above ground, although claw marks and fur on trees indicated lions sometimes tried to reach the bait. The tree climbing limitations of lions and the few kills they took suggested the majority of kills leopards cached in trees were safe from lions.

Hunting Dogs

The hunting dog population in KNP has been subject to widespread fluctuations. The hunting dog population in the southern district of the park collapsed in the 1920s and early 1930s (Pienaar 1969). One of the contributing factors was a fatal tick-borne disease. This mysterious, unidentified rickettsial infection reached epidemic proportions and all but wiped out the hunting dog population between 1927 and 1933. In the central district hunting dog populations also declined, but despite increases in impala and other prey, they did not recover like those in the southern district. Hunting dog populations in the park's northern district have not fluctuated like those in the southern district.

ECOLOGY

Hunting dogs are rare in KNP. The estimated number of hunting dogs in the park in 1964 was 335, including 150 in the northern district, 65 in the central district, and 120 in the southern district (Pienaar 1969) for an overall density of 1 hunting dog/57 km^2). The park's southern district supports the highest density (1 hunting dog/28 km^2). Large packs of hunting dogs, more than fifty individuals, have been reported in the Malelane section, one of the better habitats in the district. The high density of hunting dogs in the southern district appears to be related to the abundance of impala. In the Serengeti, Schaller (1972) estimated that there were 250 to 300 hunting dogs, or 1 hunting dog/85 to 102 km^2. If Schaller's density estimate was accurate, KNP supported 1.5 to 1.8 times more hunting dogs than the Serengeti.

I seldom observed hunting dogs during my study. During the two-year period, I observed hunting dogs 205 times on twenty occasions, for an average of 10.2 hunting dogs/observation. Of these, 154 were recorded on twelve occasions during wet seasons, and 51 on eight occasions during the dry seasons. I assumed at least two packs periodically used the leopard study areas. One large pack inhabited an area south of the Sabie River. I occasionally observed this pack along the Naphe, Doispan, Nwaswitshaka, and Lower Sabie roads (see table 10.7). The other, smaller, pack used an area north of the Sabie River. Although it is possible that the home ranges of these two packs overlapped, one observation suggested that the Sabie River may have formed a distinct boundary between the packs.

November 22, 1974. 6:00 P.M. to 7:00 P.M. Sabie River Study Area. Four hunting dogs travel together in a southwesterly direction along an abandoned road between Skukuza and the Sand River. When the dogs reach the Sabie River causeway, they become extremely nervous. The dog in the lead stops to drink at a pool of water and later all the dogs drink. After several minutes the dogs get on the tarmac road and nervously mill

TABLE 10.7 *African Hunting Dogs Observed in and near the Leopard Study Areas,* 1973–1975

			Hunting Dog Packs				
			Pack No. 1			Pack No. 2	
			(South of Sabie River)			(North of Sabie River)	
Year	Month	Day	Location	Number	Day	Location	Number
1973	Aug.	12	Naphe Rd.	2			
	Dec.	10	Lower Sabie Rd.	12			
1974	Jan.	3	Nwaswitshaka Rd.	13	9	Mutlumuvi Bridge	8
		19	Naphe Rd.	14			
		28	Lower Sabie Rd.	15			
	May	15	West of Skukuza	8			
	June	16	Nwaswitshaka Rd.	9			
		17	Doispan Rd.	4	18	Road to abattoir	1
	July	25	Naphe Rd.	10			
	Oct.	11	Naphe Rd.	24			
	Nov.				22	Sand River Rd.	4
1975	Jan.	19	Lower Sabie Rd.	20	7	Sand River Rd.	10
	Feb.				20	Sand River Rd.	10
	March	27	Doispan Rd.	16	10	Sand River Rd.	8
	May	11	North Naphe Rd.	12			
	June				26	Road to abattoir	5
Average number of wild dogs per observation				12.2			6.6

around looking across the causeway and in the direction they had just come. Finally, after fifteen minutes, the dogs retreat, disappearing in the direction they had come.

It appeared that the dogs were reluctant to cross the Sabie River. If these were the same dogs I saw on two occasions north and on five occasions south of the Sand River, they probably crossed the Sand River causeway, which was shorter than the one over the Sabie River.

Hunting dogs apparently are not territorial, because five packs in the Serengeti used the same 250 km^2 of plains (Schaller 1972). Hugo Van Lawick and Jane Van Lawick-Goodall (1971) observed a larger pack chase smaller packs away from their breeding grounds, and a smaller pack immediately left an area when a larger pack returned. At other times they observed strange dogs join packs and then depart. Thus large packs may split into several smaller packs and hunt in separate areas. Later, when the packs meet, individuals recognize each other and are friendly.

The movements of hunting dogs in the leopard study areas were extensive. The maximum distance between locations of observations of hunting dogs in pack No. 2 was 10 km, averaging 4.4 km between five observation locations. Long periods also elapsed between observations of hunting dogs. The average number of days between observations (n = 13) of pack No. 1 was fifty-three days, with a range of from 1 to 120 days. Pack No. 2 was seen less often (n =

6), with an average of 88 days between observations and a range of 18 to 160 days. The periods between observations suggested that both packs traveled extensively. If the maximum distance between locations was representative of home range sizes, then circular home ranges were at least 452 km² and 79 km², for packs 1 and 2, respectively.

The mobility of hunting dog packs is dependent on the mobility of the young. Schaller (1972) reported that one pack hunted an area of at least 160 km² for 2.5 months; another used an area of 210 km² in 3 weeks while the pups were confined to a den; and a third used an area of about 110 km². After the pups left the dens, the ranges of two packs of the hunting dogs expanded to 620 and 710 km².

I was unable to determine the sex of most of the hunting dogs I observed because they were in dense vegetation, traveling rapidly, or observed at a distance. Only on one occasion, February 20, 1975, was I able to accurately sex nine of ten hunting dogs in pack No. 2 near the Sand River causeway. There were five males, four females, and one of unknown sex. Pienaar (1969) reported that of twelve pups dug from a communal warren along the Naphe Road, eight were males and four were females. Of 237 hunting dogs destroyed in carnivore control operations in the park between 1954 and 1960, 60% were males and 40% were females. Estes and Goddard (1967) observed one pack in the Ngorongoro Crater that consisted of six males and one female. In the Serengeti, Schaller (1972) reported more males than females in nineteen packs, equal sex ratios in six packs, and more females than males in eight packs. Of 327 dogs more than six months old, 58% were males and 42% females. Of young pups in four litters at dens there were twenty-two males, seventeen females, and seven unidentified. In contrast, Leyn (1962) found more females than males (68:32) among wild dogs in Kagera National Park.

I did not observe small pups with hunting dogs in the leopard study areas. Hunting dogs in KNP breed in communal dens and raise their young together (Pienaar 1969). One such den was reported along the Naphe Road. On October 11, 1974, I discovered twenty-four hunting dogs lying on a large termite mound along the Naphe Road 12 km southwest of Skukuza. Ten of the dogs were adults; fourteen were pups about four to six months old. Pups begin to accompany the adults when they are about two and a half months old, and by nine months they are about the same size as adults (Schaller 1972).

I did not know if the fourteen pups I observed represented one litter. Pienaar (1969) reported several litters had twelve pups at a communal den along the Naphe Road. Litters at four dens had four, ten, fourteen, and sixteen pups (Schaller 1972); others varied from four to eleven (Kuhme 1965). An average of 7.8 pups in seventeen litters was recorded in a South African zoo (Brand 1963). If the pups I observed were four to six months old, they would have been born sometime between April and June. Although hunting dog pups can be born at any month of the year, in the Serengeti a birth peak in March and April coincided with abundant prey on the plains (Schaller 1972).

The scarcity of young hunting dogs in the leopard study areas suggested that mortality rates were high. Young pups were observed in pack No. 2 for two years and in pack No. 1 only in late 1974. Pack No. 1 declined from twenty-four in October 1974 to twenty in January 1975, sixteen in March 1975, and twelve in May 1975. The loss of twelve individuals in seven months was an extremely high mortality rate. On several occasions I saw injured or sick individuals in pack No. 1. In March 1975 two males had drooping ears and open lesions all over their bodies, and at least three other dogs had open lesions, particularly on their faces. The open lesions exposed inflamed tissue beneath the skin. Two months later, only twelve individuals were observed, but one large droop-eared male and remaining pack members appeared in better physical condition. On February 20, 1975, one of ten individuals in pack No. 2 was limping badly. Schaller (1972) reported a high loss rate in one pack he observed; several that died exhibited symptoms of distemper and one pup was diagnosed with distemper.

The average pack size for pack No. 1 was 12.2 with a range of 4 to 24. Before fourteen young were observed in this pack, the maximum number of adults observed was fifteen on January 28, 1974. Three other observations in December 1974 and January 1975 revealed twelve, twelve, and fourteen individuals. On one occasion only four hunting dogs were seen, but others could have been nearby in dense vegetation. Pack No. 2 averaged 6.6 hunting dogs with a range of 1 to 10. This pack's size was more consistent, perhaps because pups were never observed. Pienaar (1969) reported that the average pack size in the southern district during 1967 and 1968 was eleven. Average pack sizes in other African parks were 8.8 in Kafue National Park (Hanks in Schaller 1972), 11 in Kagera National Park (Leyn 1962) and 9.2 (Kruuk and Turner 1967) and 9.9 (Schaller 1972) in the Serengeti. Maximum reported pack size in KNP was forty in the Malelane district south of the leopard study areas (Pienaar 1969).

Impala are the principal prey of hunting dogs in KNP (Pienaar 1969). Impala, kudu, waterbuck, and reedbuck made up 96% of prey taken by hunting dogs from 1954 to 1966 (see table 10.8). The highest predation rate on impala occurred from February through April, when 23% of the impala killed by hunting dogs were young-of-the-year. The lowest predation on impala occurred during August-October, at the end of the dry season and before lambing. Pienaar (1969) estimated each wild dog killed 35 head/year in KNP, but Estes and Goddard (1967) and Schaller (1972) estimated kill rates of fifty to fifty-eight animals per dog each year for smaller prey.

I saw hunting dogs kill or feed on three occasions. On January 28, 1974, at 6:00 P.M., I observed fifteen hunting dogs along the Lower Sabie Road 0.6 km east of Skukuza. The dogs ran toward a herd of impala 80 to 100 m away. The

TABLE 10.8 *Species of Recorded Wild Dog Kills in Kruger National Park, 1954–66*

Prey Species	Number	Relative Frequency
Impala	2,389	87.0
Kudu	125	4.6
Waterbuck	73	2.7
Reedbuck	38	1.4
Bushbuck	29	1.1
Wildebeest	11	0.4
Nyala	8	0.3
Warthog	8	0.3
Sable antelope	6	0.2
Zebra	4	0.1
Eland	3	0.1
Mountain reedbuck	2	—
Tsessebe	1	—
Others	48	1.8

SOURCE: Pienaar 1969.
Others include gray and red duiker, steenbuck, Sharpe's grys-buck, klipspringer, suni, bushpig, and ostrich.

impala started running when the dogs were 30 m away, but the dogs rapidly singled out and captured a small calf. By 6:20 they had consumed the entire calf, leaving only a few bloody spots and rumen contents on the ground. By 6:30 the pack was traveling again. On the evening of May 15, 1974, I saw eight hunting dogs along the Sabie River lying near the remains of a six-month-old female impala, which they had just eaten.

I observed nine hunting dogs kill a wildebeest calf several kilometers west of the NRSA on June 16, 1974. The dogs had already surrounded the calf when I arrived on the scene at 4:30 P.M. After my arrival the dogs eviscerated the calf while it was still standing and then pulled it down. After pulling it down, the dogs began to feed, pulling chunks of skin and flesh from the calf as I looked on. By 5:00 about one-third of the calf was eaten. Of nearly eight hundred kills of hunting dogs reported by Pienaar (1969) in KNP, only one was a wilde-beest—a calf killed sometime between May and July.

The hunting success of hunting dogs is high, especially if their quarry is a calf. Schaller (1972) reported that hunting dogs were successful 75% of the time when they attempted to capture young wildebeest and 95% of the time with young Thompson's gazelle. Their overall success rate was 89%, similar to the 85% success rate reported for hunting dogs in the Ngorongoro Crater by Estes and Goddard (1967). Once hunting dogs left their den or rest area, the average time required to capture prey in the Serengeti was thirty minutes. Wild dogs will harass leopards in KNP if given the opportunity. Packs have been seen pursuing leopards with the obvious intent of killing them, and leopards are periodically robbed of their kills by wild dogs (Pienaar 1969).

Cheetahs

According to Pienaar (1969), cheetah were probably never abundant in KNP, and numbers have been declining since the period 1926–46. He attributed the decline to two reasons: the indiscriminate way in which these comparatively slow-breeding predators were destroyed in the park before 1960, and, perhaps more importantly, the progressive deterioration of suitable hunting habitat through brush encroachment. Brush encroachment is suspected of inhibiting cheetahs from capturing their favorite prey, impala.

ECOLOGY

During my two-year study I observed cheetahs in or near the study areas on only three occasions. My only observation of cheetah in the SRSA was of an adult and a cub.

> December 6, 1973. 5:00 P.M. to 5:40 P.M. While driving along the firebreak road south of the Sand River, I observe an adult cheetah and cub 200 m away standing in an area of short grass and *Acacia nilotica* trees. The cheetahs watch a herd of impala about 80 m away. The impala are watching the cheetahs but are not visibly alarmed. The adult sits and four minutes later lies down, still watching the impala. The cub, who has been about 10 m from the adult, joins it after ten minutes. Fifteen minutes later, after the impala alarm call, the cheetah turns away from them and lies down. When I leave at 5:40, the cub and adult are still watching the impala.

The cheetahs were in a large open area where impala concentrated during the wet seasons. I estimated that the cub was six to ten months old. Cubs usually accompany females for fifteen to seventeen months before becoming independent (Schaller 1972). I passed by the area where I observed the cheetahs almost daily for nineteen months but never saw them again, so I assumed she was just passing through the area. Perhaps the leopard study area was only a part of the female's home range. In Nairobi Park cheetah occupied permanent home ranges that varied from 75 to 102 km^2; in Serengeti National Park a female and two cubs used an area at least 65 km^2 (Schaller 1972).

Although cheetahs were not observed in the NRSA, I observed two adjacent to the study area. On December 10, 1974, a cheetah had just finished feeding on a freshly killed impala near the Manyahule waterhole. Two tawny eagles and two bateleur eagles were circling overhead when I first arrived. After resting twenty-five minutes in the nearby shade, the large male cheetah calmly walked away into the brush. This was 4 km west of the study area and the cheetah left, traveling west. I saw another cheetah of unknown sex on July 16,

1975, along the Naphe Road about 100 m from the study area. After lying in the grass for 1.7 hours, the cheetah walked away from the road and study area.

These observations and the fact that I seldom observed cheetah anywhere in the southern district of the park suggested that neither study area was intensively used by resident cheetahs. Although cheetah may have used the study areas on occasion, their use of the area was brief and infrequent. Pienaar (1969) estimated that there were only 263 cheetah in KNP in 1964 with approximately 63, 110, and 90 in the northern, central, and southern districts, respectively. This would mean an average of about 1 cheetah/39 km^2 in the southern district and 1 cheetah/72 km^2 for the entire park.

Leopards and cheetah periodically interact in the park. In 1964 a large leopard caught a full-grown male cheetah, weighing 45 kg, along the Naphe Road near the study area. The leopard killed the cheetah by strangulation and hoisted its carcass into a marula tree near the road. On another occasion a large cheetah was observed chasing a young leopard for a considerable distance (Pienaar 1969).

Lesser Predators

LESSER MAMMALIAN PREDATORS

In most instances, smaller mammalian predators do not directly compete with leopards for food, but they sometimes harass them, scavenge their prey, or become prey. Because of these roles, they were considered significant to leopards in the study areas.

Black-backed jackals.   Black-backed jackals, common throughout the park, were occasionally observed in both study areas. They have apparently experienced periodic fluctuations similar to those observed in spotted hyena and wild dog populations, and they seem to be especially numerous in the badly trampled and overgrazed areas of the park (Pienaar 1969). Prey such as rodents, hares, and ground nesting birds are believed to be more vulnerable to jackal predation in overgrazed areas because protective vegetative cover is minimal. Pienaar (1969) reported that black-backed jackals in KNP feed mainly on carrion, hares, squirrels, mongooses, rodents, the eggs and chicks of ground nesting birds, lizards, tortoises, insects, wild fruits, and other vegetable matter. Of 138 large prey remains fed upon by black-backed jackals between 1954 and 1966, the majority (87%) were impala calves. Other prey included steenbuck (5%), wildebeest calves (4%), and oribi (1%) (Pienaar 1969).

Although I periodically observed black-backed jackals and often heard them at night, I had no way to estimate their numbers. Home ranges of a similar species, the golden jackal, were estimated to be 2.6 to 5.2 km^2 and 10.4 to 23.4 km^2/pair in the Ngorongoro Crater and Serengeti, respectively (Lawick and

Lawick-Goodall 1971). The higher density of smaller mammals and birds in the SRSA, suggest that black-backed jackal density was probably also higher there than in the NRSA. If densities were similar to those observed in the Ngorongoro Crater, about 1 pair of jackals/4 km^2, at least four pairs may have lived within the SRSA.

Black-backed jackals were never observed scavenging around the kills of leopards. I did observe jackals scavenging with hyenas at a hippopotamus carcass on two occasions, frequently observed jackals scavenging at a garbage dump, and I periodically saw single jackals traveling or hunting in short grass areas in both study areas. Jackals probably avoided leopards because of the danger. A 10-kg jackal (Pienaar 1969) would have little defense against a full grown leopard. However, on at least 1 occasion in the park, 8 black-backed jackals near Rabelais dam molested a leopard at a wildebeest kill to such an extent that the leopard abandoned the kill (Pienaar 1969).

African civets. African civets were common in the leopard study areas, especially in the SRSA. These large nocturnal viverrids were seldom observed, but they were frequently captured when traps were initially set at ground level for leopards. Eight captures in the SRSA and two in the NRSA suggested a greater abundance in the SRSA (see table 10.1). Tracks of civets in the dusty firebreak roads were also frequently observed in the SRSA. Nine of ten captures of civets were during the dry season, and of these, seven were captured in August and September. One civet was apparently attracted by odor alone (from a previous bait) and was captured in an unbaited trap. Civets are solitary and nocturnal. During the daytime, they conceal themselves in tall grass, thickets, or old aardvark or porcupine burrows. At night they feed on a variety of prey including birds, birds' eggs, small game, lizards, rodents, frogs, snails, slugs, insects, berries, fruits, and young shoots of bushes (Dorst and Dandelot 1970). That they also fed on carrion was evident by their attraction to bait in the leopard traps.

Leopards occasionally fed on civets in the SRSA. The remains of a civet were once discovered in a tree where young M2 had fed. On another occasion a leopard I was observing killed a civet and carefully plucked much of its fur before eating it.

Genets. Genets were frequently captured in leopard traps, and their tracks were often observed in the study areas. Twice as many genets (37) were captured in the SRSA as in the NRSA (19) (see table 10.1). Because genets can climb trees, they were captured throughout the study period in leopard traps. The capture success of genets suggested there were about twice as many in the SRSA as in the NRSA. The greater density of cover and higher prey densities were undoubtedly factors contributing to higher viverrid densities in the SRSA. Genets occasionally were eaten by leopards. I found the remains of three genets in the feces of leopards and once watched a leopard unsuccessfully stalk a genet. Genet remains were also found in the feces of a leopard in Matopos National Park (Smith 1977).

Ratels.   Ratels, or honey badgers, are reported to be courageous and bold, showing little fear even when confronting larger animals (Dorst and Dandelot 1970). They inhabit a wide variety of habitats ranging from open, dry savannas to dense forests and are primarily nocturnal. Ratels are omnivorous, feeding on small animals and insects, and are said to be especially fond of insect larvae and honey.

I observed five ratels on four occasions and once captured an emaciated ratel in a trap set for a leopard. Four of the free-roaming ratels, as well as the captured ratel, were in the NRSA. Four of five ratels were seen during the day. Ratels were traveling alone on three occasions and traveling as a pair on another. The captured ratel and four of the other five were in grass dominated habitats. Only one ratel was seen in the SRSA; it was attacked by a leopard during the night as it was digging into a bank beside the road.

Mongooses.   Banded, dwarf, and slender mongoose were regularly observed in the leopard study areas. The white-tailed mongoose was seen only on two occasions, once in each study area. Three packs of banded mongooses were regularly seen in both study areas, and I estimated there were at least three packs per study area. Along the Nwaswitshaka River one pack frequently used a small koppie that probably served as a focal point for their activities. Groups of dwarf mongooses were usually seen in areas with many burrows and termite mounds. Slender mongooses had the widest habitat distribution. I saw these weasel-like viverrids in dry savanna as well as in the dense riparian zones.

The smaller mongooses feed on insects and insect larvae, although small rodents, birds, birds' eggs, reptiles, spiders, and fruits and berries are also consumed (Dorst and Dandelot 1970). Most of the mongooses I saw were either traveling or basking in the sun near burrows or other protected places. Although I occasionally saw mongooses scratching in the grass or leaves, I was unable to determine what they were eating. Mongooses occasionally became the prey of leopards (Dorst and Dandelot 1970), and I found remains of prey in the feces of leopards that appeared to be mongoose but could not identify the species. Grobler and Wilson (1972) reported finding the remains of a slender mongoose in the feces of a leopard from Matopos National Park.

RAPTORS AND VULTURES

Large raptors and vultures are common in KNP including the leopard study areas. Because they prey upon some of the smaller mammals, birds, and reptiles that are also eaten by leopards or scavenge from leopard kills, they too play a role in the predator community.

Eagles and other birds of prey. Of five species of eagles periodically observed in the leopard study areas, the most frequently observed were the tawny and bateleur. Tawny eagles were twice captured in leopard traps, and bateleur eagles were often seen soaring above the study areas. Both species scavenge as well as hunt their own prey, and their scavenging habits may explain why they were more frequently observed than other eagles in the study areas. On two occasions in the SRSA, I saw martial eagles feeding on leguans but did not consider them abundant. Black-breasted snake-eagles were seen infrequently, primarily in the more open habitats in the NRSA, where on two occasions I saw them feeding on snakes. Fish eagles were common along the Sabie River, especially during dry seasons when the water was low and clear.

Tawny eagles were seen at three (5%) of the kills of leopards. One tawny eagle sat in a tree 10 m from a kill while the leopard, avoiding nearby tourists, hid in a thicket 30 m away. After waiting twenty-five minutes, the eagle flew to the kill and fed. Three minutes later another tawny eagle joined the first and both fed for fifty-two minutes. They then flew to a nearby tree and watched the carcass. The leopard did not chase the eagles away.

After another leopard left its kill in a tree, a tawny eagle soared over several times, landed within 3 m, watched the carcass for twelve minutes, and then landed on it to feed. Several minutes later another tawny eagle joined the first. Once, after hearing something nearby, both eagles flew away in panic but returned after ten minutes and continued feeding. I estimated at least 1 kg of flesh was consumed by the tawny eagles in less than an hour. One of these kills was well hidden in a tree, so some eagles were apparently quite adept at locating kills of leopards. However, based on the low number of kills of leopards that tawny eagles found, they did not appear to seriously compete with leopards.

I saw few other large birds of prey. I once watched a secretary bird capture a small unidentified snake in an open meadow west of the NRSA, but suspected the cover was too dense for them over most of the study areas. Although spotted eagle owls were frequently heard at night during certain months, I observed these large owls during the day on fewer than ten occasions. All the owls I saw were in large ebony trees along the Nwaswitshaka or Sabie Rivers. I also observed smaller owls on fewer than ten occasions and was able to identify three of them as pearl-spotted owls.

Vultures. The most abundant species of vulture in KNP is the white-backed vulture (Braack 1983), and they are the species most frequently observed feeding on carcasses. Other less-common species, include the Cape, white-headed, Egyptian, hooded, and lappet-faced vultures.

Kills of leopards seldom attracted vultures in the study areas. Most carcasses were so well concealed by overhead foliage that vultures apparently did not see them. Vultures were present at only five (7%) of the leopard kills I observed, and of those, three were cached in trees that had sparse foliage. Vultures fed on the carcasses of prey only if the leopard was gone, and I never saw more than three vultures at a kill. On one occasion vultures plucked out the eyes and ate a small portion of flesh from an impala carcass while the leopard was away. My observations suggested that only a few kills of leopards were discovered by vultures, and they were so well guarded by the leopards that the vultures consumed little flesh.

REPTILES

Of the reptiles, crocodiles are considered to be the most important predators in KNP (Pienaar 1969). Other common large reptiles, such as the tree and river leguans, may occasionally become prey of leopards. Remains of snakes have been found in leopard feces (Smith 1977), and poisonous snakes may periodically kill leopards (Turnbull-Kemp 1967).

Crocodiles.   Crocodiles were the most important reptilian predator in the SRSA, and on several occasions, when the river was in flood, they came at least 1 km up the Nwaswitshaka River. Crocodiles not only took kills from leopards, on one occasion, perhaps two, they killed and ate leopards in the study areas. In KNP, where crocodiles prey on mammals as large as giraffe and buffalo, the bulk of mammalian prey appears to be impala, waterbuck, kudu, and bushbuck (Pienaar 1969). Adult crocodiles, however, probably prey more heavily on aquatic reptiles such as the hinged terrapin than on mammals. Individuals measuring 4.0 to 4.3 m are common in the park, and crocodiles exceeding 4.6 m and estimated to weigh more than 454 kg have been encountered in the Levubu and Olifants Rivers in the northern district. Crocodiles have probably taken more human lives in the park over the years than all mammalian predators and poisonous snakes combined.

Crocodiles in the Sabie River had favorite pools, rocks, and islands where they sunned themselves. These places were used regularly by the crocodiles during the dry season, but during wet seasons crocodiles could be observed anywhere along the river. When the water was high and frequently muddy, crocodiles were probably able to capture mammalian prey easier than during dry seasons. When floodwaters backed up into smaller tributaries, crocodiles followed. On several occasions I saw them lying motionless in the water waiting for impala to drink. Once, I observed a large crocodile in a small tributary of the Sabie River with an adult male impala in its jaws, and several times I frightened crocodiles into the Sabie River as I walked along its banks.

Leopards living along the Sabie River occasionally lost their kills to crocodiles. On December 7, 1973, I discovered a female leopard in a tree with a freshly killed impala. The next morning I saw the leopard dragging the carcass toward the flood swollen Sabie River. Later in the day I followed the drag marks and discovered that a crocodile had taken the carcass from where the leopard had hidden it on the ground under a tree. Tracks of the crocodile, drag marks, and impala hair indicated the remaining 25% to 50% of the carcass was taken by the crocodile. There was no sign of a struggle with the leopard as the crocodile dragged the carcass 10 m through wet sand into the muddy river. Crocodiles probably successfully stole kills of leopards only if such kills were left on the ground near rivers and the leopard was unable to defend them.

Snakes. Although Pienaar (1969) reported many large pythons in KNP and indicated that pythons were sometimes killed by leopards, I saw only one large python, probably the same individual, on two occasions. This huge reptile was at least 4 m long and about 15 cm in diameter. On both occasions the python was observed in a patch of reeds where a female leopard (F11) had hidden her cubs. The python was coming out of a burrow the first time I saw it on November 8, 1974. Thirteen days later I saw it only 10 m from the burrow, in the dense patch of reeds. The female leopard lost her cubs sometime during the thirteen-day period, and the python was a prime suspect.

This large python and three smaller ones were observed during the wet seasons. On December 27, 1973, I saw a 2.3 m python on the road to Lower Sabie and a 1.3 m python on January 18, 1974, in the same vicinity. On November 12, 1974, I saw a 1.6 m python on the Paul Kruger Road bordering the eastern end of the NRSA. Although mammalian prey as large as bushbuck and reedbuck ewes, kudu, waterbuck, and wildebeest calves have been swallowed by pythons in the park, impala and duiker are taken most frequently (Pienaar 1969). Turnbull-Kemp (1967) reported a mature leopard was once disgorged by a large python. I did not frequently see smaller snakes in the park. Black mambas were seen on at least three occasions in the NRSA, including one about 3 m long near the Nwaswitshaka River waterhole. Cobras were seen on at least two occasions in the SRSA. Once, on March 29, 1974, a cobra attempted to feed on the bait inside a leopard trap along the Sabie River. I saw puff adders at least fifteen times, night adders at least ten times, and boomslangs on at least two occasions. On about twenty other occasions I encountered snakes 1 to 2 m long but was unable to identify them because of the dense vegetation.

Summary

1. Other major predators inhabiting the leopard study areas, in order of decreasing abundance, were spotted hyenas, lions, hunting dogs, and cheetahs.
2. Spotted hyenas were observed 112 times, including 37 at kills of leopards, in and near the leopard study areas. Hyenas were also captured thirteen times in traps set for leopards.
3. Hyenas usually associated in small groups, averaging 3.6 and 3.0 hyenas/group in the Sabie and Nwaswitshaka River study areas, respectively.

4. Estimated average densities of hyenas were 1 hyena/1.1 to 2.1 km^2 in the Sabie River study area and 1 hyena/2.9 km^2 in the Nwaswitshaka River study area.

5. Impala, especially newborn lambs, appeared to be important prey of spotted hyenas in the leopard study areas.

6. At least two hyenas were killed and eaten by a radio-collared male leopard, and one was killed by lions in the Sabie River study area.

7. Hyenas scavenged at 51% of the kills of leopards I discovered. At most kills (68%) only one hyena was present. Two were present at 8 kills (29%) and three at only one kill (14%). Most kills of leopards were safe from hyenas, but a hyena was able to take an impala carcass away from a subadult female leopard before she could cache it in a tree.

8. More than three hundred lion observations were recorded 101 times, and lions were captured eleven times in traps set for leopards in and near the leopard study areas.

9. One hundred sixty-three lion observations (54%) were recorded from roads open to tourist traffic and 138 (46%) from roads closed to tourists. One hundred sixty-seven (55%) were recorded during wet seasons.

10. Lions in and near the Sabie River study area associated in groups averaging 3.7 lions, but only one pride of lions regularly used that study area. This pride included eleven individuals (three adult males, four females, three small cubs, one unidentified). The estimated density, excluding cubs, was 1 lion/4 to 8 km^2.

11. Limited evidence suggested impala were an important prey of lions in the Sabie River study area, but this pride also killed mature giraffe on at least two occasions.

12. Many of the observations of lions in the Nwaswitshaka River study area were singletons (43%), and of these, 59% were adult males.

13. The average group size of lions in the Nwaswitshaka River study area was 2.3 lions. Two prides, one of at least seven adults and subadults and the other of five adults, used the Nwaswitshaka River study area. In the NRSA the estimated density of lions, excluding cubs, averaged 1 lion/10.2 km^2.

14. Lion group and pride sizes were greater, prides contained more adult males, and females were seen with cubs more often in the Sabie River than in the Nwaswitshaka River study area. These differences were probably related to the densities, seasonal distribution, size, and habits of prey in the study areas.

15. Only one (2%) of fifty-five kills of leopards were taken by lions. Lions apparently were unable to detect or appropriate most kills leopards cached in trees.

16. African hunting dogs were seen on only twenty occasions in and near the leopard study areas. An average of 10.2 hunting dogs were seen per observation and most (75%) were seen during wet seasons.

17. Only two packs of hunting dogs were identified in both of the leopard study areas. One pack had twenty-four dogs, including fourteen pups; the other contained no more than ten adults.

18. Hunting dogs in the leopard study areas preyed upon impala and a young wildebeest. Previous data indicated that impala were the primary prey of hunting dogs in the southern district of Kruger National Park.
19. Cheetah were rarely observed (three occasions) near the leopard study areas. A female and cub were seen in the Sabie River area, and an adult male was once observed feeding on an impala near the Nwaswitshaka River study area.
20. Black-backed jackals, African civets, genets, ratels, and mongooses were frequently seen or captured in the leopard study area. They probably did not seriously compete with leopards for prey, but they sometimes became prey of leopards.
21. Eagles and vultures were common in the leopard study areas. Although tawny eagles fed on about 5% of the leopard kills and vultures fed on 7%, most kills of leopards were so well concealed in trees these scavengers were unable to find them.
22. Crocodiles were abundant in the Sabie River and sometimes moved short distances up the Nwaswitshaka River when it was in flood. They occasionally took kills from leopards near the Sabie River and on one, perhaps two occasions, killed and ate radio-collared leopards.

PART THREE
Synthesis

11
The Adaptable Leopard: A Synthesis

In the past the leopard was the world's most widespread large solitary felid. Today the leopard is still the most enduring large felid found throughout the Old World. The reasons for this persistence, in regions where other carnivores have been exterminated, vary and are the subject of this synthesis.

Ecological Adaptations

The most conspicuous attributes of leopards are their ability to live in different environments and their flexible diets. Common leopards (*Panthera pardus*) are ecological generalists rather than specialists. They are not specialized, like the cheetah, to pursue swift prey in open habitats. They are not specialized like the lion, whose size and sociality allow it to capture large prey and protect kills from formidable predators and scavengers. Unlike their distant Asian cousins, common leopards are not restricted like clouded leopards to dense tropical forest or like snow leopards to rugged, high-altitude terrain. Common leopards can survive by feeding on large or small prey; they can be efficient scavengers; and they are not adverse to preying upon domestic stock, a behavior that brings them into direct conflict with humans.

HABITAT ADAPTABILITY

The evolutionary history of the *Panthera* has been an initial adaptation away from warm, moist biotopes, and, specifically for the leopard, a gradual invasion of dry, tropical areas (Kleiman and Eisenberg 1973). The once widespread distribution of the leopard, the behavioral strategies leopards use to survive under climatic extremes, and the leopard's use of available stalking and protective cover suggest there are few natural habitats where leopards could not exist today given adequate prey and protection from humans. The ability of leopards to live in different environments can be appreciated by considering the climatic extremes that leopards can tolerate.

The distribution of leopards at high altitudes and northern latitudes attests to its ability to survive in extremely cold and harsh environments. In Africa leopards or signs of them, have been observed at 4,572 m on the southern slopes of Mount Kibo and at 4,205 m on Mount Mawinzi, the second highest peak of Mount Kilimanjaro (Guggisberg 1975). Although leopards do occasionally venture up to the snow line on these African mountainous peaks, the reported discovery of a leopard frozen in ice on Mount Kenya is apparently untrue. According to Guggisberg, in 1926 a Dr. Reusch discovered the frozen carcass of a leopard at 5,638 m the rim of Kilimanjaro's Kibo Crater. A later photograph of the famous leopard, however, revealed it to be lying on top of a rocky pinnacle—not embedded in the ice. This observation suggests that life at 5,638 m is very difficult—even for a leopard.

In high elevation habitats leopards in summers apparently are seldom found above 3,000 m, more likely below 2,400 m. Schaller (1979) reported that in the Himalayan Mountains, common leopards were found in forests up to within a few kilometers of snow leopard habitat, and he once observed a leopard unsuccessfully stalk a goral at 2,600 m. In Nepal common, or forest, leopards were also reported in valleys within the range of snow leopards (Hillard 1989). Guggisberg (1975) reported leopards probably do not ascend higher than 3,000 m in the Himalayas and that in the Caucasus, leopards were observed to spend summers in alpine meadows at about 2,500 m.

Leopards, like cougars, probably occupy high altitude environments only during summers. Cougars spent 82% of their time above 1,800 m in summer, but only 9% in winter (January through May) (Seidensticker et al. 1973). By descending they avoided deep snow and preyed on overwintering ungulates. In Kashmir, Ward (in Guggisberg 1975) observed that although leopards did not mind some snow, they did not winter at high altitudes. Leopards, again like cougars, probably cannot travel easily through deep snow. Cougar home range boundaries in winter are affected by snow depths, and cougars, to avoid breaking new trails in deep, difficult snow, may occasionally walk in one anothers' tracks or in trails made by humans (Seidensticker et al. 1973). Bobcats also have difficulty traveling through snow more than 15 cm deep (Marston

1942), and in Idaho bobcats sometimes stayed for three or four days in caves and rock piles during snowstorms to avoid breaking trails in deep drifting snow (Bailey 1972). Bobcats also traveled on well-packed trails made by jackrabbits rather than breaking their own trails through deep snow.

Although little is known about leopards in high-altitude mountainous habitats, in winters they probably prey on ungulates that migrate to lower elevations. In the Himalayas, prey of leopards such as goral are usually found at elevations between 2,500 m to 3,000 m, and seldom above 4,000 m. Markhor are found at 4,000 m in summers, they winter below 2,200 m (Schaller 1977). Cougars follow elk and mule deer in winters to lower elevations, valley bottoms, or southern-facing slopes (Hornocker 1970; Seidensticker et al. 1973). The few reported observations of leopards at high altitudes and our knowledge of the behavior of the cougar, suggest that deep snow probably forces leopards to lower elevations in winters. At lower elevations traveling is easier and prey more available. At high altitudes, rodents, which might be leopard prey in summers, would be hibernating in winter.

Harsh climatic extremes also occur at high northern latitudes within the leopard's range in Asia. At their most northerly distribution in eastern Siberia (52° N. latitude), the Siberian tiger and Amur leopard apparently share similar habitats (Prynn 1980). The Siberian tiger spends the summer above 1,200 m, but winters at lower elevations (Guggisberg 1975). In the winter temperatures in this eastern Siberian habitat can drop to −30 to −35°C, and snow several feet deep may persist throughout the winter.

Some physical adaptations of Amur leopards to harsh, cold environments include a larger body size, long soft winter coats, and long furry tails (Prynn 1980). Like the Siberian tiger, thick body fat is probably also laid down on the belly and flanks during winter. Large, solid-black rosettes on a light background of pale straw yellow, tan-beige, or light gray also appear to be common among leopards living in northern environments. This color combination, similar to that of the snow leopard, apparently helps the leopard blend into a background of snow during the winter. The large body size of the northern Amur leopard, up to 76 kg for males (Stroganov 1969), also probably helps reduce heat loss. The Amur leopard's preference for cliffs and rocky broken terrain in eastern Siberia is apparently a behavioral strategy for coping with winters; caves and rock piles can provide temporary escape from deep snow and reduce heat loss to the cold, blowing wind.

In extremely hot and dry environments, leopards use caves to escape high daytime temperatures and to reduce water loss. For example, leopards inhabiting the desertlike escarpments in Namibia, where annual rainfall averages 175 mm, often use caves as lairs for feeding and rearing cubs (Brain 1981). One of these caves examined by Brain extended 46 m into the rock. Another cave apparently had been used by leopards for at least forty years because, during that period, 104 leopards were captured outside the cave entrance. The use of caves by leopards is apparently an important survival strategy in some desert environments. Use of caves would greatly reduce water loss during the hot daylight hours and protect young from other predators. Bobcats in the dry desertlike plateau of southern Idaho frequently used caves to escape from the heat during the day in hot summer months and to rear their young (Bailey 1979). Even on the hottest summer days caves were much cooler and the humidity higher than outside.

In deserts where caves are absent, leopards use another strategy to escape heat and moisture loss. In the southern Kalahari Desert, where the annual precipitation is less than 250 mm and surface temperatures may reach 70°C, leopards spend the hot daylight hours in porcupine and aardvark burrows or crawl under the dense foliage of the witgat tree *(Boscia albitrunca)* (Bothma and Le Riche 1986). These trees, which are also used by lions to reduce water loss (Eloff 1973), have branches down to the ground, forming an enclosed, tent-like canopy against the hot dry air.

Leopards that inhabit deserts are not dependent on free water to drink but will drink water if it is available. Leopards probably obtain water from the tissues of their prey and adjust their activity and movements to coincide with periods of cool temperatures (Bothma and Le Riche 1986). When water was available in the Kalahari Desert, male leopards drank water, on the average, only once every 3.6 nights or every 52.2 km the leopard moved. Mean distance between known successive drinks was 21.1 km with a minimum of 2.0 km and maximum of 38.6 km (Bothma and Le Riche 1986). The mean known period without water for male leopards was 2.8 days. A female with cubs went at least fifteen days and traveled 201.4 km without a drink of water.

Some of the leopard's other strategies to conserve water and energy were apparent during my study. These strategies included: (1) resting during the daytime, (2) increasing activity at night, (3) reducing daytime activity during cold, dry seasons compared to the warmer, moist wet seasons, (4) increasing activities on cool, overcast days and decreasing them on hot, clear days, (5) using cool shade during the hottest hours of the day, and (6) for females, except those with dependent cubs, conserving more energy and water by reducing overall activities compared to males. These and perhaps other adaptations have enabled leopards to successfully exploit extremely hot and dry deserts such as the Namib, Kalahari, northern Sahara, Sinai, and Arabian, providing that caves, burrows, or some vegetation is present to provide relief from the hot daytime temperatures.

One important component of leopard habitat is adequate stalking cover. For leopards, stalking cover can either be provided by vegetation or irregular terrain. The denser the vegetation or more rugged the terrain, the better the habitat for leopards. But where stalking cover is sparse, leopards can use darkness for concealment. In the Kruger National Park study areas, leopards were generally more active during the night than the day, and they ventured into open, brackish flats with little vegetative cover more often at night than during the day. The relationship between stalking cover and use of darkness by felids has been studied for the African lion (Elliot et al. 1977). Lionesses generally avoided hunting during the day if the vegetation or topographic irregularities were less than 0.3 m high because visual clues were most important to their prey.

Leopards in my study areas often remained in dense riparian cover during the day, venturing forth after dark into the more open *Acacia*-dominated areas. Leopards' peak seasonal use of dense riparian vegetation occurred during the late dry season when vegetative cover was sparse. Peak use of the open *Acacia*-dominated areas occurred during the middle of the wet seasons when vegetative cover was the greatest.

Leopards, like cougars and some smaller solitary felids, can take advantage of, and may even prefer, rocky, broken terrain. In rugged terrain felids can greatly increase their search area by climbing onto and visually scanning surrounding areas from ridges, tops of cliffs, and rock outcrops. More than 95% of the locations of cougars were associated with timbered or rocky broken areas (Seidensticker et al. 1973). Bobcats also fed more frequently on the less abundant cottontail rabbit than the plentiful jackrabbit because cottontails inhabited rocky terrain (Bailey 1972).

Another favorable aspect of rugged terrain for leopards is that terrain and clumped vegetation can restrict the number of escape routes available to prey. The steepness of the slope can increase felid hunting success. For gazelle, the probability of capture by lions increased sharply if the gazelles fled upslope, even with slope angles less than 20° (Elliot et al. 1977). The probability of capture of wildebeest and gazelle also increased, the greater the flight angle of the prey. Capture probability increased for wildebeest fleeing at angles more than 80° from the approach of lions. In gazelles, probability of capture rose sharply for flight angles greater than 100°.

Although leopards are tolerant to climatic extremes, the majority of habitats they occupy occur within more amenable climatic zones. In sub-Saharan Africa, a diversity of vegetation types occupied by leopards, ranging from extremely moist rain forests to hot and dry deserts, represent nearly a tenfold difference in annual rainfall and include at least 35 degrees of latitude difference (see table 11.1). Most of the vegetation zones in sub-Saharan Africa (see table 11.2) were probably once occupied by leopards until human development intensified at the turn of the century. Many of the leopard's habitats have been fragmented into small blocks, developed for agriculture or livestock grazing, deforested, or the natural prey decimated. Nevertheless leopards continue to persist in some of these habitats. For example, a few leopards still occur in the developed Cape and Karoo Shrubland Zone but only within the rugged mountainous coastal forest of the southwestern Cape Province in the Republic of South Africa (Norton and Lawson 1985; Norton and Henley 1987; Norton et al. 1986).

TABLE 11.1 *Diversity of Environments occupied by Leopards in Sub-Saharan Africa*

Major Vegetation Type [1]	Location	Annual Rainfall in Milimeters	Main Prey [2]	Sources [3]
Lowland rainforest	Tai N. P., Ivory Coast	1,600–3,200	Forest duikers[a]	(1)
	Ituri Forest, Zaire	1,700	Okapi (?)[b]	(2)
	Makokou, Gabon	1,700	unknown	(3)
Dry savanna	Serengeti N. P., Tanzania	814	Thomson's gazelle[c]	(4)
	Tsavo N. P., Kenya	451–572	Impala[c]	(5)
Woodland	Kruger N. P., South Africa	554	Impala[a]	(6)
	Matobo N. P., Zimbabwe	500	Hyraxes[a]	(7)
Cape and Karoo shrubland	Cape Province, South Africa	500–770	Hyraxes[a]	(8)
Kalahari savanna	Kalahari-Gemsbok N. P., South Africa	250	Springbok[c]	(9)
Namib Desert–Karoo semidesert shrubland	Hakos Mountains, Namibia	175	Hyraxes[e]	(10)

[1] Simplified from White 1983 and Tucker et al. 1985.
[2] Determination of major prey: [a] = fecal analysis, [b] = radio-collared okapi, [c] = kill remains, [d] = tracking leopards in sand, [e] = prey remains from cave.
[3] SOURCES: (1) = Hoppe-Dominik 1984, (2) = Hart and Hart 1989, (3) = Charles-Dominique 1977, (4) = Kruuk and Turner 1967, Schaller 1972, Bertram 1982, (5) = Hamilton 1976, (6) = Pienaar 1969, This study, (7) = Grobler and Wilson 1972, Smith 1978, (8) = Norton and Lawson 1985, Norton and Henley 1987, Norton et al. 1986, (9) = Bothma and Le Riche 1984 and 1986, Mills 1984, (10) = Brain 1981.

The woodland (includes *Miombo*) and dry savanna (includes Somali-Masai Arid) vegetation zones, characteristic of much of sub-Saharan Africa, continue to be important habitat for leopards. With its diversity of prey species, density of vegetation, and topographic irregularities, these two vegetation zones, which extend more than 30 degrees of latitude and include both dry and wet climatic regimes, may still support the greatest number of leopards in Africa.

One of the most natural, largest, and least-known vegetation zones occupied by African leopards is the lowland rain forest located primarily in the Zaire (Congo) River Basin. Little ecological information, including their status, distribution, and abundance, is available for leopards from this habitat. In the Ituri Forest region of northeastern Zaire, in "mixed moist semi-evergreen forest" where annual rainfall is about 1700 mm, leopards are present and preyed upon

TABLE 11.2 *Wildlife Habitats in Sub-Saharan Africa (1986)*

Vegetation Formation	Original Area (km²)	% Remaining	Protected Areas (km²)	% Protected
Dry forests				
Upland montane	790,712	37.2	21,494	7.3
Woodland	5,896,200	42.4	427,467	17.2
Other	1,556,896	41.0	64,004	11.3
Moist forests	4,699,704	39.7	132,457	7.1
Savanna/grassland	6,954,875	40.8	296,957	10.5
Scrub/desert	176,600	97.8	17,361	10.1
Wetland/marsh	61,700	70.9	2,370	5.4
Mangroves	87,870	44.6	1,120	2.9

Adapted from Harmon 1990.

radio-collared okapi (Hart and Hart 1989). Leopards also occur farther west in equatorial rain forests in Gabon, where annual rainfall again averages 1700 mm (Charles-Dominique 1977; White 1992). Further west, leopards still occupy the Tai National Park in the West African rain forests of the Ivory Coast, where they feed on forest duikers, rodents, and monkeys (Hoppe-Dominik 1984). But this park (Kingdon 1989) and much of West Africa's rain forest have been severely altered by man (Matthiessen 1991; Stuart et al. 1990).

FEEDING ADAPTABILITY

The most thoroughly documented attribute of leopard ecology is their versatile feeding habits. At least ninety-two species of prey have been documented in the leopard's diet in sub-Saharan Africa alone (see table 11.3). In Kruger (Pienaar 1969; this study) and Serengeti National Park (Bertram 1982), leopards preyed upon at least thirty-two and thirty-one species, respectively. The range of prey taken by leopards varies from large ungulates, small carnivores, and rodents to arthropods such as crickets and scorpions.

Several patterns emerge from a review of the leopard's feeding strategies (see tables 11.3, 11.4, and 11.5): (1) mammals form the most important class of prey in the leopard's diet, (2) ungulates are the most important order of mammal in the leopard's diet, (3) most prey of leopards are terrestrial and are captured on the ground, (4) when leopards capture arboreal prey, with perhaps an exception in rain forests, the prey species, such as vervet monkeys and baboons, also spend considerable time on the ground, and (5) leopards are opportunistic feeders and will not hesitate to take any prey that is vulnerable.

That mammals are the most important prey of leopards, rather than birds, reptiles, or other classes of animals, is understandable. Most birds and reptiles are unavailable to leopards because of their arboreal or aquatic habits and escape strategies. Most are also too small to benefit leopards energetically, compared to the variety of mammals available as prey in leopard habitats.

Among the mammals, ungulates probably dominated many prey communities during the evolution of leopards. Felids began evolving in Africa in the late Miocene when artiodactyls began dominating the African landscape (Maglio 1978). When the *Panthera* began evolving in Africa, from the early to mid-Pleistocene, the family Bovidae had probably already reached a maximum in its diversity of species (Maglio 1978). During the last million years, the world's climate repeatedly swung from wet to dry and from warm to cool. During warm wet periods in Africa, forests spanned the equator. During dry periods, deserts connected the present-day Sahara and Kalahari Deserts (Kingdon 1989). The majority of Africa's staple species become exceptionally adaptable (Kingdon 1989) during such climatic changes. Leopards probably preyed upon the many varieties of ungulates that were available during this period in the earth's history, not only in Africa but perhaps in Europe and Asia as well.

TABLE 11.3 *Prey Species in Diets of Leopards in Sub-Saharan Africa*

Location	Data Source	Mammals								Aves	Reptiles	Fish	Arthropods	All species	
		Artiodactyla	Perissodactyla	Rodentata	Primates	Carnivora	Lagomorpha	Hyracoidea	Others[1]					This Study	Different Species
Zambia	(1)	11	1	3	2	3	1	—	—	—	1	1	—	22	22
Tanzania	(2)	7	1	1	1	2	—	1	—	4	1	—	—	18	12
S. Africa	(3)	18	2	2	2	3	1	—	1	1	1	—	—	31	13
Tanzania	(4)	7	2	—	1	4	—	—	1	1	1	—	—	15	2
Zimbabwe	(5)	7	1	4	1	1	2	1	1	2	3	—	1	24	6
S. Africa	(6)	4	1	2	1	3	—	—	—	1	—	—	—	12	1
Kenya	(7)	8	1	2	2	1	1	1	1	1	2	—	3	23	4
Zimbabwe	(8)	11	1	—	—	—	—	1	—	1	—	—	—	14	0
Ivory Coast	(9)	7	2	8	7	5	—	1	2	1	—	—	—	33	25
S. Africa	(10)	7	—	2	—	4	—	—	1	—	—	—	—	14	3
S. Africa	(11)	6	—	—	1	6	—	1	1	1	—	—	—	16	3
S. Africa	(12)	7	—	3	1	1	1	1	—	3	1	—	1	18	8

DATA SOURCE: (1) = Mitchell et al. 1965, (2) = Kruuk and Turner 1967, (3) = Pienaar 1969, (4) = Schaller 1972, Grobler and Wilson 1972, (6) = This study, (7) = Hamilton 1976, (8) = Smith 1977, (9) = Hoppe-Dominik 1984, (10) = Bothma and Le Riche 1984, (11) = Mills 1984, (12) = Norton et al. 1986.
[1]Insectivora, Pholidota, Tubulidentata.

TABLE 11.4 *Cumulative Relative Frequencies in Occurrence of Prey Found in Leopard Feces in Sub-Saharan Africa*

Major Vegetation Zone[1]	Data Source	Prey Remains (%)									
		Mammals							Aves	Reptiles	Arthropods
		Atiodactyla	Rodentata	Other[2]	Lagomorpha	Carnivora	Primates	All			
Lowland rainforest	(1)	37	16	5	—	6	20	94	1	—	—
Woodland	(2)	62	7	—	—	3	8	96	2	2	—
Woodland	(3)	21	17	32	8	1	3	82	5	2	2
Woodland	(4)	31	8	50	3	1	1	94	1	—	—
Woodland	(5)	24	11	56	3	1	—	95	4	—	—
Dry savanna	(6)	24	22	12	6	—	1	65	17	5	11
Cape and Karoo	(7)	51	2	42	—	—	—	99	—	—	—
Shrubland	(8)	68	—	32	—	—	—	100	—	—	—
Shrubland	(9)	25	5	65	2	—	2	98	1	—	1
Shrubland	(10)	48	—	51	—	—	1	100	—	—	—
Shrubland	(11)	33	12	30	3	—	—	78	12	3	6
Shrubland	(12)	75	—	16	—	—	—	91	9	—	6

DATA SOURCE: (1) = Hoppe-Dominik 1984, (2) = This study, (3) = Grobler and Wilson 1972, (4) = Smith 1978 [Game Park study area], (5) Smith 1978 [Eastern study area], (6) = Hamilton 1976, (7) = Manson 1974 cited in Norton et al. 1986, (8) = Stuart 1982 cited in Norton et al. 1986, (9) = Norton et al. 1986 [Cedarberg Wilderness study area], (10) Norton et al. 1986 [Gamka study area], (11) = Norton et al. 1986 [Jonkershoek study area], (12) = Norton et al. 1986 (Wemmershoek study area].

[1] Simplified from White 1983 and Tucker et al. 1985.

[2] Insectivora, Pholidota, Tubulidentata, Hyracoidea.

TABLE 11.5 *Prey Species Remains in Leopard Scats*

| Vegetation Zones | Source Data | Types of Mammalian Prey (%) | | | Number of Scats |
		Terrestrial	Partially Arboreal[1]	Arboreal	
Lowland rainforest	(1)	47	5	30	215
Woodland	(2)	77	3	0	200
Woodland	(3)	93	1	0	247
Woodland	(4)	95	0	0	91
Woodland	(5)	72	8	0	94
Dry savanna	(6)	64	1	0	51
Cape and karoo	(7)	95	0	0	54
Shrubland	(8)	98	2	0	129
Shrubland	(9)	99	1	0	59
Shrubland	(10)	100	0	0	24
Shrubland	(11)	100	0	0	25

[1]Species spending some time on the ground (except for four lowland rain forest species of carnivores, includes only vervet monkeys and chacma baboons).
DATA SOURCES: (1) = Hoppe-Dominik 1984, (2) = Grobler and Wilson 1972, (3) = Smith 1978 [Game Park study area], (4) = Smith 1978 [Eastern study area], (5) = This study, (6) = Hamilton 1976, (7) = Manson 1974 cited in Norton et al. 1986, (8) = Norton et al. 1986 [Cedarberg Wilderness study area], (9) = Norton et al. 1986 [Gamka study area], (10) = Norton et al. 1986 [Jonkershoek study area], (11) = Norton et al. 1986 [Wemmershoek study area].
(Data are the estimated cumulative, relative frequencies of occurrence in scats.)

Most studies of leopard food habits, regardless of continent or habitat, indicate that ungulates are the leopard's principal prey (see table 11.4). Leopards preyed primarily on wild goats in the Himalayas Mountains (Schaller 1977), on seven species of ungulates in the Kalahari Desert (Bothma and Le Riche 1986), upon impala or gazelles in brushland and woodland habitats (Kruuk and Turner 1967; Schaller 1972; this study), and on forest antelopes in the tropical rain forest (Hoppe-Dominik 1984).

Most prey of leopards are terrestrial and are, thus, captured on the ground. Even in tropical rain forest habitats where arboreal species of all leopard prey occurred most frequently in the leopards' diet (38.9%) (Hoppe-Dominik 1984), terrestrial prey were still taken more often (61.1%). Among mammalian prey, even some arboreal species, such as the colobus, mangabey monkey, and palm squirrel, apparently spend at least some time on or close to the ground (Kingdon 1974; Struhsaker 1975; Kingdon 1989). At least one primatologist (Struhsaker 1975) doubted that leopards would be effective predators on arboreal prey at heights greater than 10 m in tropical rain forests. Baboons and vervet monkeys are the most common primates that regularly occur in the leopard's diet (see table 11.5). In one recent study in the Manovo-Gounda-Saint Floris National Park, Central African Republic, baboons (olive baboons) occurred more frequently (22%) among kills ($n = 23$) of leopards than elsewhere in Africa. Nevertheless ungulate kills were still found more frequently (56%) (Ruggiero 1990). Baboons and vervet monkeys spend considerable time on the ground, but flee to, or rest in, trees and rocky cliffs to escape from leopards and other predators (Isbell 1990; this study).

Leopards, like other inquisitive felids, are opportunistic predators. I observed opportunistic hunting on several occasions during my study when genets, civets, and ratels elicited a stalking or attack response from leopards merely because they happened to be nearby. Some leopards may also develop a particular taste for prey not normally found in the leopard's diet. For example, one male leopard in my study area, perhaps out of desperation for food, apparently developed a taste for spotted hyenas. Given an opportunity, leopards will prey on unusual species. Some unusual prey items reported in leopard diets include European storks, secretary birds, vultures, pythons, and fish (Mitchell et al. 1965; Kruuk and Turner 1967; Pienaar 1969; Schaller 1972).

Leopards, unlike other large felids, will readily scavenge and feed on decomposing flesh. Carcasses of large prey killed by other leopards or other predators, or prey dying from other causes are potential sources of food. This aspect of leopard feeding behavior was noted by several early writers (Corbett 1947; Stevenson-Hamilton 1947; Turnbull-Kemp 1967) who attributed scavenging to the leopard's overall success as a large carnivore.

Despite their opportunistic feeding behavior, leopards seldom take humans as prey. This behavior may reflect the leopard's basic fear or avoidance of humans, at least during daytime (this study). Avoidance of humans by leopards may have evolved relatively recently. Some early hominids may have been the victims of leopard predation (Brain 1981), but early hominids may have benefited from leopards by scavenging from leopard kills (Cavello 1990). When leopards do become man-eaters, most man-killing apparently begins by accident (Turnbull-Kemp 1967). Exceptions include man-eaters that began feeding on cadavers or began man-eating because of their injuries. "On a whim of the moment, in a fleeting error of judgment, or in a second of panic-stricken surprise a human is simply and easily overpowered" (Turnbull-Kemp 1967). These accidents usually involve attacks on small children, or crouching or bent-over adults. Erect adult humans are seldom attacked. For example, in early 1987, a leopard reportedly killed ten people near the Pokhara Valley about 160 km west of Katmandu, Nepal (Anchorage Daily News 1987). All the victims were children between the ages of three and thirteen. This pattern of attacks on humans is similar to that of another wild felid, the North American cougar. Most (64%) of the victims of cougar attacks are children five to nine years of age, many of whom are attacked when alone (35%) or with other children (43%) (Beier 1991). Unexperienced yearlings and desperate underweight cougars are most likely to attack humans.

The size of prey most important to leopards is difficult to assess because of biases in the methods of diet analysis. Fecal analysis suggests leopards favor smaller prey. However, the remains of smaller prey are easier to identify than those of large prey. Another bias using fecal analysis to determine diet is that leopard feces are seldom randomly sampled. Feces collected from the home range of one individual that happens to favor small prey or from the home

TABLE 11.6 *Leopard Prey Based on Prey Remains Found in Leopard Scats*

Vegetation Zones	Source Data	Size Classes of Prey (%)				Number of Remains
		(0–5 kg)	(5–10 kg)	(10–23 kg)	(23+ kg)	
Lowland rainforest	(1)	45	23	26	6	201
Woodland	(2)	78	0	17	5	294
Woodland	(3)	65	0	23	11	235
Woodland	(4)	74	0	19	5	90
Woodland	(5)	31	0	3	65	86
Dry savanna	(6)	88	0	8	3	177
Cape and Karoo	(7)	73	0	27	—	158
Shrubland	(8)	51	0	49	—	78
Shrubland	(9)	67	0	27	6	33
Shrubland	(10)	25	0	28	47	32

DATA SOURCES: (1) = Hoppe-Dominik 1984, (2) = Grobler and Wilson 1972, (3) = Smith 1978 [Game Park study area], (4) = Smith 1978 [Eastern study area], (5) = This study, (6) = Hamilton 1976, (7) = Norton et al. 1986 [Cedarberg Wilderness study area], (8) = Norton et al. 1986 [Gamka study area], (9) = Norton et al. 1986 [Jonkershoek study area], (10) = Norton et al. 1986 [Wemmershoek study area].
Data are the estimated cumulative, relative frequencies of occurrences of remains of species, classes, or orders of prey of approximate known size in scats of leopards.

range of a young leopard, which may be more apt to take small prey, would bias the data in favor of small prey. Leopards feeding on larger prey may also deposit their feces near kill sites, and unless these sites are sampled, they would be underrepresented in collected fecal samples.

Locating kills of leopards to determine leopard diet would bias the data toward large prey because leopards rapidly and entirely consume small prey. Using leopard movements to locate kills of radio-collared leopards would also favor large prey because leopards are stationary for detectable periods only at kill sites of large prey.

Studies of leopard food habits based on fecal analysis indicate between 31% and 88% of the remains are from small prey weighing less than 5 kg (see table 11.6). In contrast, nine of eleven studies based on detecting kills indicate 45% to 89% of the prey species weigh from 23 kg to 84 kg (see table 11.7). These apparently contradictory conclusions may appear when studies are conducted in the same areas if fecal analysis and kill information are compared. In Matopos National Park, Zimbabwe, fecal analysis indicated more than 70% of the prey remains in leopard scats were from prey weighing less than 5 kg; kill data indicated more than 80% of the leopards' prey weighed more than 23 kg (Grobler and Wilson 1972). In my study areas the differences between sizes of prey taken by leopards were not as great. Fecal analysis indicated more than 30% of leopard prey weighed less than 23 kg, but the kill data indicated nearly 90% of the prey weighed more than 23 kg. Leopards appear to obtain most of their food from ungulates in the 23-kg- to 84-kg-size class (see table 11.7). However, leopards are probably taking small prey between their kills of larger prey.

Although leopards are capable of killing prey three to four times their own weight and, according to one analysis (Packer 1986), rank third among the large felids (the cougar and snow leopard rank first and second) in their ability to kill prey larger than themselves, large prey is seldom killed by leopards (see table 11.6). When species of prey killed by leopards occur in large-size classes, it invariably includes calf or juvenile ungulates. For example, among ungulates weighing more than 84 kg killed by leopards, all eland and wildebeest kills reported in Zimbabwe were calves (Grobler and Wilson 1972); all hartebeest, wildebeest, and gemsbok killed by leopards in the southern Kalahari Desert were calves (Mills 1984); ten of twelve other large ungulates killed in Zimbabwe were calves (Smith 1978), as were five of seven other large ungulates in the Kalahari Desert (Bothma and Le Riche 1984).

The occasional killing of large prey by leopards probably attests to the leopard's strength and ability to kill such large prey rather than its prey preferences. The ability to successfully kill large prey undoubtedly comes with repeated attempts, large prey abundance, and age and experience of the individual leopard. Killing large prey probably includes some risk for leopards. Potential injuries may include punctures and bruises from horns, hooves, tusks, and teeth; broken bones; and even death. Avoiding potentially dangerous large prey is probably to a leopard's advantage, particularly if smaller, less dangerous prey are available. Cougars, which also kill large prey relative to their own weight, do so at some risk and periodically incur injuries (Hornocker 1970).

In Africa the larger ungulates are usually the prey of either larger social felids such as the lion, or other social carnivores such as spotted hyenas and wild dogs. In Asia the tiger fills the same ecological niche by taking large ungulates (Seidensticker 1976a). If the recent phylogeny of felids derived by genetic studies is correct (O'Brien et al. 1987), leopards may have evolved later than present-day lions and tigers. Leopards may have evolved in response to a variety of medium-size prey that were not as significant as prey to larger felids because of energetic hunting and capture costs.

The capability of leopards to flourish and successfully reproduce while feed-

TABLE 11.7 *Leopard Prey Based on Kills of Leopards*

Location	Source Data	Size classes of prey %				Number of Kills
		(0–23 kg)	(23–84 kg)	(84–296 kg)	(296 +)	
Zambia	(1)	19	48	17	0	96
Tanzania	(2)	0	60	11	0	55
S. Africa	(3)	4	83	9	<1	7,465
Tanzania	(4)	0	81	10	0	164
Zimbabwe	(5)	7	60	13	13	15
S. Africa	(6)	2	89	0	0	55
Kenya	(7)	17	50	10	3	30
Zimbabwe	(8)	16	45	26	5	38
S. Africa	(9)	10	10	20	5	20
S. Africa	(10)	9	65	10	0	80

DATA SOURCE: (1) = Mitchell et al. 1965, (2) = Kruuk and Turner 1967, (3) = Pienaar 1967, (4) = Schaller 1972, (5) = Grobler and Wilson 1972, (6) = This study, (7) = Hamilton 1976, (8) = Smith 1978, (9) = Bothma and Le Riche 1984, (10) = Mills 1984.
Data are the estimated cumulative, relative frequencies of occurrences of species among observed kills.

ing primarily on small prey (less than 5 kg) is questionable. Although small prey regularly appears in the diet of leopards, small prey has dominated the diet of leopards in only two study areas thus far (Grobler and Wilson 1972; Smith 1978; Norton et al. 1986). Unless small prey is exceptionally abundant, it may be difficult for female leopards to successfully rear one to three young on a diet of small prey alone. If the energetic costs of rearing young leopards is comparable to that of the female cougar feeding on mule deer (Ackerman et al. 1986), the energetic costs could vary from 6, 9, to 12 kg/day for one, two, to three cubs, respectively. If a female leopard depended for food entirely on rock hyraxes weighing 2.5 kg to 3.5 kg each (Kingdon 1974), she would have to kill two to five hyraxes each day for a least nine to twelve months to successfully feed herself and rear one to three young. This would mean killing a total of from 540 to 1,825 rock hyraxes. Small prey population densities would have to be very high for leopards to maintain reproduction.

In the two study areas where leopards frequently fed on small prey, leopard reproduction appeared to have been low. In three years of following tracks of leopards in Matopos National Park, Zimbabwe, where leopards fed primarily on hyraxes, Smith (1978) reported tracks of leopards together only 37 (5%) of 730 times and did not specifically mention the presence of females with cubs. In the mountains of the southwestern Cape Province in South Africa, where leopards also fed frequently on rock hyraxes, leopard densities were extremely low because of low prey densities and persecution by farmers (Norton and Lawson 1985; Norton and Henley 1987). Females with cubs were not identified nor mentioned in any of the several southern and southwestern Cape Province leopard study areas.

In most habitats occupied by African leopards, medium-size ungulates weighing between 23 kg and 84 kg make up most of the leopard's diet. In habitats where medium-size prey are scarce or absent, such as the isolated mountain ranges in the southeastern Cape Province of South Africa, leopards apparently can subsist predominately on small prey. Although leopards are capable of killing large prey three to four times their own body weight and are capable of subsisting on small prey, it may be difficult for leopards to success-fully rear young under such prey conditions. Leopards living in such precar-ious, prey-limited environments are probably more susceptible to population declines or extirpation than leopards living in areas where medium-size ungu-lates dominate the prey community.

COEXISTENCE WITH OTHER LARGE FELIDS

Leopards also have a wide geographic distribution because they have been able to coexist with other large felids and carnivores. One reason for this coexistence is that leopards can usually escape from other carnivores by climbing trees. Leopards reduce spatial and food competition with other felids by using habi-tats not favored by their larger cousins and by eating different-size prey. Although information on the relationships of leopards and other felids is scarce, it is sufficient to provide some insight into this aspect of leopard ecology.

Importance of escape and protective cover. Escape cover is critical for leopards living among other large, dangerous predators. Leopards are usually able to escape from the attacks of other carnivores because of their smaller size, agility, and ability to climb trees, rocks and cliffs. Where trees occur, leopards can normally escape from spotted hyenas and wild dogs because these predators cannot climb. Climbing trees is probably also successful in most encounters with lions. I once observed a leopard climb a tree to escape from lions; others have also documented that behavior (Schaller 1972; Bertram 1982). Lions were reported to kill six leopards in KNP from 1936 to 1946 and 1954 to 1966 (Pienaar 1969).

The escape strategies used by leopards to avoid lions may contribute to their higher densities, compared to cheetahs, in multipredator communities. Cheetahs, unlike leopards, are unable to escape into trees or cache their kills in trees; instead, they must flee upon the approach of lions or hyenas (Schaller 1972). The main predators on cheetahs appear to be lions and hyenas. Predation on cheetahs by other predators, including leopards, also appears to be an important mortality factor (Eaton 1974). Cheetah cubs are especially vulnerable, and Eaton hypothesized that predation on cheetah cubs is the most important natural limiting factor in cheetah populations.

In the Serengeti National Park, mortality among free-living cheetah cubs was high and caused primarily by predation (Laurenson et al. 1992). Cub mortality even in lairs (dens) was high; only ten of thirty-six litters (27.7%) emerged from the lair. Lions were seen killing two litters and predation was suspected in nine other cases. Female cheetahs also abandoned between two to five litters when they had to move long distances (18 km round-trips) from their lairs to locate herds of Thompson's gazelle. Cheetah cubs were still subject to high mortality once they left the lairs and accompanied the females. Cub numbers were reduced by 52.8% from when they were first seen outside the lair until approximately three months of age. The principle cause of mortality during this period was predation, probably lions and spotted hyenas. One 9.5-month-old cheetah was wounded by a leopard and finished off by a spotted hyena (Burney 1980).

Protective cover is a habitat requirement for leopards to successfully rear young. Leopards often use inaccessible rocky outcrops (koppies) for dens (Adamson 1980); they may also use thick vegetative cover, caves (Brain 1981), or burrows of aardvarks (Bothma and Le Riche 1984) for dens. Leopard cubs are vulnerable to predation because female leopards must often be away from their cubs (Seidensticker 1977; this study). Female leopards are not always present at their dens to defend their cubs, and after the cubs are able to follow the female, trees again become important escape routes. Female 6 sent her small cub up a nearby tree when I approached their kill.

The caching of prey carcasses in trees where it can be eaten at leisure, is another effective other-predator-avoidance behavior of leopards. That this be-havior is a response to the presence of other predators is shown in areas where leopards alone are the dominant predator. In Wilpattu National Park, Sri Lanka, where leopards are the largest predator, they rarely place their kills in trees. Instead, they merely feed on the carcasses at the kill site or drag them into nearby cover. Jackals, the only other significant predator in the park, were not abundant, and the sloth bear did not prey on large game animals (Eisenberg and Lockhart 1972). Usually the mere presence of other predators among leop-ards is enough to cause leopards to cache their kills in trees. Prey-caching behavior by leopards occurs throughout Africa as well as in some parts of Asia where tigers still exist (Seidensticker 1976a).

Habitat separation.   The use of different habitats is another strategy of leop-ards to avoid confrontations with other large felids. Insufficient escape cover may limit the use of some habitats by leopards if other large felids are present. Examples include open plains and perhaps some desert areas where trees and rock outcrops are scarce. Even where trees occur, leopards may sometimes be caught by surprise by lions before they can escape into a tree. Such encounters may be one reason that leopards appear to use denser riparian forest rather than more open savannas in some areas where lions are present (Schaller 1972).

A strategy leopards seem to use to coexist with tigers is to seasonally use habitats less used by tigers. In Chitwan National Park, Nepal, when a female tiger used tall grass areas, a female leopard with an overlapping home range used a nearby forest area. When the tiger used the forest habitat more often, the leopard shifted its use to the tall grass area (Seidensticker 1976a). The leopard also used open burns more frequently than the tiger, but walked along roads infrequently and used different trails and crossings from the tiger. Later, as the population density of tigers increased, they began killing leopards, which were forced into the park with the tigers because of surrounding leopard habitat destruction outside the park (McDougal 1988). In Kanha National Park, India, leopards were not permanent residents where tigers were numerous. Instead, leopards used disturbed habitats near villages at the periphery or outside the park (Schaller 1967).

Differential use of prey.   Leopards also coexist with other large felids by feed-ing on different, usually smaller-size prey. In Kruger National Park, leopards feed on a greater variety of species (primarily smaller prey) than lions. When there were overlaps in prey species (wildebeest, waterbuck, etc.) leopards took primarily young individuals; lions took adults (Pienaar 1969). In Kruger Na-tional Park, leopards (1) fed more on smaller prey than all other predators, (2) fed on prey primarily in the 23 kg to 84 kg weight class, compared to lions, which fed on prey primarily in the 84 kg to 296 kg weight class, (3) fed more on brush-loving species (bushbuck and warthog) in the 23 kg to 84 kg weight class

TABLE 11.8 *Comparative Diets of Major Predators in Kruger National Park, 1936–66.*

Weight Class Prey	Frequency of Occurrence (%) of Kills				
	Lion	Cheetah	Leopard	Wild Dog	Spotted Hyena
0–23 kg					
Other[1]	0.3	3.1	4.5	1.8	1.6
23–48 kg					
Impala	19.6	67.9	77.7	87.0	58.8
Bushbuck	0.3	1.1	3.9	1.1	2.7
Reedbuck	0.3	5.3	2.2	1.4	0.5
Warthog	1.9	0.6	1.4	0.3	0.5
Mountain reedbuck	0.02	—	0.1	0.1	—
84–296 kg					
Waterbuck	10.6	6.7	3.9	2.7	12.8
Greater kudu	11.0	6.8	2.9	4.5	10.2
Wildebeest	23.7	5.0	1.3	0.4	11.2
Nyala	0.1	—	0.4	0.3	—
Tsessebe	0.4	0.7	0.2	0.04	—
Sable	1.5	0.4	0.1	0.2	—
Roan	0.3	0.1	0.04	—	—
Zebra	15.9	1.8	1.2	0.1	0.5
296 + kg					
Eland	0.5	0.1	0.2	0.1	—
Buffalo	9.3	0.1	0.2	—	1.1
Giraffe	4.0	0.2	—	—	—

SOURCE: Pienaar 1969.
[1]Includes grey and red duiker, steenbuck, Sharpe's grysbuck, klipspringer, suni, bushpig, chacma baboon.

than did cheetahs and wild dogs, which fed more on a grass-preferring species (reedbuck), and (4) fed on less prey in the 84 kg to 296 kg class than lions, cheetahs, and spotted hyenas (see table 11.8).

In Serengeti National Park, only two of twenty species of prey were shared by both leopards and lions, and in neither case in large numbers (Bertram 1982). Lions commonly took adult large prey such as zebra, wildebeest, and kongoni; leopards took medium-size prey such as impala and Thompson's gazelle. Leopards also took a wider variety of prey than lions. Leopards preyed upon at least thirty-one species (Bertram 1982); lion kills included only twenty-two species (Schaller 1972). In the Kalahari Desert, where there were fewer prey and leopards and lions fed on the same ungulates, leopards again took young individuals and lions adults (Bothma and Le Riche 1984).

Leopards and tigers sharing the same habitat also appear to use different prey. In Chitwan National Park, Nepal, leopards most often killed smaller prey in the 25 kg to 50 kg weight class (wild pig, axis deer, and muntjac), and tigers killed prey primarily in the 50 kg to 100 kg weight class (sambar and hog deer) (Seidensticker 1976a). Where tigers are absent or have been exterminated, leopards feed on a wider variety of prey (Muckenhirn and Eisenberg 1973) and may even become more numerous (Seidensticker 1986).

The relationships between leopards, tigers, and smaller felids in the rain forests of southeast Asia (Seidensticker 1986) exemplifies some of the previously discussed interfelid relationships. In extensive tracts of monsoon forest and savanna in Thailand/Burma (Myanmar), where clouded leopards, Asian golden cats, and marbled cats are absent and do not compete with leopards for smaller arboreal and terrestrial prey, the habitat is good for leopards. However, in Borneo where these smaller felids occur, leopards are absent. In Java, where these smaller felids are absent, leopards have apparently expanded their range into the rain forest, but only after tigers were eliminated. In Sumatra and in the Sundarbans mangrove forest, where tigers occur, leopards are absent.

Leopards, then, coexist with other larger, potentially dangerous felids and other carnivores by climbing trees to escape attacks, by using protective cover to rear their young, by using habitats less used by other larger felids, and by preying on more and smaller species than lions and tigers. Smaller arboreal felids may compete with the leopard for smaller prey, especially in rain forest habitats where much of the prey is arboreal. Leopards appear to do best in environments with an abundance of ungulate prey in the medium-size range (23 kg to 84 kg) where competition with other felids, both large and small, is minimal.

Behavioral Adaptations

Leopards, like the majority of the 253 living species of carnivores (Wilson 1975) and the vast majority of felids, are solitary. As solitary predators, leopards are able to efficiently utilize an area's resources, locate or avoid each other as needed, and successfully rear young; they have a simple, practical social system for the use of space. Leopards maintain this system by effective communications. The leopard's simple but effective land tenure system is proposed as another reason leopards have been successful throughout their range.

AN EFFICIENT, FUNCTIONAL LAND TENURE SPACING SYSTEM

The following analysis is based primarily on observations obtained during this study. The land tenure system of leopards can be visualized as three distinct, superimposed layers of home ranges of leopards. Each layer contains a mosaic of home ranges of leopards of the same sex and similar social rank. The two major social ranks of leopards include: (1) residents that are sexually mature, usually older individuals that do most, if not all, of the breeding (dependent offspring are included with the females because they share her home range), and (2) floaters (Barash 1977) that are usually younger leopards in excess of the number of leopards that can bred and successfully rear young in a given area. Leopards that are floaters do not contribute their genes to the next generation unless they can change their social status and become resident. These floaters can be: (1) young independent leopards that have not yet dispersed from their natal ranges, (2) leopards from other regions temporarily passing through the area (transients) either in search of vacancies among the residents or challenging a resident for the right to breed, or probably less often (3) leopards of any age that have lost their dominant, resident social status for some reason, such as injury, illness, or fighting. Leopards interact socially with other leopards of the same sex within their own social class as well as in the other social class. These interactions may range from subtle modes of communication, such as scent marking and vocalizations, to outright aggression.

The first and most important layer of home ranges in the land tenure system of leopards is that of the resident female leopards. Each resident female, and eventually her offspring, may occupy exclusive to overlapping home ranges depending on habitat quality, but females seldom associate with each other. They seem, instead, to avoid each other by using scent marks and vocal signals to inform each other of their current whereabouts. Prior rights to breed and rear young in an area appear to be acknowledged and respected by later-arriving females.

A second layer within the land tenure system of leopards contains much larger and generally less overlapping home ranges of resident male leopards. Their home ranges are in effect superimposed over the home ranges of the resident females. Because resident males apparently compete with each other for the right to breed with the resident females in the area, their home ranges overlap less than those of females. Resident females are usually occupied rearing offspring, so resident males must constantly patrol their home ranges to assess the changing reproductive status of the resident females. Males must be available to breed immediately with females if they are receptive. Males probably determine the reproductive status of a female by the scent she leaves in her marking fluid or her urine (Brahmachary and Dutta 1981, 1987, 1991), her calls, and her response to their calls. Resident males must also patrol their home ranges to advertise their presence to other resident and floater males, who are constantly searching for opportunities to expand their range, occupy vacancies, and breed with the resident females.

The third layer of home ranges within the leopard land tenure system is occupied by floaters of both sexes. It may be seen as being superimposed over the other two layers. A resident female's offspring remain in her range until they become independent. Then, via exploratory movements, these offspring, males earlier than females (Le Roux and Skinner 1989), gradually extend their natal ranges beyond that of their mothers into the home ranges of adjacent leopards.

Resident female leopards appear to establish their home ranges within the best available habitat. Their home ranges probably contain the greatest abundance of available prey, protective cover to rear young, and escape cover to flee from other predators. The size of a resident female's home range is probably determined by the locations of den sites and the minimum amount of prey needed by the female and her offspring during the leanest time of the year. Where prey densities are low, female home ranges will be large, and where prey is abundant home ranges will be smaller. Home range size may also depend on the locations of suitable places to hide and rear offspring.

In communities of large predators, survival of the resident female leopard and her offspring may be dependent on her ability to escape from other large predators. Large trees, dense vegetation, rocky cliffs, and koppies all provide needed protective cover for a female leopard rearing young (Stevenson-Hamilton 1947, Schaller 1972, Adamson 1980, Brain 1981, this study); they are important and necessary habitat components of her home range. In the Nwaswitshaka River leopard study area, the home ranges of resident females were oriented along the river and its major tributaries. Prey densities were the highest there, and resident females spent much of their time within this vegetation zone.

The intense utilization of the best available habitat by breeding, resident female leopards, rather than by resident male leopards, is to the advantage of the females and their offspring. The resident female spends most of her time caring for offspring, and she must be able to obtain sufficient food with the least expenditure of energy. This is especially critical during the period when she leaves her young behind to hunt for prey. In prey-poor habitats, females probably spend more time away from their cubs, thus exposing them to danger from other predators.

If suitable habitat is already occupied by a resident female leopard, newly arriving females have at least two options. One is to continue to search for suitable habitat, unoccupied by a resident female, where opportunities to breed may be higher. Another option is to remain in or near their natal range where their opportunities to successfully breed and rear young may be limited. In the Londolozi Game Reserve, near Kruger National Park, two female leopard offspring remained well within their natal ranges for at least thirty months (Le Roux and Skinner 1989). If a nonresident female is bred by the resident male, her chance of successfully rearing young may be low because she and her offspring must compete with the resident female, which may be her mother, for protective den sites and food. Some resident female leopards may eventually have to evict their own female offspring. When one female leopard had two chance encounters with her thirty-month-old female offspring, little aggression was displayed (Le Roux and Skinner 1989). However, fighting between female leopards does occasionally occur and may even result in death (Le Roux and Skinner 1989). It is probably to the resident female's advantage to avoid other leopards, especially other female leopards. Such avoidance can be facilitated by scent marking, vocalizations, and remembering areas where past encounters have occurred with other leopards. Avoidance and recognition of prior rights keep females separated from each other and other leopards in time and space.

Resident male leopards probably attempt to breed with as many female leopards as practical, regardless of the female's social status. Resident male leopards must, therefore, actively exclude other potentially breeding males from their area. Active exclusion of other males is accomplished by constant scent marking, vocalizations, direct confrontations, and on rare occasions, actual fighting. Resident males must frequently visit and make their presence known throughout their home ranges. They have to be familiar with all the females in their area because female reproductive status changes as offspring mature and become independent. In environments where there is no single peak breeding period, as in my leopard study areas, females may become receptive to breeding over an extended period of time. Resident male leopards probably have to periodically assess the female's reproductive status in such environments because they cannot rely on physical clues such as photoperiod and temperature.

The spatial consequences of the reproductive strategy of resident male leopards is that they occupy large areas. The maximum size of their home ranges is probably determined by the size of the females' home ranges rather than prey conditions. In high-quality habitats where the resource requirements of resident female leopards can be met in relatively small areas, males will be able to include many females within their own large home range and still exclude other males. In poor quality habitats, where the home ranges of females are also large, fewer females will be included in a male's home range. A threshold may occur in some poor quality habitats and an area may only support a single male and female resident leopard.

The maximum size of a resident male leopard's home range is probably also influenced by his mobility, his aggressiveness, and the environmental characteristics that influence the effective life span of his scent marks. Male leopards must be physically capable of traveling great distances in short periods of time to regularly visit and mark important places, such as key trails along boundaries of their home ranges. They must also be aggressive in order to exclude other males through scent marks and vocalizations.

Because this reproductive strategy of male leopards probably selects for a larger body size than females, males require more resources and appear to have several advantages over females for obtaining food. Because of their larger size, male leopards are probably more capable of killing larger prey and defending kills and scavenged carcasses. Scavenging from large carcasses can be dangerous in multipredator communities. Female leopards were never observed scavenging in my study areas.

Dispersal of subadult male leopards from their natal home ranges and their influx into surrounding areas appears to depend upon the availability of prey and population density of leopards. An abundance of prey in the natal area might encourage a stay-at-home floater strategy, particularly if exploratory movements indicate the surrounding areas are already occupied by resident males. Several of the subadult males in my study areas exhibited this behavior. In a nearby area, a subadult male leopard dispersed from his natal home range at the age of fourteen months, returned for two days at the age of nineteen months, and then left again (Le Roux and Skinner 1989). To be successful in this strategy, the floater would have to remain inconspicuous, to avoid the chance of eliciting an aggressive response and eviction by the resident male. Ways to remain inconspicuous are to avoid the resident male and avoid vocalizing and scent marking—all behaviors exhibited by subadult males in my leopard study areas.

Subadult males that are floaters can physically and sexually mature while they are avoiding the resident male. This maturation may increase their chance of later assuming resident status if the resident male dies or is injured. Avoiding the resident male may not be difficult for a floater because the resident male is constantly advertising his presence.

Subadult female leopards do not appear subject to the same social pressures regarding reproduction as males. They do not disperse as early in life nor as widely as subadult males and may not compete with each other for access to males. If prey and cover are abundant, females do not seem to intensively compete with each other for those resources. Because subadult female leopards seem to be more sedentary and less likely to disperse over wide areas, dispersing subadult males are important to leopard populations because they ensure inbreeding is kept to a minimum. The reproductive strategies of females address the problems of successfully rearing young; those of males reflect the competition for breeding with females.

Fighting among leopards was rare in the leopard populations I studied. Although some leopards had scars and wounds, I could not determine their origin. However, one leopard was severely bitten by a ratel; another was apparently wounded while capturing an impala; and another leopard killed dangerous hyenas and ate them, which suggested that leopards are periodically wounded while capturing prey. Actual fighting among highly specialized carnivores is not advantageous to their physical well being and survival (Ewer 1968, Hornocker 1969). Corbett (1947) remarked that fights between leopards were unusual because they invariably kept to their own areas. He described one fight between two leopards as having two rounds of five and ten to fifteen minutes, respectively, with a third much shorter round. Then, the apparent intruding male leopard—the man-eating leopard of Rudraprayag—left the area and returned to its previously used area.

Fighting among three radio-collared male leopards in Tsavo National Park, Kenya, was suspected but not confirmed (Hamilton 1976). All suspected fights involved the same male leopard fighting with adjacent radio-collared males. Exploratory movements, mating with a female, smaller size, and reaction to encounters with other males all strongly suggest this particular leopard was a younger male attempting to establish itself as a resident in an area already occupied by two resident males leopards. This hypothesis is strengthened by information that during the only observed encounter between the two established resident male leopards, no fighting occurred. But both resident males apparently fought with the smaller male. In the Londolozi Game Reserve, an encounter between two apparently unrelated female leopards and between an adult female and a male leopard resulted in at least brief fights, with both individuals suffering superficial injuries in the latter encounter (Le Roux and Skinner 1989). Fighting among leopards in a small population in the Judean Desert was also observed (Ilany 1990; appendix C).

Observations of leopards fighting each other can be summarized as follows: (1) fighting among leopards appears uncommon; (2) male leopards may fight more frequently than females, but fights between females do occur; (3) fights are usually of brief duration; and (4) fights seldom result in the death of either of the combatants. Apparently most fights between males occur when one male is attempting to establish himself in an area already occupied by a resident male. Fighting may also be more intense or prolonged if competing males are of the same size or exhibit equivalent aggressive behavior. In leopard populations where resident males occupy home ranges for long periods and, thus, have a distinct advantage over newcomers because of their familiarity with the surroundings and each other, little fighting appears to occur. In leopard populations with a high turnover rate of resident males, fighting may occur more frequently.

A solitary way of life appears to be the most efficient way of locating and using medium-size to small, relatively sedentary, but dispersed, prey. This way of life manifests itself among leopards as a mosaic of overlapping home ranges, with the best habitat occupied by females to ensure their survival—and that of their offspring. The size of the home range of females is apparently related to the quality of the habitat. Important habitat components for females are abundance, distribution, and availability of both prey and protective cover. The size of males' home ranges depends on the size of the area they can regularly monitor, thus excluding other males from having reproductive access to the females. The number of females within a male's home range is dependent on habitat quality. In high-quality habitats that will support more females per unit area, males will be able to include more females within their exclusive area. In lower quality habitats that support fewer females per unit area, males can only include a few females within their large exclusive areas. Females are a necessary component of a successful male's home range.

Overlap between home ranges of leopards varies with the abundance and distribution of prey and cover, and the size of the home ranges. This land tenure system allows young leopards to remain in the same area used by their parents as long as food is abundant and they do not compete reproductively with their parent or other resident leopards. When resources become scarce or when the young leopards become sexually mature, they face increasing social pressure from the residents to disperse. Dispersing may increase the chances of survival for young leopards, minimize the chances of inbreeding, and allow leopards to occupy vacancies in occupied habitats or to colonize new habitats. Although fighting may be one way that resident leopards keep other leopards from occupying their home ranges, such instances appeared to be uncommon in my study areas. Instead of fighting, leopards used various modes of communication to space themselves and thereby avoided injuries.

EFFICIENT MODES OF COMMUNICATION

Leopards use three modes of intraspecific communication: visual, auditory, and olfactory. Visual signals are perhaps the least used because leopards are often most active in darkness or in dense cover. Visual signals are probably very effective during close encounters and may, as in the case of scrapes, enhance the effectiveness of olfactory signals.

Auditory signals are probably used more often by leopards than visual signals, but are effective only in special circumstances because of their short duration. Too frequent vocalizations may attract other dangerous predators. Leopards appear less vocal in areas where lions are abundant (Schaller 1972, this study) and more vocal where lions or other large predators are less abundant or absent (Eisenberg and Lockhart 1972, Hamilton 1976).

Olfactory signals seem to be the most important mode of communication among leopards. Olfactory signals transmit through darkness and around obstacles; they have great energetic efficiency, potential for long-range transmission, and long life spans; they can be renewed; and they have the capacity to transmit information after the sender has gone (Wilson 1975). In addition, highly specific information can be encoded in the signal by altering its chemical composition. And olfactory signals can be used to mark specific places.

A chemical analysis of the scent of leopard cubs (adults were not available) and tigers (Brahmachary and Dutta 1981, 1987, 1991) suggests that leopard scent has many of the same properties as tiger scent. This scent contains "tigeramine," a phenylethylamine, and other amines that are believed to be a chemical blueprint of the species and perhaps individuals. Tigeramine is a high molecular weight (121) pheromone whose long-lasting effects may result from being fixed in lipids from the anal glands mixed with the urine. The odor of this scent lasts at least ten days and persists even after rainfall. Zoo tigers could perceive the odor of scent from a distance of 0.3 m to 0.6 m. Various amines, lipids, and hormones in tiger scent are believed to characterize the scent of tigers with the following information: (1) species, (2) individual, (3) sex, and (4) status of sexual condition.

My study suggested that resident leopards did the majority of vocalizing and scent marking. Sexually immature and nonbreeding leopards seldom vocalized or scent marked. Observations of other large carnivores indicate that established residents or dominant individuals are responsible for most of the socially oriented or spacing signals. Schaller (1972) did not see male lions spray urine until they were 3 1/4 to 3 1/2 years old and believed most lions did not scent mark until they were almost two years old. The youngest observed lion (a female) that scraped was twenty-two months old. Among wolves, lone wolves and wolves outside established packs never howled in response to artificial calls; they either remained silent or moved away when they heard the calls of other wolves (Harrington and Mech 1979). Lone, nonbreeding wolves rarely scent marked, and they defecated and urinated infrequently in roads and trails used by established packs (Rothman and Mech 1979).

The advantages of primarily resident leopards vocalizing and scent marking are (1) it provides newcomers, or floaters, with a means to determine if an area is already occupied by resident leopards; (2) it provides site specific information needed by nonresidents or floaters to avoid resident leopards on a short-term basis; (3) it permits simultaneous use of an area by both residents and floaters; and (4) it allows resident leopards to confine most of their signals to narrow zones and toward other residents, rather than broadcasting signals throughout their home range.

Leopards apparently have the flexibility to use various signals in different environments. Urine spraying by leopards is frequently seen in warm, moist environments (Eisenberg and Lockhart 1972; this study). The leopards I observed sprayed urine more frequently during wet than dry seasons. In cool, dry conditions marking with feces or with visual scrapes, with or without a small amount of scent (urine and anal scent) appears more common (Bothma and Le Riche 1986). In arid, desiccating environments scrapes alone, or with only a small amount of scent, would be more effective and conserve water. Effectiveness is further enhanced in dry, open environments by the addition of feces to the scrape, a strategy apparently used by leopards in the dry Karchat Hills of Pakistan; by snow leopards in open, windblown environments (Schaller 1979); and by bobcats in rocky, arid, and windblown habitats (Bailey 1972). Leopards in my study areas deposited feces on roads more often during dry than wet seasons. Bobcats used their feces to demarcate dens and travel routes in an arid environment (Bailey 1972), and cougars scraped at frequently used sites in a relatively low-precipitation mountainous region (Seidensticker et al. 1973).

Different signals are probably used by leopards to provide specific types of information. Vocalizations may indicate that (1) home range is currently occupied by a resident leopard; (2) home range is still occupied by a specific individual (this assumes that individual recognition is possible); (3) an individual is sexually receptive for courtship or mating; or (4) a leopard is moving through a common travel route or zone of home range overlap. Examples from my study areas of each of the above included (1) leopards vocalizing at random in their home ranges, (2) my ability to distinguish several individuals solely by their calls, (3) the union of resident male M14 and resident female F11 after F11 was attracted to M14's call, and (4) F12 and her mother avoiding F4 by negatively responding to F4's calls when she passed near their kill.

The response of leopards to the vocalizations of other leopards may depend on their social status. In mammals, resident status is characterized by great self-confidence, aggressiveness, and familiarity with the area the resident inhabits (Ewer 1968). Floater, or transient, status is characterized by insecurity, submissiveness, and unfamiliarity with the area. When a resident leopard hears a call that it recognizes as that of a neighbor, it may ignore it or respond to it with its own call. The same call heard by a floater may evoke uneasiness, promote silence, and cause the individual to move out of the area to avoid contact with the resident. Resident lions seldom move toward roars from other lions (Schaller 1972), but trespassers into another lion pride's territory may alter their movements in response to roars (Schenkel 1966). Similarly, packs of wolves howl at kills, but lone wolves without social status seldom howl (Harrington and Mech 1979).

Site specific information is provided by scent marks and scrapes of leopards along trails, near kills, and perhaps along boundaries of home ranges. Resident mammals scent mark familiar routes because it helps them maintain home range familiarity (Ewer 1968). The home ranges of leopards are characterized by a network of travel routes (game trails, dry stream bed, roads) that the leopard regularly uses. Because residents and floater leopards may use the same network of trails, they constantly send and receive signals as they move about the area. Floaters using common travel routes can assess that resident leopards are already occupying an area and may leave little of their own scent behind as they use these common trails. Residents may only make subtle adjustments in their movements in response to the scent marks of their neighbors, but floaters may drastically alter their movements in response to the same scent mark.

In summary, as a solitary predator utilizing large areas and often moving in darkness or in dense cover, leopards have efficient modes of communication. Primarily used to keep leopards separated in space and time, their signals can also be used to enhance encounters. Visual signals, which are the least used, are probably the most important at close range. Auditory signals are used more frequently than visual ones, but must be used with caution because they can alert prey and other dangerous predators. Olfactory signals seem to be the most important mode of communication among leopards. These signals are long-lasting, site specific, and can probably provide specific information about the individuals that leave them. Resident leopards seem to use auditory and olfactory signals more frequently than nonresident leopards, but their signals are probably detected and evaluated by nonresident leopards. Leopards also appear to have the flexibility to use one or more visual, auditory and olfactory signals to increase signal efficiency in different environments.

TABLE 11.9 *Comparative Ecological and Behavior Characteristics of Large Solitary Felids*

Characteristic	Leopard	Tiger	Snow Leopard	Jaguar	Cougar
ECOLOGICAL					
Habitats	Brushland, forests	Forests, tall grass	Mountains, alpine areas	Rainforest, swamps	Forest deserts
Principal prey	Antelope, small prey	Large, medium deer	Sheep, goats, small prey	Peccary, small prey	Deer, elk, peccary
Ranges of home range sizes[1]:					
Males (km^2)	16.4–388	44.7–60.2	11.7–22.7	28–168	41–453
Females (km^2)	5.6–487	12.8–19.7	19.7–38.9	10–70	31–373
Ranges of densities (#/100 km^2)	0.6–16.4	0.2–4.7	0.2–10.0	1.6–6.6	1.0–2.3
BEHAVIORAL					
Males territorial	Yes	Yes	No ?	Sometimes ?	Sometimes
Females territorial	Sometimes	Sometimes	No ?	Sometimes ?	Sometimes
Common scent marks	Urine/Scrapes	Urine/Scrapes	Scrapes	Scrapes ?	Scrapes
Spacing vocalizations	Rasp call	Roar	Yowl ?	Pulsed roar	None
Fighting common	No	No	No	Unknown	Sometimes

[1] Radiotelemetry studies only.

Comparison of Leopards to Other Large Solitary Felids

Leopards share many ecological and behavioral characteristics with other large solitary felids because they often are utilizing similar ecological niches (see table 11.9). These niches are usually forested habitats or forest edges (Eisenberg 1986) characterized by closed dense vegetation and inhabited by relatively small or medium-size, diverse, dispersed prey. More open niches, characterized by grassy savannas and larger or herding prey, were exploited by felids that developed different strategies. These strategies, exhibited by cheetahs and lions, included cursorial pursuit of prey, coordinated group hunting, and protection of highly visible kills from other social scavengers and predators. In contrast, leopards and other solitary felids rely on camouflage for concealment, employ stealth, and take advantage of cover to stalk and capture by surprise. These felids developed individual hunting tactics because their prey is dispersed in closed environments, and close visual coordination of position and movements such as group hunting tactics are difficult to employ.

One of the most primitive, larger, still-living, and least known solitary felids to exploit a closed-environment-dispersed-prey niche is the clouded leopard. Clouded leopards probably split off the felid evolutionary tree from five to seven million years ago (Collier and O'Brien 1985; O'Brien et al. 1987), or about three million years before the common leopard is recognized as a species. Clouded leopards appear to be especially adapted to extremely moist, hot, tropical forests and have several traits (long bushy-tail and flexible ankle joints) that suggest arboreal ability (Kitchener 1991). Where they occur together, common leopards probably compete for prey with, but dominate, clouded leopards. In Thailand clouded leopards avoided using trails and roads used by tigers and Asiatic leopards (Rabinowitz 1989, 1990, 1991; Rabinowitz and Walker 1991). In Borneo, where leopards and tigers are absent or scarce, clouded leopards spent considerable time on the ground (Rabinowitz et al. 1987). The common leopard appears much more habitat adaptable and takes a greater variety of prey than the clouded leopard. Whereas common leopards hunt primarily on the ground, clouded leopards seem to be more adept at taking arboreal prey in dense canopy forest (Rabinowitz 1989, 1990, 1991; Rabinowitz and Walker 1991). Not enough is known about clouded leopard social organization to compare it to that of the common leopard.

Snow leopards apparently split off the felid evolutionary tree about three to four million years ago (Collier and O'Brien 1985; O'Brien et al. 1987), or one to two million years before the common leopard. Today, snow leopards inhabit a harsh, windy, seasonally cold, high altitude environment that common leopards avoid. Common leopards have been encountered at about 3,650 m within snow leopard habitat in Nepal (Hillard 1989). But usually when distribution of these two leopards overlaps, common leopards are more likely to be found at lower slopes or valley bottom, which are more densely vegetated (Schaller 1977). Snow leopards seem to prefer deeply dissected terrain broken by cliffs, ridges, and gullies (Jackson and Ahlborn 1988a, 1988b, 1989). In some areas of Nepal, however, both species of leopards may periodically use the same areas. Common or forest leopards are considered to be much more aggressive than snow leopards (Hillard 1989), and larger. Both species may periodically take the same prey, but one of the favorite prey of snow leopards, the blue sheep, is seldom taken by common leopards. Where they occur together in mountainous habitat, common leopards seem to take wild goats and domestic livestock, whereas snow leopards take blue sheep, marmots, and livestock (Schaller 1977).

Common and snow leopards apparently use similar methods of communication. In one area in Nepal, snow leopards, like common leopards elsewhere, frequently scraped along commonly used trails. The intensity of scraping increased during the mating season when nearly half of the scrapes were remarkings of existing scrapes (Ahlborn and Jackson 1988). Like leopards, both sexes of snow leopards scent marked by repeatedly visiting and marking the same rocks, some of which may have been marked by several generations of snow leopards. Snow leopards apparently yowl loudly to locate each other during the mating season. But they do not have a rasping call comparable to that of common leopards to advertise their presence.

In the only snow leopard area intensively studied thus far, the leopards did not display a land tenure system comparable to that of other leopards that have been studied thus far. Rather than maintaining home ranges with little to moderate overlap, at least four snow leopards shared the same general centers of activity to which they frequently returned (Jackson and Ahlborn 1988a, 1988b, 1989). But, like common leopards, snow leopards staggered their visits to commonly used areas and spaced themselves at least 1.6 km apart on the same day.

Some of the differences in the land tenure system noted in the snow leopard study may have been related to the ages of the monitored snow leopards. Most were young individuals and young snow leopards may remain with the female for nearly two years (Jackson and Ahlborn 1988a, 1988b, 1989). These particular snow leopards may have shared the same area because they all grew up in the same vicinity, were related, or had not yet dispersed.

The nature of the terrain and prey distribution also influenced snow leopards' use of space. Because of the steep gorges and high prey-poor ridges and peaks, snow leopards had to share common travel corridors. The preferred-use areas they shared supported high densities of blue sheep, the snow leopard's principal prey. Exclusive use of home ranges may be difficult where prey is seasonally clumped. This pattern was exhibited by snow leopards and by some leopards in my study areas. After leopards have been studied in several harsh environments and snow leopards have been studied in more amenable environments, more similarities may be discovered in their social organizations.

Leopards once shared a substantial portion of their range with tigers and have many common ecological and behavior characteristics. Both felids are solitary and inhabit diverse habitats but prefer dense vegetation; they may feed on the same prey. However, in some situations, tigers appear more likely to prefer warm, moist, dense grass-dominated or forest vegetation, whereas the leopard is apt to be found in more open, drier habitat types (Schaller 1967; Seidensticker 1976a; Sunquist 1981). Tigers also can take larger prey that is too large to be taken by leopards.

Tigers typically compete with, but dominate, leopards where they occur together (Seidensticker 1986) because of their similar ecological attributes but disparate size. Leopards do well in areas where tigers are scarce or absent. But where tigers occur, leopards have to switch to perhaps less-preferred or smaller prey, alter their behavior to avoid tigers, and reduce use of or avoid habitats favored by tigers (Schaller 1967; Seidensticker 1976a).

Leopards and tigers appear to have similar land tenure systems based on a common theme of females competing with each other for space to hunt prey and feed their young, and males competing with each other for access to estrous females (Eisenberg 1986). To achieve spacing, both species use comparable modes of communication, although vocalizations of the leopard and tiger differ. Male and female leopards and tigers spray urine or scent, scrape, and occasionally use feces as signals in a comparable manner, although the details have been much better described for the tiger (Smith et al. 1989). Female leopards, like female tigers (Smith et al. 1987), compete with each other for prey and for sites to rear young. Male leopards, like male tigers (McDougal 1977; Sunquist 1981), compete with each other for the right to breed with females. The result of this intrasexual competition is generally a land tenure system with nonrelated, resident females occupying exclusive to slightly overlapping ranges where prey and rearing sites are evenly distributed and prey densities high. Where prey are clumped and densities low, overlapping ranges occur.

Adult breeding males generally hold exclusive ranges, but the size of their range and number of breeding females they overlap probably varies with environmental quality. Variations in this general theme may occur in harsh or perhaps human-altered environments. When leopards share an area, they still space themselves in time and are rarely close together. Inconsistencies to this generalized system may appear if the status of individuals are unknown. For example, some females with highly overlapping ranges may either be current or former offspring in the process of establishing themselves as breeding residents near or within their natal ranges. Other females and males may be nonbreeding floaters awaiting vacancies to occur in the land tenure mosaic.

Data from thirty-six breeding tigers (fourteen males, twenty-two females) gathered between 1972 and 1989 in the Royal Chitwan National Park, Nepal, provided many insights into the tiger land-tenure system, breeding success, and cub survival. Future long-term studies of leopards will likely reveal many similarities between them and tigers. All breeding female tigers remained in the same territories throughout their reproductive lives, an average of 6.1 years. Two were eventually driven from their territories by other females. Once driven from their territories, these two never reproduced again but lived 13.5 and 15.5 years in the wild. Male tigers did not become reproductively successful until an average age of 4.4 years (range 3.4 to 6.8 years). The average productive life of male tigers was only 2.8 years, only half that of females, and ranged from 7 months to 6.3 years. The breeding territories of males were established by direct takeovers and often expanded after fights with adjacent males. A male tiger that had seven females within his breeding territory for four years fathered twenty-seven offspring to dispersal age. Other males displaced from their breeding territories in less than one year did not father any offspring. When the study area supported a stable population of resident male tigers (1976–77), 90% of the cubs in the area survived and no infanticide was observed. When resident males died and new males took over their territories, cub survival declined to 33% because infanticide became widespread.

In the Western Hemisphere jaguars and cougars are ecological equivalents to the leopard. Although similar in appearance to leopards, jaguars are heavier and stockier. They occupy many comparable habitats. Jaguars appear to favor dense, moist forest habitats, often near water, and their principal prey is apparently peccaries and capybaras (Guggisberg 1975; Schaller and Crawshaw 1980; Rabinowitz and Nottingham 1986; Crawshaw and Quigley 1991). Unlike leopards, jaguars spend considerable time in or near water. Like leopards, jaguars are not highly arboreal, but do use trees as resting places. Because jaguars are the dominant predator in carnivore communities, they do not have to cache kills in trees to protect them from larger predators and scavengers. For the same reason, jaguars may not be as agile as leopards; they probably seldom have to escape into trees for protection from other predators.

Although larger than leopards, jaguars generally take smaller prey. The ratio of maximum prey size to female weight indicates that female leopards may take prey up to 4.0 times their own weight, compared to only 2.3 for female jaguars (Packer 1986). In Belize jaguars preyed primarily on dense populations of small, sedentary prey such as armadillo, pacas, and red brocket deer (Rabinowitz and Nottingham 1986). But in the Panatal of Brazil, larger prey such as cattle and capybara were taken (Schaller and Crawshaw 1980).

Land tenure systems of jaguars appear to be comparable to that of leopards. In an area where cattle made up the majority of prey biomass, the home ranges of neighboring female jaguars overlapped; male ranges were quite discrete, although they were occasionally intruded upon by neighboring males and transients (Schaller and Crawshaw 1980). In a natural, prey-rich area, the home ranges of female jaguars did not overlap, but those of males did (Rabinowitz and Nottingham 1986). In a seasonally flooded area where jaguars preyed upon large, highly mobile prey, the home ranges of female jaguars again overlapped (Crawshaw and Quigley 1991). These variations in jaguar land tenure systems appear comparable to those observed among leopards. For jaguars (Crawshaw and Quigley 1991) and leopards (this study), the complexity of the habitat; sites to rear young; the density, size, distribution, and mobility of prey; and prey availability between seasons determines the importance and limitations of marking or defending an exclusive area against conspecifics of the same sex. Regardless of the land tenure system that develops, both jaguars and leopards apparently are capable of sharing areas by remaining separated in time and space. Fighting is rare.

Jaguars and leopards use comparable modes of communication, the types and frequencies of signals varying with the environment. In some areas, jaguars seldom vocalize or scent mark (Schaller and Crawshaw 1980); in others, scent marking by scrapes and feces is common (Rabinowitz and Nottingham 1986), especially in areas of home range overlap. Presence of humans and degree of persecution, habitat and precipitation patterns, density of conspecifics, and sizes of home ranges are some variables that influence modes of communication among both species.

From the standpoint of adaptability, the cougar, or puma, may have more in common with the leopard than the jaguar. Like the leopard, the cougar has adapted to a wider variety of habitats and prey communities than the jaguar. Cougars are found in South and North America, in habitats ranging from tropical forests, woodland savannas, deserts, and temperate forests, to seasonally snow-covered mountains and coniferous forest. Like the leopard, the cougar's range was once even more widespread than it is today. Given legal protection, the cougar, like the leopard, also adapts well to human presence (Beier 1991), primarily because it is secretive and seldom seen.

The diet of cougars, like that of leopards, shows great variation in prey species and size. In Peruvian tropical rainforest, the main prey of pumas were small cricetid rodents, pacas, and agoutis (Emmons 1987). In Chile pumas fed primarily on introduced European hares and ungulates such as guanacos, guemuls, and domestic sheep (Yanez et al. 1986). In North America pumas in an arid environment fed primarily on desert mule deer, peccaries, lagomorphs, rodents, and porcupines (Leopold and Krausman 1986). Further north, in seasonally snow-covered, mountainous habitat, wapiti (elk) and mule deer were the most important prey (Hornocker 1970). The principal prey of an isolated population of *Felis concolor coryi* using mixed swamp and hammock forest in the Florida Everglades (Belden et al. 1988), was wild hog, white-tailed deer, raccoon, and armadillo (Maehr et al. 1990).

Unlike leopards, cougars are capable of taking much larger prey, compared to their own body weight. For female cougars the maximum of the ratio of prey size to female weight is 6.2; the mode is 2.4. For female leopards the maximum is 4.0; the mode is 1.1 (Packer 1986). When feeding on large prey such as wapiti, however, cougars primarily (53% of kills) took young less than a year old. Large male wapiti 5.5 years or older made up less than 5.0% of the wapiti kills (Hornocker 1970). This selection of prey size by cougars is similar to leopards, which also take mainly younger individuals of the larger prey species.

The land tenure system of cougars, like that of leopards, demonstrates a flexibility dependent on the density and sizes of their home ranges and the sizes, densities, seasonal distribution, and mobility of their prey. In the Idaho Primitive Area, where the seasonal distribution of large, mobile prey varied, female cougars shared home areas, but those of neighboring males were exclusive (Seidensticker et al. 1973). This spatial organization of cougars was also found in similar environments in Utah (Hemker 1982) and Wyoming (Logan et al. 1986). In California, however, where nonmigratory prey densities were higher, male cougars shared overlapping areas and the areas of two females were exclusive (Hopkins et al. 1986).

The environmental and population variables that determine land tenure systems in leopards and cougars are not fully understood. But two patterns often appear regardless of the degree of home range overlap among males or females of both species. First, when sharing common areas, leopards and cougars temporally separate themselves by direct avoidance or with the use of urine and scent spraying, scrapes, feces, and vocalizations. Second, this spacing, or mutual avoidance, tends to reduce aggressive encounters and fighting. Although both male and female leopards may sometimes fight or even kill each other, fighting is apparently infrequent. Likewise, although fighting occurs among cougars, especially between males (Ashman 1975; McBride 1976; Logan et al. 1986), fighting seems to be infrequent between females. Future studies may discover that fighting occurs more frequently in unexploited, isolated populations of both leopards and cougars.

The frequent removal of established resident male cougars through hunting may increase the activities of transient males and the amount of competition for space (Logan et al. 1986). If the communication systems normally maintained by resident males disappear after they are removed, the remaining males may fight more often. Frequent removal of established breeding males may increase infanticide among tigers (Smith and McDougal 1990). Such human-caused effects merit investigation and consideration in the conservation of large felids.

12

The Conservation of Leopards

MUCH of the world's remaining leopard habitat occurs in countries with enormous social, economic, and environmental problems. In many African countries wildlife conservation programs face pressures from rapidly expanding human populations. The consequences of these expanding populations often include overcultivation, overgrazing, and desertification. The growing of cash crops instead of food crops and bad irrigation projects also cause environmental problems. But some claim these are related more to economic decisions than to human population pressures (Timberlake 1986; Anderson and Grove 1989). When one considers the magnitude of such problems, the difficulties associated with conserving wildlife or preserving a single endangered or threatened species, such as the leopard, appear almost unsurmountable.

To complicate matters further, many people in rural areas live in poverty and depend on wildlife for food or the little cash they can get from the trade in wildlife (Allen 1980). It has been suggested that the key to avoiding environmental bankruptcy in Africa is to recognize that (1) the needs of the rural African who lives off the land must be addressed, (2) the environmental degradation has been caused by unwise governments and development policies , and (3) programs can no longer be controlled by foreign governments and outside experts (Timberlake 1986; Anderson and Grove 1989). Despite the risk of seeming like an "outside expert," in this chapter I discuss some issues that should be considered in efforts to ensure the conservation of leopards.

Conservation Issues

For convenience I grouped the issues associated with leopard conservation into two general categories: (1) factors affecting leopard populations and (2) assumptions assessing the leopards' population status.

Factors Affecting Leopard Numbers and Distribution

Leopard populations, like wildlife populations throughout the world, are exposed to the ever-increasing pressures from the planet's rapidly expanding human population. Of 1,036 species of vertebrates listed in the IUCN's *Red Data Books* during the 1970s, when the leopard was first classified as endangered, 67% were threatened by the loss or contamination of their habitats and only 37% from overexploitation (Allen 1980). The problems faced by leopards are typical of those affecting many other threatened or endangered species of wildlife.

Habitat changes.   Loss of habitat is the most serious long-term threat to leopards and their prey. According to one recent estimate (Harmon 1990), an average of approximately 50% of the original habitats in Africa have already been lost (see table 11.2). Some of the greatest losses have been in forest and savanna grassland types, which probably supported the highest densities of leopards. Habitat changes in Africa are caused by the intensification of agriculture, spread of semisubsistence agriculture, livestock grazing, and deforestation (Myers 1976). Specifically, these habitat changes include the desertification of the Sahel zone south of the Sahara, pastoral expansion in the northern and agricultural expansion in the southern Guinea/Sudan woodland zone south of the Sahel zone, loss of tropical rainforest in the equatorial region, and the expansion and intensification of agriculture and livestock grazing in eastern and southern Africa.

The famines in the 1980s and early 1990s in Ethiopia, Chad, Somalia, and Mozambique, the worst in African history, and the shortages of food in at least thirteen other African countries within the leopard's range exemplify the basic problems facing wildlife conservation. The continent's last remaining secure leopard habitat may be in the extremely dry region of southern Africa (Kalahari Plateau) and the extremely wet region in the Zaire River Basin, regions where we have little information on leopard populations. Because of climatic extremes in both regions, human population densities are still relatively low compared to the rest of Africa.

Habitat changes affect leopards in several ways. In addition to predator removal, intensive agriculture greatly reduces, or eliminates, most of the leopard's large terrestrial prey. The larger ungulates are often eliminated leaving only small prey or species able to coexist with or elude humans. This greatly reduces the carrying capacity of an area for leopards, despite the fact that some leopards may be able to survive in such human-altered environments. The development of drought-resistant strains of maize could also open up previously unsuitable agricultural lands, which are usually wildlife habitat, to agriculture (Myers 1976). Although increased use by livestock may not be as devastating to the habitat as agriculture, grazing is often accompanied by the loss of natural prey because of competition with livestock and degradation of the range.

If the tsetse fly *(Glossina spp.)* is eradicated from the majority of its 10-million-square-kilometer range, about 37% of the continent (Murray and Njogu 1989), as proposed by the United Nations Food and Agricultural Organization, tsetse-free arid regions are likely to be overstocked with domestic livestock (Coe 1980). This will result in further degradation of wildlife habitats including that of the leopard and its prey. Even if livestock replaces natural prey, it is unlikely that livestock owners will tolerate livestock killing, especially when they can take matters into their own hands.

The clearing of forests could, in certain circumstances, potentially result in better leopard habitat because natural secondary forest growth can provide more stalking cover and prey for leopards than mature forest. However, at high levels of human use, prey of leopards is eliminated during cutting programs (Wilke et al. 1992). If enough escape cover remains, some individual leopards may survive in highly developed areas by preying on domestic animals. But extremely high densities of leopards, such as those speculated by Eaton (1977b) to occur within the city limits of Nairobi, Kenya, are probably atypical in developed areas.

The periodic reports of leopards in various habitats do not necessarily equate to viable breeding populations. Occasional leopards sighted within or near metropolitan areas may not be indicators of general leopard abundance but merely temporary dispersers far from viable breeding populations. Several leopards persisting in small, fragmented islands of habitat do not mean that leopards are plentiful everywhere.

Natural events, such as Africa's widespread, prolonged droughts, can also influence the quality of leopard habitat. In one drought-prone region, the Sahel, debate centers on whether the disastrous droughts over the past several decades are merely periods in a natural long-term cycle or whether the Sahel's climate is, indeed, progressively getting drier—and the extent to which these climatic changes have been magnified by the activities of humans (Timberlake 1986). Even in large protected areas such as Kruger National Park, evidence exists of a long-term (eighty to one hundred years) natural pattern of alternate below-average and above-average rainfall periods (Pienaar 1985). Severe droughts can have drastic effects on the park's abundance of wildlife. My studies of leopards coincided with a particularly wet rainfall cycle (1971 to 1979) when habitat and prey conditions were ideal. A similar study of leopards conducted during a period of prolonged drought, such as that from 1961 to 1970 or after my study in the 1980s and early 1990s, would probably reveal significantly different prey and leopard populations.

Where species, including leopards and their prey, have already been reduced to small isolated populations in fragmented habitats by the activities of humans (a "first strike"), a drastic natural event such as a severe, prolonged drought could be enough to exterminate a population, at least locally. The scenario of a first strike event followed by additional decimating factors to explain the extinction of some species has gained favor among some paleontologists (Raup 1991).

Poisoning.   Leopards, especially young or disabled individuals, are particularly vulnerable to poisoned baits because scavenging can be an important alternative feeding strategy. To aggravate this problem, poisoning is probably most widespread in areas where natural prey populations are already low because of agricultural or livestock use. Leopards are probably less attracted to baits if natural prey is abundant. According to Myers (1976) the principal poisons used in Africa today are the chlorinated hydrocarbons such as DDT and toxaphene, which are widely used to control ticks in livestock. As more areas become available to livestock grazing and agriculture, the use of poisons will probably increase. Myers believed that poisoning is a greater threat to leopards than are traps, snares, or guns and is capable of extirpating leopards from sizable areas in a short time.

Livestock depredation.   The best documented information on the incidence and significance of leopard predation on livestock is that of Hamilton (1981) for Kenya. Admitting the information was incomplete and that actual incident rates were probably higher, he reported that only 29% of the reporting game stations in the country regarded the leopard as a major stock raider; 32% reported it was a minor offender; and 38% had no problems with leopards. In most areas, other predators such as lions and hyenas were considered a greater problem than leopards. Leopards were unpopular because they preferred sheep, goats, occasionally cattle (usually calves), dogs, and chickens, but they were not considered major predators on adult cattle. Problem leopards were either shot or live-trapped and translocated.

The number of leopards reported shot in Kenya between 1957 and 1959, before leopards were routinely translocated, averaged 17.3 per year. This increased to 45.3 per year between 1960 and 1962. Thereafter, the average number of leopards shot annually was fifteen to twenty. The number of leopards translocated between 1957 and 1980 varied from 4 to 30 per year, with an average minimum of 13.5 per year from 1977 to 1980. Because many leopards killed in defense of livestock were unreported, Hamilton estimated the total number of leopards controlled each year by killing and trapping combined was 113. Most stock-raiding leopards were males by a ratio of 1.6 males per female.

The most serious incidences of livestock killed by leopards in Kenya occurred in densely settled agricultural lands, ranching districts, and in the pastoral Masai districts. Large ranches can tolerate the loss of ten to twenty sheep annually to leopards, but small livestock owners can be devastated by the loss of several individuals. On large ranches total annual loss of livestock to leopards probably seldom exceeded 0.5% per year because many ranches still retained substantial herds of wild prey such as impala and Thompson's gazelle. Although large ranches usually reported instances of livestock killing, nomadic herdsman were more likely just to eliminate the predator, often by poisoning.

Although stock-raiding leopards may be treated similarly in other countries, there are several differences between Kenya and other African countries. The first two differences suggest greater impact on leopards outside of Kenya, the third less impact than in Kenya. First, Kenya has a relatively highly organized game management organization; other countries do not. Owners of stock in other countries, therefore, may take matters into their own hands by killing leopards themselves rather than relying on game officials to kill them. Second, the number of firearms, especially military automatic weapons, in some African countries is probably much higher than in Kenya, which again may encourage stock owners to handle problem leopards on their own. Third, in contrast to the rapid rate of agricultural expansion in Kenya, slower expansion in other countries with lower population levels and growth rates may reduce the impact of livestock grazing on leopards. However, even if Kenya is atypical among African countries today because of its rapid agricultural development, other countries will eventually have similar stock-raiding problems with leopards as their human populations expand.

Commercial trade in skins. The alarming trend in imports of leopard skins into the United States in the late 1960s, as well as the worldwide trade in leopard skins, was the primary reason for listing the leopard as an endangered species by the United States in March 1972 and for including it in Appendix I of the Convention of International Trade in Endangered Species (CITES) in 1975. Commercial trade was, and probably continues to be, the driving economic force behind the illegal hunting (poaching), trapping, and snaring of leopards. When the leopard was listed as an endangered species, the heavy flow of leopard skins into the United States immediately ceased, and there have been few legal imports since 1972 (U.S. Department of the Interior 1982). The impact of CITES on the total number of leopards killed in Africa, once estimated to approach 50,000 per year, is more difficult to determine. In some areas, such as Mozambique (which exported about twelve hundred skins annually between 1966 and 1971), the trade declined significantly after the ban on U.S. imports (Myers 1976). The ban apparently had similar effects on the number of leopard skins reported exported from other countries. According to Myers (1976), South Africa apparently had the greatest concentration of fur dealers in Africa in the early 1970s, but merely served as a funnel for skins obtained outside the country in route to world markets. Many illegal skins were included in the imports; total numbers were estimated at fifteen hundred to twenty-five hundred skins per year between 1966 and 1972. The price per skin averaged $250 to $300, with a high of $680 per skin. In nearby Botswana, demand was still great, especially from France, Japan, and Hong Kong. Overall, the ban on international trade in skins appears to have had significant and positive effects on the numbers of leopards being killed. However, a demand for leopard skins still exists. And there are people willing and able to circumvent existing regulations, as well as the intent of international treaties and conservation laws of the majority of exporting and importing countries in the world, for purposes of monetary profit. Demand in continental Europe in 1973, after the new controls were applied, was said to surpass that of the late 1960s (Myers 1976). Any relaxation of existing regulations or amendments that might create loopholes in those regulations should be considered with extreme caution; such loopholes might once again trigger a dramatic increase in the numbers of leopards illegally killed for their skins.

LEOPARD STATUS SURVEYS

The numbers of leopards estimated for sub-Saharan Africa have ranged from 233,050 to 1,155,500 (Eaton 1977b; Teer and Swank 1978; Martin and Meulenaer 1988). One estimate (Martin and Meulenaer 1988) was 714,105 leopards (95% confidence interval 598,102 to 854,066) predicted from a computer model. Reviewers were critical of the model and concluded that it was invalid and that its predictions were too high and unreliable because its assumptions were false (Meadows 1991). The primary assumptions of the model were (1) all suitable leopard habitat is occupied by leopards; (2) leopards occur at maximum densities in all occupied habitats; and (3) numbers of leopards are correlated to rainfall, with higher numbers of leopards occurring in areas with higher rainfall (Martin and Meulenaer 1988).

The model's first assumption, that all suitable leopard habitat is actually occupied by leopards, is highly improbable but admittedly unknown. Many potential habitats outside protected or managed areas in sub-Saharan Africa no longer support natural prey populations because of human settlement, livestock grazing, and agriculture. Even with the model's assumption that modified habitat supports only one-tenth the densities of leopards as unmodified habitat, it still results in predictions of unusually high numbers of leopards. For example, the model predicted 23,472 leopards for highly developed Republic of South Africa, when it is likely that fewer than 3,000 live there (Meadows 1991). Even three thousand leopards for the country seems high if Kruger National Park, the country's largest area of high-quality leopard habitat, supports only about seven hundred leopards. In any scenario it is highly doubtful that more than 22,700 leopards occurs in the Republic of South Africa outside Kruger National Park.

The second assumption, that leopards always occur at maximum densities the habitat can support and that any relaxation in human harvest will result in a rapid recovery, has yet to be documented in the field, especially outside of protected or managed areas. Little is known about prey and leopard densities throughout most of sub-Saharan Africa outside a few parks, reserves, or game ranches. Diversity and abundance of leopard prey, habitat fragmentation and other ecological factors probably have a greater influence on leopard densities in such areas than human harvest pressures.

The third assumption, that leopard densities increase with increasing rainfall, is questionable because of differences in leopard prey species diversity, sizes, abundance, and adaptations such as the proportion of prey that is terrestrial or arboreal as rainfall increases. The rainfall and herbivore biomass relationship may be valid in general, but a herbivore biomass increase does not necessarily equate to increased leopard prey biomass. The increase in herbivore biomass can be exhibited in large herbivores (elephant, hippopotamus, buffalo) or herd-forming species (zebra and wildebeest), which provide little food for leopards. The model's relationship between leopard densities and rainfall was derived primarily from leopard densities estimated or documented in national parks (fifteen of twenty-three data points), or game reserves, research areas, and game ranches (three of twenty-three) where leopard densities would be higher than in surrounding nonprotected habitat. Many of the leopard population density estimates, especially in the high rainfall habitats, appeared to be crude estimates unsubstantiated by intensive field studies of marked individual leopards.

Leopards, like most solitary felids, are extremely difficult to census because of their secretive nature and preference for dense cover. Converting observed leopard tracks, kills, or feces into actual leopard numbers is questionable because of unknown leopard movement patterns, travel route biases, and differences in habitat quality. Some estimates of leopard densities and populations appear to have been based on assumptions undocumented by studies, or on incidental or anecdotal observations. Two aspects of estimating leopard numbers that deserve closer scrutiny are assumptions about leopard densities in rainforest habitats and the dangers of extrapolating leopard densities from small protected areas to vast unprotected areas.

Leopard densities. One assumption of those compiling leopard status information is that rainforests provide optimum habitat for leopards. Leopard densities are believed to be as high as one individual per 3 to 4 km^2 (Myers 1976; Eaton 1977b; Martin and Meulenaer 1988) because of assumed abundant prey and the scarcity of competitors such as lions, wild dogs, hyenas, and cheetahs. These presumed high densities of leopards are supposedly maintained by an abundance of primates and arboreal rodents, birds, reptiles, and fish, because ungulate biomass is admitted to be low (Myers 1976). The rainforest habitat is assumed to favor a predator that is stealthy, an expert climber; operates alone; and hunts equally well by night or day. But studies of the carrying capacity of rainforest habitats, the behavior of primates and arboreal rodents in African tropical rainforests, and knowledge of leopard hunting habitats and diets strongly suggest that rainforests may not be the tropical haven once envisioned for leopards.

Several studies published since Myers's (1976) leopard status survey indicate rainforests support relatively low prey biomass. The biomass of primates in the Kibale Forest in western Uganda was estimated at 2,217 kg/km^2 (Struhsaker 1975), the biomass of prosimians in the Congolese rainforest in Gabon at 20.7 to 27.5 kg/km^2 (Charles-Dominique 1977), and the biomass of African rainforest squirrels in Gabon at 740 kg/km^2 (Emmons 1980). These biomasses were only 0.10% to 50% of those of leopard prey in the Kruger National Park study areas. Furthermore, most of these rainforest mammals would not be available to leopards because either they spend much of their time in treetops or outermost limbs to avoid climbing predators (most forest primates and prosimians) or they rest overnight inside hollow trees with small cavities for openings (tree squirrels).

Few studies have identified the leopard as a major predator on arboreal prey. Crowned-hawk eagles were the major predators on monkeys in the Kibale Forest, and humans were considered to be the major predator on monkeys throughout Africa (Struhsaker 1975). Struhsaker reported leopards were not important predators on the red colobus monkey because they were largely ineffective as predators at heights above 10 m. Although the leopard was among the nine species of carnivores sharing habitat with the prosimians in Charles-Dominique's (1977) study area in Gabon, he believed that leopards were largely terrestrial and seldom preyed on the monkeys. Emmons (1980) reported the major predators on tree squirrels were avian (accipiters and crowned eagles), with others perhaps, such as snakes, driver ants, and viverrids, as major nocturnal nest predators. Such information strongly suggests that (1) the biomass of primates and arboreal rodents are extremely low in rainforest habitat compared to ungulate and other leopard prey in woodlands or woodland-savanna habitats, (2) many species are not vulnerable to leopard predation, (3) leopards are ineffective predators on arboreal mammals, and (4) the major predators on arboreal mammals are raptors, and perhaps snakes and viverrids, not leopards.

Little information is available on the densities or biomass of birds in African rainforests. Although a relatively large number of forest species (about two hundred) are found in African rainforests, only a few (six species) in the family Phasianidae (Amadon 1973) are ground-dwelling and large enough to provide prey for leopards. Because of the sparse understory, they are more likely to occur at lower densities in rainforests than in woodlands or savannas. The majority of forest birds are small tree dwellers and, hence, would provide little food or be unavailable as leopard prey.

Although small prey may form a significant component of the leopard's diet in some areas (Grobler and Wilson 1972; Smith 1977), the size and hunting habits of the leopard indicate that it is primarily a predator of terrestrial prey. In my study areas most arboreal and avian prey (vervet monkeys and guinea fowl) were probably captured while they were foraging on the ground.

If leopards occur at high densities in African rainforests, they are probably dependent on terrestrial prey. But estimates of terrestrial mammalian densities and biomass in African rainforests indicate a relatively low prey base. Excluding elephant, hippopotamus, and buffalo, which are relatively free from leopard predation, there are relatively few terrestrial ungulates. Bourlière (1963a) using Collins's (1959) data, estimated only 5.6 kg/km^2 for three small forest ungulates that could be potential leopard prey in a rainforest on Ghana, but later (Bourlière 1973) acknowledged the area may have been atypical because of its lack of elephant, buffalo, bongo, okapi, and giant forest hog.

In other rainforest regions, bongo, bush pig, and giant forest hog, as well as aardvark, pangolins, and porcupines, may add to the leopard prey diversity. Nevertheless, overall terrestrial mammal biomass still appears low compared to savanna and woodland habitats. In comparing biomass of terrestrial mammals in South Asia and South America, Eisenberg (1980) noted that tropical evergreen forests cannot support high biomasses of terrestrial herbivores because the bulk of productivity is locked up in the crowns of the trees, making it inaccessible to terrestrial forms. These ecological considerations suggest that leopard densities in rainforest areas would be less, not greater, compared to areas where most prey are medium-size terrestrial mammals.

Leopard densities in favorable rainforest habitats are unlikely to exceed 1 per 10 km^2 and are more likely to be nearer to 1 per 30 km^2. It would be difficult for leopards to exceed the densities documented in the Kruger National Park study areas (1 per 3.3 km^2) where leopard prey biomass varied seasonally from 2,932 to 6,186 kg/km^2. These are the highest densities documented for leopards based on known, radio-collared individuals. In Wilpattu National Park, Sri Lanka, where the leopard is the only major predator and prey densities are higher than in rainforests, the estimated density was 1 resident leopard per 29 km^2 (Eisenberg and Lockhart 1972). In African rainforests, where terrestrial mammalian leopard prey probably averages less than 500 kg/km^2, densities of leopards could be lower. Only intensive field studies of marked leopards in rainforest habitat will provide a definitive answer.

The reported high densities of one or more leopards per square kilometer in the Matopos National Park, Zimbabwe, (Eaton 1977b) is also puzzling. Apparently he based his conclusions on a report by Grobler and Wilson (1972) and later correspondence with Grobler. But neither Grobler and Wilson (1972) nor

Smith (1977), who later studied leopards three years in the same park, reported such high densities of leopards. Smith (1977) reported only 1 resident leopard per 6 km^2 in his two study areas and estimated a maximum density for the 432 km^2 park at one hundred leopards, or 1 leopard per 4.5 to 5 km^2. Until studies are conducted with marked or individually recognizable leopards, densities of 1 leopard per square kilometer remain undocumented.

Extrapolating leopard population densities.   Numbers of leopards for entire countries, even entire sub-Saharan Africa, have been estimated by extrapolating leopard densities documented primarily in parks and preserves to nonprotected areas (Eaton 1977b; Martin and Meulenaer 1988). Such estimates can be deceptively high (Schonewald-Cox et al. 1991). First, the most accurate estimates of leopard densities have been obtained in protected parks where natural prey is abundant, human impacts on the habitat are insignificant, and poaching, snaring, and trapping are controlled (Schaller 1972; Eisenberg and Lockhart 1972; Hamilton 1976; Smith 1977; this study). To extrapolate comparable densities to unprotected areas where prey is scarce, the habitat has been significantly altered by humans, and the killing of game is uncontrolled, is questionable. Second, to assume, without study, that certain densities of leopards are found in selected habitats and then to extrapolate the same densities over tens or hundreds of thousands of square kilometers, is also questionable. Third, the assumption that leopard densities are probably high in developed areas because other predators, such as lions and hyenas, have been eliminated, or because much prey is available in the form of livestock, has yet to be documented by field studies. For example, leopard densities were quite low, mortality of adult leopards was high, and vacancies remained unfilled for prolonged periods in areas surrounding Royal Chitwan National Park, Nepal, settled by humans (Seidensticker et al. 1990). Because of these factors, countrywide and continentwide estimates of leopard numbers throughout Africa must be viewed with continued skepticism.

THE STATUS OF THE SUB-SAHARAN AFRICAN LEOPARD

The leopard status reports of Myers (1976, 1986) are probably the most reliable. He lived in Africa for twenty-four years and traveled extensively throughout most African countries. Rather than repeat his country by country synopsis, or those of Eaton (1977b) and Teer and Swank (1978) who did not conduct field surveys, I merely summarize Myers's conclusions.

In West Africa the massive human population growth in sixteen countries totaling 7.5 million square kilometers, was expected to rapidly disrupt wildlife habitats. As stocks of wild herbivores were depleted for food under human pressure, leopards would become confined to small, remote, rocky areas with little human use, or to parks and preserves. Most leopards were extirpated along the coastal forest belt in West Africa by 1945. In northeastern Africa, which has mainly desert or semiarid habitat, leopards were being confined to relatively small forest habitats. Throughout much (perhaps 80% or more) of the area, wild game had been depleted leaving little natural prey. An exception was about 500,000 km^2 of southern Sudan, which may retain some of the best wildlife habitat left in Africa.

The leopard was expected to become scarce in East Africa because of expanding agriculture and livestock grazing. Leopards were expected to survive in relatively good numbers in the large parks and preserves and in isolated areas of low human population density. In Equatorial Africa the status of the leopard appeared fairly secure, primarily because of the low human population density and unsuitable nature of the land for intensive agriculture and livestock grazing. The Miombo woodland-savanna, south and east of the equatorial forest, supports low densities of wild herbivores and carnivores because of the scarcity of water. Human population expansion was expected to be restricted primarily along the floodplains intersecting the woodlands. Because of wildlife's dependence on water, human impact on the floodplains was expected to be significant, and the density of wildlife including leopards, was expected to decline. Because the region was still inhabited by the tsetse fly, which transmits sleeping sickness to cattle, the human demand for protein was expected to be reflected in increased killing of game rather than increased consumption of livestock.

Leopards in southern Africa were expected to become increasingly confined to small, remote areas or to parks and preserves because of intensified livestock grazing and agriculture. Except for some assumptions on densities of leopards, I suggest that Myers's surveys are still the best assessment of the leopard's status and future in sub-Saharan Africa.

Despite the leopard's ability to adapt to different habitats and prey, and the capacity of individual leopards, compared to other large carnivores, to alter their behavior in close proximity to humans, leopard populations have already seriously declined and will continue to decline compared to their former numbers and distribution (Myers 1986; Hamilton 1986). Martin and Meulenaer (1988) maintained that the available range for the leopard in sub-Saharan Africa is likely to decline by half within the next twenty years. Like other large carnivores, leopards can be rapidly exterminated over large areas by habitat changes, poisoning, and illegal hunting and trapping. As changes in their habitat occur, leopards will become increasingly confined, as they already have been in West and Southern Africa, to small isolated and often rocky areas throughout their ranges or to parks and preserves surrounded by developed land. In some areas, such as the tropical rainforest and miombo woodland-savanna, leopard densities may not yet be significantly influenced by man. The same applies to the larger parks and preserves. In Kenya, Hamilton (1986) believed that leopards were still widespread and possibly increasing in numbers relative to the late 1960s and early 1970s.

Conservation Strategies

Any conservation program that fails to involve the participation and cooperation of the rural people whose lives it will invariably alter is unlikely to be successfully implemented (Anderson and Grove 1989). This is probably the most important principle to be applied to wildlife conservation efforts. It is often the most difficult, especially when the human population itself is living in poverty and struggling to survive. The conservation of wildlife, including leopards, will have to consider many different strategies to be successful. Some strategies may work in some areas but not in others.

Wherever practical, one basic strategy should be to nurture, through education and environmental awareness programs, the deep-seated awareness within many cultures that recognizes the intrinsic value of all forms of life, to create a reverence for life in its many forms (Passmore 1974; Norton 1987). Some religious and philosophical systems have always viewed other forms of life as having an intrinsic or even sacred value. Others are still human-centered. It is to be hoped that the beauty, complexity, and mystery associated with other forms of life are still appreciated in most cultures. The leopard has several advantages in this regard: its beauty, strength, and cunningness are still respected by cultures throughout Africa.

Another strategy of leopard conservation would be to emphasize that conservation efforts can provide material benefits to humans. This strategy would have to be tailored to the different cultural, economic, and social conditions found throughout the leopard's range. One such approach, the World Conservation Strategy (1980), sees economic development and conservation as compatible goals, with increasing contact between developed and developing countries, the developed countries playing a significant role in fostering conservation in Third World countries.

Efforts to conserve leopards should consider at least three approaches: (1) an environmental education and awareness program primarily intended for school-age children, (2) a program devoted to protected areas such as parks and preserves, and (3) a program to conserve but utilize leopards on a sustained basis for the material benefit of rural people living outside of, or adjacent to, protected areas.

ENVIRONMENTAL EDUCATION

Environmental education could stress the intrinsic, ecological, and economic value of leopards. The intrinsic value of leopards would have to be communicated within different cultural contexts. Even where large carnivores compete with humans for livestock, carnivores can still be admired and valued for their prowess, cunning, strength, and beauty. Humankind's general admiration for these characteristics might be transferred into positive conservation benefits by making people aware that such qualities are preserved when leopards are protected. One does not necessarily have to have a high material standard of living to appreciate the value of other forms of life. A deep respect for other life forms exists in some cultures even when those forms are regularly taken for food, clothing, medicine, or religious purposes.

Environmental education should also emphasize the ecological indicator value of predators, stressing that humans, predators, and prey can coexist in harmony if an ecosystem is healthy. People can be shown that predators do not always endanger one's livestock or decimate natural prey. This approach, however, will be exceedingly difficult, if not impossible, to implement in degraded systems where not only wildlife, but human life, can barely exist. Fortunately, large areas still exist outside parks and preserves, such as in the Zaire River Basin and Kalahari Desert, where ecosystems have not yet been seriously degraded.

Environmental education should also emphasize the potential economic value of leopards, especially in the rapidly developing, nonprotected areas of the leopards' range. Consumptive or nonconsumptive uses of leopards can be emphasized. Consumptive use of leopards could include carefully controlled trophy hunting if the area is large enough to provide a sustained yield. Trophy hunting would have to be closely regulated to prevent overharvest and illegal hunting. Nonconsumptive uses of leopards could include viewing and photography.

To obtain the support of people who share their land with leopards, the potential economic value of leopards should be emphasized. One estimate placed a potential value of $29 million U.S. dollars on the exploitation of leopards throughout sub-Saharan Africa, $19 million of which could be derived from sport hunting and $10 million from trade (Martin and Meulenaer 1988). If local inhabitants receive revenue or benefits generated by trophy hunters or tourists paying to observe leopards, they may be more apt to tolerate leopards. Otherwise, they are unlikely to support wildlife conservation programs. One goal should be to allocate some of the revenues derived from wildlife into increased standards of living for the local residents. Such programs adjacent to parks and reserves have already been initiated with varying success in Kenya's Amboseli National Park (Western 1982) and in the Pilanesberg Game Reserve in the Republic of South Africa (J. L. Anderson, personal communication). Financial contributions from concerned citizens in developed countries can also help conservation efforts. Some conservation organizations in the United States include the World Wildlife Fund, African Wildlife Foundation, and World Conservation International.

THE ROLE OF PARKS AND PRESERVES

The long-term future of the leopard and other species of wildlife in Africa is probably most secure in parks and preserves. However, because parks and preserves make up less than 5% of the 20 million square kilometers in sub-Saharan Africa (Myers 1976), they must be carefully managed or their value will be lost. Parks and preserves serve as ecological yardsticks for comparison with human-altered environments; they are valuable educational and training tools; and they have aesthetic and economic value usually associated with tourism (Coe 1980). The leopard's role as a major predator influencing herbivore behavior and population dynamics, as well as its public appeal and beauty, make it a valuable component of any African park or preserve. To maintain viable leopard populations in parks and preserves, ecological, development-related, and genetic factors must be considered.

Ecological considerations.   Although protective areas such as parks and preserves have not been established exclusively for leopards in Africa, many ecological factors can influence leopard populations. Fortunately, leopards do not required special habitats because they are generalists and can exist in a variety of environments. Leopards need only adequate space, protective and stalking cover or terrain, sufficient prey, and protection from exploitation to survive.

To maintain genetically viable populations of leopards, large areas are required. The size of reserves required to support a viable population of leopards over the long run is important because (1) the smaller the reserve, the more rapidly species will become extinct; (2) the rate of extinction is more rapid in large than small species; (3) only a small fraction of vertebrates will disperse between reserves even if they are only a few kilometers apart; and (4) dispersal corridors between small reserves can easily be disrupted by land-use practices (Frankel and Soule 1981). Thus maintaining viable populations of leopards will be more difficult in small parks than in large ones. Although the carrying capacity of parks for leopards will vary among different habitats and among different prey and predator communities, it will probably be difficult, in the long run, to conserve viable leopard populations in parks less than 500 km² in size because of genetic considerations. Leopards confined to small parks are also more apt to eventually stray outside protective, artificial boundaries where they are eventually killed. The impact of human encroachment through poaching, snaring, and trapping will also have a greater impact on wildlife, including leopards, in small parks than large parks. Furthermore, confining species such as leopards inside small parks will be difficult because they can easily climb over or go under fences.

Leopards reared inside such small areas have no place to disperse but outside the park; leopards outside parks often conflict with humans (Seidensticker et al. 1990). Managers of such areas may eventually be faced with the difficult decision of culling leopards from small endangered populations to reduce opposition from local residents (Cobb 1981). One approach to address this issue would be to have a buffer zone around the park boundaries, where dispersing leopards could settle but be subject to a regulated harvest to the benefit of local residents. The practicality of such management would vary among small reserves and be dependent on the human population density, livestock density, and the ecological carrying capacity of the park and the buffer zone.

A protected area's ecological carrying capacity for leopards will vary with habitat features such as the degree of topographic relief, amount of rocky outcrops for protected den sites, and the type and density of vegetation providing stalking cover. The size, density, availability, and movement of leopard prey will also strongly influence an area's carrying capacity for leopards, as will the types and densities of other large carnivores in the area. An ideal protected area for leopards would be characterized by rugged topography with many rocky cliffs and outcrops; be covered by dense understory of vegetation; and support high-density populations of sedentary, medium-size ungulates and few other large predators. From a leopard's perspective, a poor area would be characterized by flat topography with few if any rocky outcrops, sparse vegetation, migratory prey or resident prey that are too large or too small for leopards, and other larger predators. Most protected areas inhabited by leopards probably fall between these two extremes.

Although establishing additional protected areas for wildlife will be difficult, future parks could be chosen to include features favorable to leopards. Fortunately, rugged landscapes are often candidates for potential parks because of their decreased value for agriculture, power projects, and human settlement. Intensive management may be necessary to maintain leopards in small parks or in large parks with artificial boundaries that are surrounded by developed lands. Vegetation can be altered to favor leopards and their prey by the use of fire and control of major herbivores. Fire is believed to be an important factor in maintaining tropical grasslands (West 1965 cited in Eltringham 1979); prevention of fire may result in brush encroachment (Pienaar 1969). To benefit leopards and their prey, it may be desirable to control the fire periodicity to allow brush to encroach.

Brush encroachment in Kruger National Park has been one of the main factors responsible for the sharp increase in impala populations, the primary prey of leopards. Brush also provides ideal stalking cover for leopards. Because brush encroachment results in a decline in grassland-favoring species, a park manager must consider favoring brush-loving species and their predators over grass-loving species. For example, although brush encroachment in Kruger National Park has favored leopards, it has probably been to the detriment of cheetahs in the park (Pienaar 1969). A rotational burning scheme where blocks are burned at regular intervals for a desired effect appears practical. Undesired fires are extinguished. Such programs are used in Kruger National Park (Van Wyk 1971) and Kabalega Falls and Kidepo Valley National Parks (Eltringham 1979).

Controlling large herbivores in parks may sometimes be necessary because intensive browsing by large herbivores, such as elephant, hippopotamus, and buffalo, may affect an area's capacity to support other ungulates and their predators. For example, elephants in the Luangwa Valley, Zambia, were felling mopane *(Colosphermum spp.)* trees at a rate of 4% of the standing crop (Hanks 1979), and in one area in Zimbabwe elephants were responsible over a seven-year period for killing 63% of the tree population or converting trees to shrub-like vegetation by reducing their height (Anderson and Walker 1974). Changing trees to a shrublike form in this environment may have benefited prey of leopards. High mortality of *Brachystegia* either directly by elephants or as a result of damage by elephants was also believed responsible for the decline of that woodland in the Chazarira Game Reserve in Zimbabwe (Thomson 1975).

The relationship between large herbivores, vegetation, and leopard prey can vary. Although intensive browsing by elephants in some areas may favor leopard prey, intensive grazing by other large herbivores may sometimes be detrimental. For example, overgrazing by hippopotamus in Ruwenzori National Park led to the depletion of the range, but after a hippopotamus culling program, numbers of other herbivores (Eltringham 1979), some of which were potential leopard prey, increased 50%.

Management actions that increase leopard prey populations should benefit leopard populations. The provision of permanent waterholes in Kruger National Park has been significant because water is one of the main factors in determining impala distribution (Whyte 1976). However, waterholes can also be detrimental to other species, such as zebra and wildebeest, because it disrupts their normal migratory behavior (Smuts 1978). Artificial waterholes have been responsible for the development of small, sedentary herds of zebra and wildebeest that became vulnerable to lion predation. Despite such detrimental impacts, providing artificial water sources appears to be more beneficial than detrimental to leopards because it increases the numbers, distribution, and stability of their prey populations. However, range utilization adjacent to waterholes should be monitored to prevent range destruction.

The composition of prey communities is also important to leopards. Healthy populations of medium-size sedentary prey such as impala, bushbuck, reedbuck, and duiker can support more leopards than populations of larger migratory prey such as wildebeest and zebra. A variety of small, terrestrial, species of prey also appears important for leopards, particularly young and temporarily handicapped individuals. Small prey may be critical in a leopard's life history because a young leopard may develop its hunting skills on small prey before switching to larger prey.

Other predators inhabiting parks and preserves may influence leopard distribution and abundance. Leopards coexist with tigers in Chitwan National Park, Nepal, because of the large biomass of prey, the large proportion of the ungulate prey in small size classes, and the dense vegetative structure (Seidensticker 1976a, 1976b). In Kanha National Park, India, leopards are resident only in areas infrequently used by tigers (Schaller 1967). After the decline of tigers in Sumatra, leopards became more widespread and abundant (Seidensticker 1986).

Similar relationships may influence leopard populations wherever leopards occur with other large predators. However, given sufficient prey and diverse habitats, leopards can survive in the presence of more dominant carnivores.

Development. The number, design, and location of roads and public facilities and the way tourism is conducted in parks and preserves can directly impact wildlife distribution, behavior, and abundance (Eltringham 1979). Because leopards in woodland-savanna and savanna regions appear to spend much of their time in riparian vegetation, the placement of roads and facilities along rivers could reduce the amount of habitat available to leopards. This impact may be negligible in large parks, but it could be significant in small parks or preserves. Also, because leopards frequently use roads as travel routes, strict regulations may be necessary to ensure that vehicle speeds are not excessive and that vehicles remain on designated roads. As roads become more crowded, a limit on the number of vehicles allowed in parks may also become necessary. Finally, because leopards feeding in trees are often disturbed by tourists, prohibiting vehicles from crowding around trees in which leopards have cached their kills may be necessary.

Genetics. Conservation should be considered a dynamic process, rather than a static preservation of a species, ecotype, or genotype (Frankel and Soule 1981). Viewed in this context, a crucial question is whether conditions in parks and reserves facilitate, restrict, or inhibit the process of continuing evolution. Unlike wild primeval populations, many populations in parks and preserves are, or will become, small, disconnected, and subject to inbreeding, genetic drift, and random fixation of alleles, resulting in a gradual weakening and genetic impoverishment of individuals. Of particular concern are large carnivores, which usually occur at low densities but have high community impact on ecosystems.

One of the concerns of habitat fragmentation and isolation of parks and preserves is the effects of inbreeding on small isolated populations. Detection of inbreeding depression is one of a series of difficult problems in conservation biology (Smith and McDougal 1991). The effects of an accelerated rate of inbreeding in small populations can potentially drive a population to extinction. The average cost of a parent-offspring or full sibling mating in mammals was calculated at 0.33 and was likely to be an underestimate (Ralls et al. 1988). This means that mortality was 33% higher in offspring of such matings than in offspring of unrelated parents. Generally, females should not mate with their fathers or sons. Observations of female leopards breeding with their sons in small endangered populations is therefore of great concern (see appendix C). Other effects of inbreeding may include susceptibility to viral epidemics (O'Brien et al. 1985), reduction of fecundity of inbred young that survive to reproductive age (Wright 1977), lower body weights, decreased longevity, and blindness (Laikre and Ryman 1991).

Franklin (1980) estimated that to avoid inbreeding fixation of deleterious genes a species needed to have a minimum genetically effective population size of fifty and estimated that five hundred was needed to maintain enough genetic variation for genetic adaptation. An affective population of fifty may lose 50% of its original genetic variation after only twenty-five generations, whereas a population of five hundred individuals retains about 95% of its original genetic diversity after fifty generations, or about four hundred years (Foose 1987).

Although the minimum effective population size needed to preserve genetic fitness has been estimated to be fifty, this is an "ideal" population. Ideal populations are composed of adult males and females of equal numbers that mate at random, are not subject to severe population fluctuations, have nonoverlapping generations, and have a random distribution of offspring among families. Because such an ideal population seldom occurs in nature, particularly with long-lived carnivores like leopards, the actual censused population size needed to provide an effective population of fifty is probably two to three times larger.

To maintain a minimum effective population size of fifty leopards would probably require an actual population of at least eighty to one hundred leopards in many parts of Africa. This is necessary because only a proportion of the actual population (63% in the Kruger National Park study area) will be breeding adults; the sex ratio among breeding adults may favor females about 2 to 1; only a proportion of the adult females produce young each year; generations overlap; and over two to three generations (about thirty years) populations have a high probability of declining to about 50% of their numbers. A reserve's carrying capacity for leopards may be influenced by changes in environment from droughts; disease among prey; poaching, poisoning, or other human impact; long-term changes in vegetation brought about by fire; changes in ground water; or use by elephants.

Among tigers in Royal Chitwan National Park, Nepal, it has been shown that the variance in lifetime reproductive success is a major factor contributing to effective population size (Smith and McDougal 1991). The degree of polygyny among male tigers is of particular importance resulting in the increased variance; some males fathered up to twenty-seven offspring to dispersal age while other males fathered none. The effective population size (N_e) for this observed population of 45 breeding female and 20 breeding male tigers ($N = 65$) was only 25.8 and 26.4 based on dispersal and breeding age offspring, respectively. The effective population size for male tigers was only nine, compared to thirty-one for females. Applying the Royal Chitwan National Park effective/observed population ratio throughout other tiger populations in reserves in India revealed effective population sizes of less than ten to about fifty tigers. Their analysis demonstrated the genetic problems encountered in attempting to conserve genetically viable populations of solitary felids in small, isolated reserves. Because leopards have a similar social organization and breeding system to tigers, conserving genetically viable populations of leopards in small isolated reserves will also be difficult.

The size of a park or preserve needed to support a minimum effective population size of fifty leopards versus five hundred leopards can be roughly estimated if the average number of leopards per unit area for that habitat type, prey, and predator community is known. Using documented estimates of leopard densities and park and preserve sizes, only about fourteen (21%) of sixty-eight major parks and preserves in Africa in 1975 had the potential capacity of supporting leopard populations while maintaining permanent genetic diversity. At least 41% of the parks and preserves were too small to support even a minimum effective population. The actual number of parks capable of supporting minimum and desired effective populations is probably much less because my assumptions were area-based only. Some of the parks and preserves included may not even have viable leopard populations today.

Populations of leopards below the minimum effective population size probably are, or soon will be, common in small parks and preserves. Managers of such areas can consider several actions to monitor and reduce the rate of inbreeding and loss of genetic fitness (Frankel and Soule 1981): (1) obtaining baseline information on the genetic variation within the population; (2) intensively managing the population to increase its effective size; (3) culling, if necessary, in a manner that maximizes genetic variation instead of in a manner that reduces it; (4) manipulating habitat, if necessary, to prevent loss of other species; and (5) initiating artificial migration by adding new members from other parks or captive populations.

Obtaining baseline genetic information may require that leopards are live-captured, blood samples taken, and genetic studies conducted. As an alternative, morphometric measurements can be compared using rigorous multivariate statistical analyses. Sequencing of proteins or DNA are other, more expensive, options.

The effective sizes of some small populations may be increased by maintaining equal numbers of adult male and female breeders (Frankel and Soule 1981). This may be difficult for leopard populations because of their social organization. Male leopards usually maintain large territories, with two to six females, from which other males are excluded. Effective population sizes can be increased by increasing the carrying capacity of the leopard's habitat. In some regions this may involve altering plant succession to favor brush and brush-favoring prey instead of grassland species. In mature forest areas it may involve letting natural fires burn or purposely using fire to set back forest succession to temporarily favor leopards and their prey.

If the probability of new individuals naturally entering the population is low or absent, maintaining genetic diversity in small leopard populations may eventually require artificial migration. The turnover rate of adult males in a population appears greater than that among adult females; therefore the periodic introduction of new males from distant parks or reserves may be the most logical first step. Because one male will probably breed with two to six females, periodically introducing new males into the population would be more practical and effective than introducing females. Because translocated males seldom remain in areas already occupied by resident males (Hamilton 1976), the artificial introduction of a new male will probably require the removal of the previous resident male just before the new male's introduction. The needed periodicity of such artificial introductions remains unknown.

Conservation Outside Protected Parks and Preserves

Unless incentives are provided to maintain leopards outside parks and preserves, leopards will eventually be extirpated over most of their geographic range. Natural faunas will be drastically altered as habitats disappear under the pressure of agriculture, livestock, and development. Large predators are usually the first component of wildlife communities to disappear. Under present socioeconomic conditions throughout most of the leopard's distribution, there is little hope that this trend will soon be reversed. A conservation strategy emphasizing the economic benefits of leopards should be one of the options considered for these nonprotected lands.

Viewing and photographing.   Many wildlife viewers visiting Africa would probably be willing to pay to observe or photograph leopards in natural settings outside of parks and preserves. Legal jurisdiction over the land should matter little as long as visitors feel they are in a wild leopard's domain. Landowners, be they individuals, tribes, or corporations, could benefit economically from leopards, which could be an incentive to maintain natural prey of leopards, or to accept occasional livestock losses to leopard predation.

Several techniques could increase contacts between leopards and observers. Slowly traveling at night along roads or trails that are used by leopards as travel routes is an exciting experience. Seeing a leopard at night is more rewarding and exciting, in my opinion, than in the daytime. During the day, leopards are usually resting or inactive. During the night, however, they are active, alert, and usually on the move or stalking prey. Adding to the excitement is the suddenness with which leopards appear out of the darkness. At night, leopards are truly observed in their own element.

Another fascinating experience is to hear leopards vocalize, or rasp. Most vocalizing occurs at dusk or in the early night. Once a leopard's approximate home range and travel routes are known, vantage points can be chosen where calls can be expected. Not, perhaps, as impressive as the call of a lion or tiger, a leopard's call, when heard from several meters away, is an unforgettable experience. Leopards may be induced to respond to an artificial call or recording, in the same manner as wolves (Harrington and Mech 1982). Frequent use of artificial calls may, however, eventually condition leopards not to answer.

Stalking natural kills of leopards while on foot, to observe feeding, is also an interesting experience. But it must not be allowed to disrupt their feeding behavior or increase predation rates. People who periodically travel on foot throughout leopard habitat will eventually encounter kills of leopards. They can be observed at a distance, or visited several times, provided the leopard is not disturbed. If disturbed, the leopard will probably move the kill. If other ground scavengers or predators are present, the leopard will probably cache its kill in a tree where visibility will be greater for the observers (and for the leopard). If enough personnel are available, several individuals could scout for kills. Clients could then be led directly to an area, instead of randomly searching for kills. Regardless of the method used to locate kills, the paramount concern must be to avoid disrupting the leopard's behavior.

Another technique to enhance observer-leopard contacts is to attract leopards with live prey or a carcass. The bait is securely tethered to a tree or stake to prevent the leopard from dragging it away. Leopards at baits can comfortably be observed from vehicles, blinds, or even permanent accommodations. Lights can be arranged beforehand if desired, but if they are too disturbing, leopards may be discouraged from feeding. Although baiting may unfairly take advantage of leopards' natural scavenging behavior and could alter their behavior, it would be acceptable if the practice enhanced their survival instead of their extirpation.

Hunting leopards.   A controversial alternative that may provide economic incentives to conserve leopards outside of parks and preserves in some countries is a highly regulated hunting program that removes only a small proportion of a leopard population each year. A closely regulated take of leopards may be not only practical but necessary, as some claim (Myers 1974; Eaton 1977b; Myers 1981; Hamilton 1986; Martin and Meulenaer 1988), to promote leopard and other wildlife conservation in Africa. Although I believe it will be increasingly difficult in the long run to justify maintaining wildlife populations solely on economic criteria, because of livestock and agriculture needs and development, hunting may be an effective conservation alternative for the immediate future. Whether one agrees or disagrees with the ethics of hunting leopards outside parks and preserves, hunting can probably be managed to benefit some leopard populations. A closely regulated hunting program for leopards for trophy purposes should not be confused with hunting leopards commercially for the fur trade. To prevent unregulated hunting and poaching of leopards for skins in areas opened to trophy hunting will require increased enforcement of current restrictions on the international trade of leopard skins—a difficult law enforcement task for most African countries.

Ideally, a hunting program for leopards should be only one part of a more comprehensive program designed to provide conservation-related economic benefits for local inhabitants. Properly managed, it could be combined with tourism or a game cropping operation where selected herbivores are also harvested on a sustained basis for protein or profit. Such programs are already conducted on some large game ranches in Namibia, Zimbabwe, and the Republic of South Africa and on concessioned lands in Botswana. The hunting of leopards will be best managed on large tracts of land that support ample populations of prey and leopards. After some preliminary surveys an estimate should be made of how many leopards could be removed annually without jeopardizing the population.

Smaller tracts or tracts with highly human-altered wildlife populations will be more difficult to manage because leopard populations there are likely to be low, with unpredictable annual recruitment. If hunting of leopards from such areas occurs, it must be extremely conservative and accompanied by frequent surveys. All hunting programs should be based on accurate assessments of leopard numbers and annual recruitment.

Three levels of surveys are possible depending on the availability of personnel, equipment, and funds. As a minimum, an area should be surveyed for a crude estimate (level 1) by placing baits of native or domestic species at strategic locations throughout the proposed hunting area. These baits should be put out during the dry season when they are most attractive to leopards, spaced 1 to 3 km apart, and monitored simultaneously to reduce double counting the same individuals. To be effective, an area would probably have to be baited from two to twelve months, starting with the most favorable-appearing habitats. This technique will probably give the best results in relatively open habitats such as woodlands, woodland-savannas, and savannas.

Once a leopard visits a bait, its sex, relative age, presence of cubs, and any distinguishing marks (scars, spot patterns) should be recorded. If possible, each leopard should be photographed to record its unique spot pattern, especially on the face (Miththapala 1991). If vehicles are unavailable or if leopards are intolerant of vehicles, a strategically placed blind may need to be constructed near each baiting station. Otherwise a vehicle can serve as a blind. An alternative would be to attach a triggering device between the bait and nearby camera with a flash attachment. Over time, one should be able to distinguish individuals visiting baits and eventually determine the approximate size and composition of the population.

A more thorough survey (level 2) would require capturing leopards in live-traps and photographing, or immobilizing and marking the leopard (ear tags, collar), before releasing it. Close examination of leopards would ensure accurate sex determination, relative age, and reproductive status of females. Accurate identification of previously marked individuals would increase the accuracy of the population estimate. For extremely large areas, one could stratify the area into distinct habitats and sample each accordingly. Estimates for each habitat would apply only to that particular habitat type.

A high level survey (level 3) would require the use of radio collars to monitor and accurately delimit the home ranges of live-captured and released leopards. Accurate information on preferred habitats, mortality rates, dispersal, and recruitment could be obtained with a level 3 survey. Level 3 surveys would be most costly and time consuming, but they would give the best results, especially in densely vegetated habitats.

The region surrounding a proposed leopard hunting area should be carefully evaluated to determine whether a population is completely isolated or whether leopards in adjacent areas might immigrate into the hunting area to replace removed leopards. If isolated but large enough to support a viable population and sustained hunting, a conservative hunting strategy would be essential—to ensure that harvesting did not deplete individuals faster then they can be naturally recruited into the population. If an influx of males, which bring different genes into the population, is unlikely, problems associated with inbreeding may arise, especially with small populations.

Leopard populations probably should not be hunted unless a minimum effective population size of fifty breeding adults, or at least eighty to one hundred individuals, are present and a viable population of leopards exists in adjacent areas. Hunting smaller, isolated populations may only contribute to their eventual demise. The size of an area that can support eighty to one hundred leopards will vary with habitat quality and may range from 300 km^2 in high-quality habitats to 5,000 km^2 in low quality habitats. As a very crude estimate, most proposed hunting areas should be at least 2,500 km^2 if habitat appears average and is adjacent to other areas supporting leopards.

Only male leopards should be taken by hunters until further information suggests otherwise. Males seem to be naturally replaced more rapidly than females; they have a higher natural mortality rate; and they are more apt to respond to baits for survey or hunting purposes. Because of their larger size and visible genitalia, they can be easily distinguished from females. Several options are available for estimating hunting rates of males, all of which result in relatively low hunting levels. One method assumes that all natural mortality is compensatory and replaced by hunting mortality. The other more realistic method assumes hunting and natural mortality may not be completely compensatory and may even be additive. One can also base hunting rates on the proportions and natural mortality rates of adult or subadult males in the population.

Information from the Kruger National Park leopard study areas suggests a hunting rate of 4% to 6% of a total population may be possible if one assumes complete compensatory mortality. When a hunting level of one-half the natural mortality rate is assumed, the hunting rate declines to 2% to 3% of the total population. A hypothetical population of one hundred leopards whose population composition and mortality patterns are similar to leopards in the Kruger National Park study areas are speculated to withstand a hunting kill of at least two, possibly as many as six, male leopards per year. Hunting rates will undoubtedly vary among populations. One computer model of leopard population dynamics predicted a 5% safe and a 10% maximum sustainable harvest level for leopards (Martin and Meulenaer 1988). One factor to consider is whether other, perhaps significant, forms of human-related mortality, such as poaching and poisoning, are already impacting a leopard population. These additional forms of mortality would lower the legal hunting rate.

Methods of hunting leopards vary from chance encounters, tracking, and trailing with dogs to baiting. Baiting may be the easiest to manage and the least disruptive to a population. Managers of hunting areas should have established baiting locations and be able to distinguish individual leopards well before hunting occurs. Another advantage of baiting is that it would allow managers to examine each leopard responding to a bait to ensure it is the desired individual to be taken. Drawbacks of baiting for purposes of hunting are that it is not challenging, does not require the hunter to have knowledge about the animal's habits, and merely turns the leopard into a target.

Hunted leopard populations should be closely monitored to ensure that hunting is not contributing to a population decline. Only selected locations within a hunting area should actually have leopards removed from them. Reduced hunting pressure should be enforced if hunted males are not rapidly replaced. Areas frequented by females, such as koppies and other rocky outcrops used as denning areas, should be avoided to prevent disrupting the females' habits and their unintentional killing. Actual hunting sites should be specific places where males are periodically observed or attracted to baits. Only specific baiting locations should be hunted, and then only on a rotational basis. For example, if a male was taken at one bait location, the next male removed from the hunting area should be taken at least two to three male-home-range-distances away. This would prevent creating a large vacancy among males, which could prevent or reduce female productivity. Baiting should occur even after a male has been taken to ensure that his replacement has appeared. In healthy leopard populations, males taken by hunters should be replaced within one to six months. To help maintain genetic diversity within the population and reduce the possibility that infanticide will become a significant mortality factor among cubs, newly arriving males should be allowed to reproduce for at least one to two years before being taken by hunters.

Leopard hunting programs will have to be closely regulated to prevent illegal harvest or overharvest. Participating countries should adopt laws or regulations specifically addressing the hunting of leopards. These regulations should be strictly enforced. Management techniques to accomplish these objectives include: (1) issuing only a limited number of nontransferable permits to hunt leopards based on certified or documented survey data; (2) requiring that sexes of leopards taken by hunters be verifiable by examining the skins; (3) immediately sealing all skins and accompanying skulls with a locking seal while they are in transport or being processed; (4) requiring that certified legal documents be kept with the skin or skull at all times; (5) requiring presentation of export and import permits per terms of the CITES treaty; and (6) registering all legally killed leopards with an international organization or clearinghouse. Exporting and importing countries that either will not accept or repeatedly violate these conditions should not be allowed to establish or maintain leopard hunting programs until corrective actions are taken. Guides, outfitters, and hunters violating these conditions should be penalized, in addition to having their hunting privileges revoked.

The Future

The future of the leopard is uncertain. The environmental problems and human poverty resulting from chronic overpopulation, soil erosion, drought, famine, civil wars, arms buildups, debts, corruption, and maladministration throughout much of the leopard's range make it exceedingly difficult for leopards, and other forms of wildlife, to survive. If current trends continue, leopard populations will slowly but surely continue to disappear in the highly developed, densely populated areas of the leopard's range. A few increasingly isolated populations of leopards may survive in some of the smaller protected areas. In these protected areas, leopards and their prey will have to be intensively managed, and much thought must be given to protecting a population's genetic variability.

Leopard populations will survive the longest in those areas that are inhospitable and where human population densities are low, such as the Kalahari Desert and perhaps the Zaire River Basin. With increasing human use of even these areas, leopard population densities are likely to steadily decline.

Leopard populations are probably the most ecologically and genetically secure in the few large protected areas remaining on the African continent. This assumes political, social, and economic events will not prevent governments from properly managing such areas in the future, which is uncertain. Fortunately, because of the leopard's adaptability, little must be done specifically for leopards within these large protected areas other than maintaining healthy ecosystems. Although even these large areas will eventually become isolated in a surrounding sea of development and humanity, their leopard populations should remain large enough to sustain long-term genetic viability.

Even under the most dismal scenarios of human population growth, social unrest, increasing severity of droughts, and global climatic changes, I hope there will be a few places left in Africa and in Asia where leopard populations will endure in their natural surroundings. If any large solitary felid is likely to survive in this region of the planet during the difficult times ahead, it is the leopard. The biologist-philosopher Julian Huxley maintained that we have the collective duty of preserving wild nature (Huxley 1957). He also said "To exterminate a living species, be it a lion or lammergeier, to wipe out wild flowers or birds over great tracts of country, is to diminish the wonder and the interest of the universe." Let us hope that the wonder and interest of future human generations will not be diminished and that they will continue to have the opportunity to see one of nature's most elegant creatures, the leopard, in the wild.

Appendix A

Common and Scientific Names of Wild Mammals, Birds, Reptiles, and Fish Mentioned in the Text

Class Mammalia

ORDER EDENTATA

Armadillo	*Dasypus novemcinctus*

ORDER PRIMATA

Galago (bushbaby)	*Galago crassicaudatus*
Mangebey	*Cercocebus spp.*
Chacma baboon	*Papio ursinus*
Olive baboon	*Papio anubis*
Gray langur	*Presbytis entellus*
Vervet monkey	*Cercopitheticus aethiops*
Colobus monkey	*Colobus spp.*

Order Pholidota

Pangolin	*Manis spp.*

Order Lagomorpha

Natal red hare	*Pronolagus crassicaudatus*
Scrub hare	*Lepus saxatilis*
Cape hare	*Lepus capensis*
Black-tailed jackrabbit	*Lepus californicus*
Cottontail rabbit	*Sylvilagus nuttalli*

Order Rodentia

Bush squirrel	*Paraxerus palliatus*
Giant forest squirrel	*Protoxerus spp.*
Temminks giant squirrel	*Epixerus ebii*
Sun squirrel	*Heliosciurus spp.*
Cricetid rodents	*Mesomys spp.*
Cane rat (giant rat)	*Thryonomy spp.*
American porcupine	*Erethizon dorsatum*
African porcupine	*Hystrix spp.*
Brush-tailed porcupine	*Atherurus africanus*
Springhare	*Pedetes capensis*
Capybara	*Hydrochoerus spp.*
Paca	*Agouti paca*
Agouti	*Dasyprocta punctata*

Order Carnivora

Black bear	*Ursus americanus*
Sloth bear	*Melursus ursinus*
Wolf	*Canis lupus*
Coyote	*Canis latrans*
Black-backed jackal	*Canis mesomelas*
Side-striped jackal	*Canis adustus*
Red fox	*Vulpes vulpes*
Asiatic (Indian) wild dog	*Cuon alpinus*
Wild dog	*Lycaon pictus*
Raccoon dog	*Nyctereutes procyonoides*
Raccoon	*Procyon lotor*
Short-tailed weasel	*Mustela erminea*
Mink	*Mustela vison*

Wolverine	*Gulo gulo*
Ratel	*Mellivora capensis*
Ferret badger	*Melogale moschata*
African Civet	*Viverra civetta*
Common genet	*Genetta genetta*
Rusty-spotted	*Genetta tigrina*
Crabeating mongoose	*Herpestes urva*
Slender mongoose	*Herpestes sanguineus*
Banded mongoose	*Mungus mungo*
White-tailed mongoose	*Ichneumia albicauda*
Dwarf mongoose	*Helogale parvula*
Spotted hyena	*Crocuta crocuta*
Brown hyena	*Hyaena brunnea*
Clouded leopard	*Neofelis nebulosa*
Snow leopard	*Panthera uncia*
Tiger	*Panthera tigris*
Leopard	*Panthera pardus*
Jaguar	*Panthera onca*
Lion	*Panthera leo*
Cheetah	*Acinonyx jubatus*
Marbled cat	*Pardofelis marmorata*
Cougar	*Felis concolor*
Serval	*Felis serval*
Caracal	*Felis caracal*
Lynx	*Felis lynx*
Spanish lynx	*Felis pardina*
Asian golden cat	*Felis temminckii*
Bobcat	*Felis rufus*
Wildcat	*Felis silvestris*

ORDER TUBULIDENDATA

Aardvark	*Orycteropus afer*

ORDER HYRACOIDEA

Tree hyrax	*Dendrohyrax arboreus*
Rock hyrax	*Heterohyrax brucei*
Cape hyrax	*Procavia capensis*

ORDER PROBOSCIDAE

African elephant	*Loxodonta africana*

ORDER PERISSODACTYLA

White rhinoceros	*Ceratotherium simium*
Black rhinoceros	*Diceros bicornis*
Burchell's zebra	*Equus burchelli*

ORDER ARTIODACTYLA

Hippopotamus	*Hippopotamus amphibius*
Wild pig	*Sus scrofa*
Bush pig	*Potamochoerus porcus*
Giant forest hog	*Hylochoerus meinertzhageni*
Warthog	*Phacochoerus aethiopicus*
Peccary	*Tayassu spp.*
Guanaco	*Lama guanicoe*
Muntjac (barking deer)	*Muntiacus reevsi*
Mule deer	*Odocoileus hemionus*
Red brocket deer	*Mazama americana*
Guemul	*Hippocamelus spp.*
Sambar deer	*Cervus unicolor*
Wapiti (elk)	*Cervus canadensis*
Hog deer	*Axis porcinus*
Axis deer	*Axis axis*
Okapi	*Okapia johnstoni*
Giraffe	*Giraffa camelopardalis*
Bushbuck	*Tragelaphus scriptus*
Bongo	*Tragelaphus euryceros*
Nyala	*Tragelaphus angasi*
Lesser kudu	*Tragelaphus imberbis*
Greater kudu	*Tragelaphus strepsiceros*
Eland	*Tragelaphus oryx*
Gemsbok	*Oryx gazella*
Roan antelope	*Hippotragus equinus*
Sable antelope	*Hippotragus niger*
Common waterbuck	*Kobus ellipsiprymnus*
Puku (kob)	*Kobus kob*
Mountain reedbuck	*Redunca fulvorufula*
Common reedbuck	*Redunca arundinum*
Vaal rhebok	*Pelea capreolus*
Kongoni	*Alcelaphus buselaphus cokii*
Hartebeest	*Alcelaphus buselaphus*
Sassaby (topi)	*Damaliscus lunatus*
Blue wildebeest	*Connochaetes taurinus*
Impala	*Aepyceros melampus*

Grant's gazelle	*Gazella granti*
Thompson's gazelle	*Gazella thomsoni*
Spring-buck	*Antidorcas marsupialis*
Duikers (forest)	*Cephalophus spp.*
Red duiker	*Cephalophus natalensis*
Grey duiker	*Cephalophus grimmia*
Oribi	*Ourebia ourebi*
Klipspringer	*Oreotragus oreotragus*
Steenbuck	*Raphicerus campestris*
Sharp's grysbok	*Raphicerus sharpei*
Dik dik	*Madoqua spp.*
Water buffalo	*Bubalus bubalis*
African buffalo	*Syncerus caffer*
Goral	*Nemorhaedus goral*
Wild goat	*Capra aegagrus*
Markhor	*Capra falconeri*
Blue sheep (bharal)	*Pseudois nayaur*

Class Aves

Ostrich	*Struthio camelus*
Hammerkop	*Scopus umbreta*
Marabou stork	*Leptoptilos crumeniferus*
European white stork	*Ciconia ciconia*
Hadeda ibis	*Bostrychia hagedash*
Secretary bird	*Sagittarius serpentarius*
Cape vulture	*Gyps coprotheres*
White-backed vulture	*Gyps africanus*
White-headed vulture	*Trigonoceps occipitalis*
Egyptian vulture	*Neophron percnopterus*
Hooded vulture	*Necrosyrtes monarchus*
Lappet-faced vulture	*Torgos tracheliotus*
Tawny eagle	*Aquila rapax*
Martial eagle	*Polemaetus bellicosus*
Crowned hawk-eagle	*Stephanoaetus coronatus*
Black-breasted snake-eagle	*Circaetus pectoralis*
Bateleur eagle	*Terathopius ecaudatus*
Fish eagle	*Haliaeetus vocifer*
Coque francolin	*Francolinus coqui*
Crested francolin	*Francolinus sephaena*
Natal francolin	*Francolinus natalensis*
Swainson's francolin	*Francolinus swainsonii*

Crowned guinea fowl	*Numida meleagris*
Pearl-spotted owl	*Glaucidium perlatum*
Spotted eagle owl	*Bubo africanus*
Red-billed hoopoe	*Phoeniculus purpureus*

Class Reptilia

Crocodile	*Crocodylus niloticus*
Leopard tortise	*Geochelane pardalis*
Tree leguan	*Varanus exanthematicus*
Water leguan	*Varanus niloticus*
Chameleon	*Chamaeleo dilepis*
Python	*Python sebae*
Night adder	*Causus defilippii*
Puff adder	*Bitis arietans*
Boomslang	*Dispholidus typus*
Black mamba	*Dendroaspis polylepis*
Cobra	*Naja spp.*

Class Pisces

Catfish	*Clarias spp.*
Yellowfish	*Barbus spp.*
Tiger fish	*Hydrocynus spp.*

Appendix B

Endangered and Threatened Classification of the Leopard

The leopard was listed as an endangered species by the United States in March 1972. It was added to appendix I of the Convention on International Trade in Endangered Species of Wild Fauna and Flora (CITES), which became effective in 1975. An endangered species is defined as "one in danger of extinction throughout all or a significant portion of its range." International trade of leopard products for commercial purposes is prohibited among countries that are members of CITES.

In 1980 the U.S. Fish & Wildlife Service (FWS) proposed to reclassify the leopard as a threatened rather than an endangered species in sub-Saharan Africa. A threatened species is defined as "one that is likely to become endangered within the foreseeable future throughout all or a significant portion of its range." Based on comments received, FWS modified its original proposal to reclassify leopards as threatened only in southern Africa, as opposed to sub-Saharan Africa. Authorities in west and northeastern Africa believed leopards in their countries were still endangered. As a result, leopards in Africa were reclassified as threatened south of a line running along the borders of the following countries: Gabon/Rio Muni; Gabon/Cameroon; Congo/Cameroon; Congo/Central African Republic; Zaire/ Central African Republic; Zaire/Sudan; Uganda/Sudan; Kenya/Sudan; Kenya/Ethiopia; and Kenya/Somalia.

The effect of the reclassification was that sport-hunting trophies of the leopard legally taken in accordance with the laws of appropriate countries in southern Africa may be imported into the United States without a permit issued, pursuant to the Endangered Species Act of 1973 that created and defined threatened status. However, the importer must obtain an import permit from the U.S. Management Authority of the Convention under the terms and conditions of CITES. This ruling became effective on March 1, 1982.

Leopards are still classified by the United States as endangered throughout the rest of their historic range. This includes the remainder of Africa and Asia Minor, India, Southeast Asia, China, Malaysia, and Indonesia. Commercial trade in all leopard products is still prohibited under appendix I of CITES.

Appendix C

Infanticide Among Leopards

Little was known about infanticide among leopards until an isolated population
of leopards of the subspecies *Panthera pardus jarvisi* was discovered and studied
in the Judean Desert and Negev Hills of southern Israel (Ilany 1986; Ilany 1990;
G. Ilany, personal communication).

At least three different male leopards killed a minimum of eleven leopard
cubs during a nine-year period (1981–89) (Ilany 1990). One male (Ben Aflul)
killed an eight-month-old male cub and immediately bred with its mother
(Humbaba), who gave birth to his offspring one year later. Another male
(Ktushion) killed two cubs of another female (Shlomtsion), then bred with her
to produce a male cub that survived. Ktushion also killed two cubs of the other
female (Humbaba) in 1983. She, however, bred with the other male (Ben Aflul).
Humbaba and Ben Aflul produced a male cub that survived. A seven-month-
old male cub of the female Shlomtsion was also killed by the adult male
(Ktushion), as was three of her other cubs. Another adult male leopard (Hordos), son of female Shlomtsion and male Ktushion, killed two three-month-old
cubs that were born to his own mother from her mating with another male
(Amrafel). One female (Humbaba) was not able to successfully rear a cub since
1983 because all her cubs were killed by male leopards. All the male leopards in
this small population competed for the right to father their offspring with one
female (Shlomtsion). Not a single adult leopard was added to this population

from 1984 to 1989 primarily because of cub mortality from infanticide (Ilany 1990).

Two instances of female leopards breeding with their own sons were observed in the Judean Desert (Ilany 1990). The female Shlomtsion bred with her son (Hordos). They had a male cub (Esarchadon), which was killed by another adult male leopard (Ktushion). The other female (Humbaba) was also once observed copulating with her son (Amrafel).

Other unique observations of leopards were recorded in this desert population (Ilany 1990). An older adult female leopard (Humbaba) fought with and was defeated by a younger female leopard (Shlomtsion). The younger female eventually expelled, by fighting and chasing, the older female (accompanied by her son Amrafel) from a substantial portion of the older female's home range. The older female (Humbaba) was observed expelling her nine-month-old daughter (Tsruyah) from the daughter's natal home range. Later, after becoming independent, Tsruyah temporarily assumed a substantial portion of her mother's home range. But after several months the old female chased the younger female away; she was never seen in the Judean Desert again. A younger male leopard (Ben Aflul) was also believed to be responsible for expelling and perhaps killing an old male leopard (Pre-cambrian).

These and other unpublished observations by Giora Ilany of Judean Desert leopards increase our knowledge of leopard ecology and behavior. They also warn us of potential threats and dangers as we attempt to conserve small, increasingly isolated populations of leopards, as well as other felids and carnivores.

References

Ables, E. D. and J. Ables. 1970. Home range and activity studies of impala in northern Kenya. *Trans. N. Amer. Wildl. Conf.*, 34: 360–71.

Ackerman, B. B., F. G. Lindzey, and T. P. Hemker. 1986. Predictive energetics model for cougars. In S. D. Miller and D. D. Everett, eds., *Cats of the World: Biology, Conservation, and Management*, pp. 333–52. Washington, D.C.: National Wildlife Federation.

Acocks, J. P. H. 1953. Veld types of South Africa. *Bot. Surv. S. Afr.*, 40: 1–128.

Adamson, J. 1980. *Queen of Sheba: The Story of an African Leopard.* New York: Harcourt Brace Jovanovich.

Ahlborn, G. G. and R. M. Jackson. 1988. Marking in free-ranging snow leopards in West Nepal: A preliminary assessment. In H. Freeman, ed., *Proc. 5th Intl. Snow Leopard Symp.*, pp. 35–49. Bellevue, Wash.: Snow Leopard Trust.

Akeley, C. and M. L. J. Akeley. 1931. *Lions, Gorillas, and Their Neighbors.* London: Stanley Paul.

Allen, R. 1980. *How to Save the World: Strategy for World Conservation.* London: Krogan Page.

Altmann, S. A. and J. Altmann. 1970. *Baboon Ecology.* Chicago: University of Chicago Press.

Amadon, D. 1973. Birds of the Congo and Amazon forests: A comparison. In B. J. Meggers, E. S. Ayensu, and W. D. Duckworth, eds., *Tropical Forest Ecosystems in Africa and South America: A Comparative Review*, pp. 267–77. Washington, D.C.: Smithsonian Institution Press.

Anchorage Daily News. 1987. World news: Leopard kills 10 children in Nepal. Thursday, January 22.

Anderson, D. and R. Grove. 1989. *Conservation in Africa: People, Politics and Practice.* New York: Cambridge University Press.

Anderson, G. D. and B. H. Walker. 1974. Vegetation composition and elephant damage in the Sengwa Wildlife Research Area, Rhodesia. *J. South Afr. Wildl. Manage. Assoc.* 4: 4–14.

Anderson, J. L. 1972. Seasonal changes in the social organization and distribution of impala in the Hluhluwe Game Reserve, Zululand. *J. Sth. Afr. Wildl. Mgmt. Assoc.*, 2(2): 16–22.

———— 1975. The occurrence of a secondary breeding peak in the southern impala. *E. Afr. Wildl. J.*, 13: 149–51.

Aschoff, J. 1966. Circadian activity pattern with two peaks. *Ecology*, 47: 657–702.

Ashman, D. 1975. Mountain lion investigations. *Fed. Aid Wildl. Restor. Rep.*, Proj. W48–6. Nev. Fish and Game Dept.

Badino, G. 1975. *Big Cats of the World.* New York: Bounty Books.

Baikov, N. A. 1927. Zuerovyi promysel v Manchzhurii (Fur trade in Manchuria). *Okhotnik.* 2. Moscow. (in Stroganov 1969).

Bailey, T. N. 1972. Ecology of bobcats with special reference to social organization. Ph.D. diss., University of Idaho.

———— 1974. Social organization of a bobcat population. *J. Wildl. Manage.* 38: 435–46.

———— 1979. Den ecology, population parameters, and diet of eastern Idaho bobcats. In *Proc. Bobcat Res. Conf., National Wildlife Federation Sci. and Tech. Series*, no. 6, pp. 62–69.

Bailey, T. N. and M. G. Hornocker. 1973. A preliminary study of the wolverine (*Gulo gulo*) in the Hungry Horse Reservoir area, Montana. Unpublished Report. 32 pp. University of Idaho, Moscow: Idaho Cooperative Wildlife Research Unit.

Baker, J. R. 1969. Trypanosomes of wild mammals in the neighborhood of Serengeti National Park. *Symp. Zool. Soc. London*, 24: 147–58.

Barash, D. P. 1977. *Sociobiology and behavior.* New York: Elsevier

Barnett, S. F. and D. W. Brocklesby. 1969. Some piroplasms of wild mammals. *Symp. Zool. Soc. London*, 24: 159–76.

Beier, P. 1991. Cougar attacks on humans in the United States and Canada. *Wildl. Soc. Bull.*, 19: 403–12.

Belden, R. C., W. B. Frankenberger, R. T. McBride, and S. T. Schwikert. 1988. Panther habitat use in southern Florida. *J. Wildl. Manage.*, 52: 660–63.

Berrie, P. M. 1973. Ecology and status of the lynx in interior Alaska. In R. L. Eaton, ed., *The World's Cats.* Vol. 1, *Ecology and Conservation*, pp. 4–41. Winston, Ore.: World Wildlife Safari.

Bertram, B. C. R. 1975. The social system of lions. *Scientific American*, 2: 54–61.

———— 1978. *Pride of Lions.* New York: Scribner's.

———— 1982. Leopard ecology as studied by radio-tracking. *Symp. Zool. Soc. London*, 49: 341–52.

Bigalke, R. C. 1974. Ungulate behaviour and management, with special reference to husbandry of wild ungulates on South African ranches. In V. Geist and F. Walther, eds., *The Behaviour of Ungulates and Its Relation to Management.* IUCN publ. 24, 2: 830–52.

Borner, H. 1977. Leopards in western Turkey. *Oryx*, 14(1): 26–30.

Bothma, J. Du P. and E. A. N. Le Riche. 1984. Aspects of the ecology and the behaviour of the leopard *Panthera pardus* in the Kalahari desert. *Koedoe Suppl.*, 27: 259–79.

———— 1986. Prey preference and hunting efficiency of the Kalahari Desert leopard. In S. D. Miller and D. D. Everett, eds., *Cats of the World: Biology, Conservation, and Management*, pp. 389–414. Washington, D.C.: National Wildlife Federation.

——— 1990. The influence of increasing hunger on the hunting behaviour of the southern Kalahari leopards. *J. Arid. Environ.*, 18: 79–84.

Bourlière, F. 1963a. Observations on the ecology of some large African mammals. In F. Howell and F. Bourlière, eds., *African Ecology and Human Evolution*, pp. 43–54. Chicago: Aldine.

——— 1963b. Specific feeding habits of African carnivores. *Afr. Wildl.*, 17(1): 21–27.

——— 1965. Densities and biomass of some ungulate populations in eastern Congo and Rwanda, with notes on population structure and lion/ungulate ratios. *Zool. Afr.*, 1(1): 199–207.

——— 1973. The comparative ecology of rain forest mammals in Africa and tropical America: Some introductory remarks. In B. J. Meggers, E. S. Ayensu, and W. D. Duckworth, eds., *Tropical Forest Ecosystems in Africa and South America: A Comparative Review*, pp.279–92. Washington, D.C.: Smithsonian Institution Press.

Braack, L. E. O. 1983. *The Kruger National Park*. Cape Town: C. Struik.

Brahmachary, R. L. and J. Dutta. 1981. On the pheromones of tigers: Experiments and theory. *Am. Nat.*, 118: 561–67.

——— 1987. Chemical communication in the tiger and leopard. In R. L. Tilson and U. S. Seal, eds., *Tigers of the World: The Biology, Biopolitics, Management, and Conservation of an Endangered Species*, pp. 296–302. Park Ridge , N.J.: Noyes Publications.

——— 1991. The marking fluid of the tiger. *Mammalia*, 55(1): 150–52.

Brain, C. K. 1981. *The Hunters or the Hunted? An Introduction to African Cave Taphonomy*. Chicago: University of Chicago Press.

Brand, C. J., L. B. Keith, and C. A. Fischer. 1976. Lynx responses to changes in snowshoe hare densities in central Alberta. *J. Wildl. Manage.*, 40: 416–28.

Brand, D. J. 1963. Records of mammals bred in the National Zoological Gardens of South Africa during the period 1908–1960. *Proc. Zool. Soc. London*, 140: 617–59.

Brander, A. D. 1923. *Wild Animals in Central India*. London: Edward Arnold.

Brereton, J. L. G. 1971. Inter-animal control of space. In A. H. Esser, ed., *Behavior and Environment: The Use of Space by Animals and Men*, pp. 69–91. New York: Plenum Press.

Brown, J. L. and G. H. Orians. 1970. Spacing patterns in mobile animals. *Ann. Rev. Ecol. Sys.* 1: 239–62.

Burney, D. A. 1980. The effects of human activities on cheetah (*Acinonyx jubatus*) in the Mara region of Kenya. M.Sc. thesis., Univ. of Nairobi.

Caro, T. M. and D. A. Collins. 1987. Male cheetah social organization and territoriality. *Ethology* 74: 52–64.

Cavello, J. A. 1990. Cat in the human cradle. *Nat. Hist.*, February 1990: 52–60.

Charles-Dominique, P. 1977. *Ecology and Behaviour of Nocturnal Primates: Prosimians of Equatorial West Africa*. New York: Columbia University Press.

Cheng, T. 1964. *The biology of animal parasites*. New York: W. B. Saunders.

Child, G. 1964. Growth and aging criteria of impala, *Aepyceros melampus*. *Occ. Papers. Nat. Mus. S. Rhod.*, 4(27): 128–35.

Child, G., H. H. Roth, and M. Kerr. 1968. Reproduction and recruitment patterns in warthog *(Phacochoerus aethiopicus)* populations. *Mammalia*, 32: 6–29.

Child, G., P. Smith and W. Von Richter. 1970. Tsetse control hunting as a measure of large mammal population trends in the Okavango Delta, Botswana. *Mammalia*, 34(1): 34–75.

Child, G. and V. J. Wilson. 1964. Delayed effects of tsetse control hunting on a duiker population. *J. Wildl. Manage.*, 28(4): 866–68.

Clark, P. J. and F. C. Evans. 1954. Distance to nearest neighbor as a measure of spatial relationships in populations. *Ecology*, 35(4): 445–53.

Cloudsley-Thompson, J. L. 1969. *The Zoology of Tropical Africa*. New York: W. W. Norton.

Cobb, S. 1981. The leopard problems of an overabundant, threatened, terrestrial carnivore. In P. A. Jewell, S. Holt, and D. Hart, eds., *Problems in Management of Locally Abundant Wild Mammals*, pp. 181–91. New York: Academic Press.

Coe, M. J. 1980. African wildlife resources. In M. E. Soule and B. A. Wilcox, eds., *Conservation Biology: An Evolutionary-Ecological Perspective*, pp. 273–302. Sunderland, Mass.: Sinauer.

Coe, M. J., D. H. Cumming, and J. Phillipson. 1976. Biomass and production of large African herbivores in relation to rainfall and primary production. *Oecologia*, 22: 341–54.

Cole, M. M. 1961. *South Africa*. London: Methuen.

Collier, G. E. and S. J. O'Brien. 1985. A molecular phylogeny of the Felidae immunological distance. *Evolution*, 39: 473–87.

Collins, W. B. 1959. *The Perpetual Forest*. Philadephia: Lippincott.

Corbett, J. E. 1946. *Man-eaters of Kumaon*. New York: Oxford University Press.

—— 1947. *The Man-eating Leopard of Rudraprayag*. New York: Oxford University Press.

Cowan, I. M. 1947. The timber wolf in the Rocky Mountain National Parks of Canada. *Can. J. Res.*, 25: 139–74.

Crabtree, I. D. 1973. *Sabi Sand Wildtuin: Warden's Report for the Year Ended 30th April, 1973*. Circular no. 452.

—— 1974. *Sabi Sand Wildtuin: Warden's Report for the Year Ended 30th April 1974*. Circular no. 483.

Crandall, L. S. 1964. *The Management of Wild Animals in Captivity*. Chicago: University of Chicago Press.

Crawshaw, P. G., Jr., and H. B. Quigley. 1991. Jaguar spacing, activity, and habitat use in a seasonally flooded environment in Brazil. *J. Zool. Soc. London*, 223: 357–70.

Crook, J. H. 1970. Social organization and the environment: Aspects of contemporary social ethology. *Anim. Behav,*. 18: 197–209.

Cross, H. 1964. Observations on the formation of the feeding tube by *Trombicula splendens* larvae. *Acarologia, Fasc. Hors. Series* 6: 155–61.

Dasmann, R. F. and A. S. Mossman. 1962. Road strip counts of estimating numbers of African ungulates. *J. Wildl. Manage.*, 26(1): 101–4.

Davis, J. W. and R. C. Anderson. 1971. *Parasitic Diseases of Wild Mammals*. Ames: Iowa State University Press.

Dorst, J. and P. Dandelot. 1970. *A Field Guide to the Larger Mammals of Africa*. London: Collins.

Eaton, R. L. 1970a. Hunting behavior of the cheetah. *J. Wildl. Manage.*, 34(1): 56–57.

—— 1970b. Group interactions, spacing and territoriality in cheetahs. *Z. F. Tierpsychol.* 27(4): 481–91.

—— 1974. *The Cheetah: the Biology, Ecology, and Behavior of an Endangered Species*. Behavioral Science Series. New York: Van Nostrand Reinhold.

—— 1977a. Reproductive biology of the leopard. *Zool. Garten*, 47: 329–51.

—— 1977b. *The Status and Conservation of the Leopard in Sub-Saharan Africa*. Tucson: Safari Club International.

Edey, M. 1968. *The Cats of Africa*. New York: Time-Life Books.

Eisenberg, J. F. 1980. The density and biomass of tropical mammals. In M. E. Soule and

B. A. Wilcox, eds., *Conservation Biology: An Evolutionary-Ecological Perspective*, pp. 35–55. Sunderland, Mass.: Sinauer.

———— 1986. Life history strategies of the felidae: Variations on a common theme. In S. D. Miller and D. D. Everett, eds., *Cats of the World: Biology, Conservation, and Management*, pp. 293–303. Washington, D.C.: National Wildlife Federation.

Eisenberg, J. F. and M. Lockhart. 1972. An ecological reconnaissance of Wilpattu National Park, Ceylon. *Smithsonian Contrib. Zool.* 101: 1–118.

Elder, W. H. and N. L. Elder. 1970. Social groupings and primate association of the bushbuck *(Tragelaphus scriptus)*. *Mammalia*, 34(3): 356–62.

Ellerman, J. R. and T. C. S. Morrison-Scott. 1966. *Checklist of Palearctic and Indian Mammals 1758 to 1946*. Oxford: Alden Press.

Elliot, J. P., I. M. Cowan, and C. S. Holling. 1977. Prey capture by the African lion. *Can. J. Zool.*, 55: 1811–28.

Eloff, F. C. 1959. Observations on the migration and habits of the antelopes of the Kalahari Gemsbok Park. *Koedoe* 2:(Pt. 1): 1–29 and (Pt. 2): 30–51.

———— 1973. Ecology and behaviour of the Kalahari lion. In R. L. Eaton, ed., *The World's Cats*. Vol. 1, *Ecology and Conservation*, pp., 90–126. Winston, Ore.: World Wildlife Safari.

Eltringham, S. K. 1979. *The Ecology and Conservation of Large African Mammals*. Baltimore: University Park Press.

Emlen, J. T., Jr. 1957. Defended area?—A critique of the territory concept and of conventional thinking. *Ibis*, 98: 352.

Emmons, L. H. 1980. Ecology and resource partitioning among nine species of African rain forest squirrels. *Ecol. Monogr.*, 50: 31–54.

———— 1987. Comparative feeding ecology of felids in a neotropical rainforest. *Behav. Ecol. Sociobiol.*, 20: 271–83.

Etkin, W. 1967. *Social Behavior from Fish to Man*. Chicago: University of Chicago Press.

Estes, R. D. 1967. Predators and scavengers: Stealth, pursuit and opportunism among carnivores on Ngorongoro Crater in Africa. Parts 1 and 2. *Natl. Hist.*, 76(2): 20–29; 76(3): 38–47.

Estes, R. D. and J. Goddard. 1967. Prey selection and hunting behavior of the African wild dog. *J. Wildl. Manage.*, 32(1): 52–70.

Ewer, R. F. 1968. *Ethology of Mammals*. London: Elek Science.

Fairall, N. 1969. The use of the eye lens technique in deriving the age structure and life table of an impala *(Aepyceros melampus)* population. *Koedoe*, 12: 90–95.

Ferrar, A. A. and B. H. Walker. 1974. An analysis of herbivore/habitat relationships in Kyle National Park, Rhodesia. *J. South Afr. Wildl. Manage. Assoc.*, 4: 137–47.

Field, C. R. and R. M. Laws. 1970. The distribution of the larger herbivores in the Queen Elizabeth National Park, Uganda. *J. Appl. Ecol.*, 7: 273–94.

Fisher, J., N. Simon, and J. Vincent. 1969. *Wildlife in Danger*. New York: Viking Press.

Fisler, G. F. 1969. Mammalian organizational systems. *Contrib. Sci. Los Angeles Co. Mus.*, 167: 1–32.

Fitzpatrick, R. 1907. *Jock of the Bushveld*. London: Longman.

Foose, T. J. 1987. Species survival plans and overall management strategies. In R. L. Tilson and U. S. Seal, eds., *Tigers of the World: The Biology, Biopolitics, Management, and Conservation of an Endangered Species*, pp. 304–16. Park Ridge, N.J.: Noyes Publications.

Foster, J. and M. Coe. 1968. The biomass of game animals in Nairobi National Park, 1960-1966. *J. Zool. Soc. London*, 155(4): 413–25.

Frädrich, H. 1974. A comparison of behavior in the Suidea. In V. Geist and F. Walther, eds., *The Behaviour of Ungulates and Its Relation to Management*. IUCN publ. 24, 1: 133–43.

Frame, G. and L. Frame. 1981. *Swift and Enduring: Cheetahs and Wild Dogs of the Serengeti*. New York: E. P Dutton.

Frankel, O. H., and M. E. Soule. 1981. *Conservation and Evolution*. Cambridge: Cambridge University Press.

Franklin, I. R. 1980. Evolutionary change in small populations. In M. E. Soule and B. A. Wilcox, eds., *Conservation Biology: An Evolutionary-Ecological Perspective*, pp. 135–49. Sunderland, Mass.: Sinauer.

Fretwell, S. D. 1972. *Populations in a Seasonal Environment*. Princeton: Princeton University Press.

Gertenbach, W. P. D. 1980. Rainfall patterns in the Kruger National Park. *Koedoe*, 23: 35–43.

———— 1983. Landscapes of the Kruger National Park. *Koedoe*, 26: 9–121.

Goodwin, H. A. and C. H. Holloway. 1972. *Red Data Book*. Vol. 1, *Mammalia*. Morges, Switzerland: IUCN.

Graham, A. 1966. East African Wild Life Society cheetah survey; extracts from the report by Wildlife Services. *E. Afr. Wild. J.*, 4: 50–55.

Graupner, E. D. and O. F. Graupner. 1971. Predator-prey interrelationships in a natural big game population. *Final Rep.* Nature Conservation Division, Subproj. TN.6.4.1/4F. Pretoria, Rep. So. Afr.: Transvaal Provincial Administration.

Green, H. 1951. The bighorn sheep of Baniff National Park. *Bull. Canada Dept. Resources and Devel, Ottawa*, pp. 1–53.

Grobler, J. H. and V. J. Wilson. 1972. Food of the leopard, *Panthera pardus* (Linn.) in the Rhodes Matopos National Park, as determined by faecal analysis. *Arnoldia Rhod.*, 5(35): 1–9.

Guggisberg, C. A. W. 1975. *Wild Cats of the World*. New York: Taplinger.

Haffner, K. Van, G. Rack, and R. Sachs. 1969. Verschiedene Vertreter der Familie Linguatulidae (Pentastomida) als Parasiten von Saugetieren der Serengeti (Anatomie, Systematik, Biologie). (Dead representatives of the Family Linguatulidae [Pentastomida] as parasites of mammals in the Serengeti [Anatomy, Systematics, Biology]). *Mitt. Hamburg. Zool. Mus. Inst.* 66: 113–17.

Hamilton, P. H. 1976. The movements of leopards in Tsavo National Park, Kenya, as determined by radio-tracking. M.Sc. thesis., University of Nairobi.

———— 1981. The leopard *Panthera pardus* and cheetah *Acinonyx jubatus* in Kenya: Ecology, status, conservation and management. Unpublished report for the U. S. Fish and Wildlife Service, The African Wildlife Leadership Foundation, and the Government of Kenya.

———— 1986. Status of the leopard in Kenya, with reference to sub-Saharan Africa. In S. D. Miller and D. D. Everett, eds., *Cats of the World: Biology, Conservation, and Management*, pp. 447–58. Washington, D.C.: National Wildlife Federation.

Hanby, J. and D. Bygott. 1982. *Lions Share: the Story of a Serengeti Pride*. Boston: Houghton Mifflin.

Hanks, J. 1979. *The Struggle for Survival the Elephant Problem*. New York: Mayflower Books.

Harmon, D. 1990. Wildlife and habitat. In *1990–91 World Resources: A Report by the World Resources Institute*, pp. 121–40. New York: Oxford University Press.

Harrington, F. H. and D. L. Mech. 1979. Wolf howling and its role in territory maintenance. *Behav.*, 68: 207–49.

—— 1982. An analysis of howling response parameters useful for wolf pack censusing. *J. Wildl. Manage.*, 46: 686–93.

Hart, J. A and T. B. Hart. 1989. Ranging and feeding behaviour of okapi (*Okapia johnstoni*) in the Ituri Forest of Zaire: Food limitation in a rain-forest herbivore. In P. A. Jewell and G. M. O. Maloiy, eds., *The Biology of Large African Mammals in Their Environment. Symp. Zool. Soc. London*, 61: 31–50.

Hawthorne, V. M. 1971. Coyote movements in Sagehen Creek Basin, northeastern California. *California Fish and Game* 57: 154–61.

Hemker, T. P. 1982. Population characteristics and movement patterns of cougars in southern Utah. M.Sc. thesis, Utah State University

Hemmer, H. 1976. Fossil history of living Felidae. In R. L. Eaton, ed., *The World's Cats*, 3(2): 1–14. Seattle: Carnivore Research Institute, Burke Museum.

Hendrichs, H. 1970. Schätzungen der Huftierbiomasse in der Dornbuch-savanna nörlich and westlich der Serengetisteppe in Ostafrika nach einem neuen Verfahren und Bemerkungen zur Biomass der anderen pflanzenfressenden Tierarten. (An estimation of the ungulate biomass in the thornbush-savanna in the northern and western Serengeti Plains in East Africa by a new method and comments on the biomass of other herbivorous animal species.) *Saugetierk. Mitt.* 18(3): 237–55.

Henschel, J. R. and J. D. Skinner. 1990. The diet of the spotted hyenas *Crocuta crocuta* in Kruger National Park. *Afr. J. Ecol.* 28: 69–82.

Hillard, D. 1989. *Vanishing Tracks: Four Years Among the Snow Leopards of Nepal*. New York: William Morrow.

Hirst, S. M. 1969a. Road strip census techniques for wild ungulates in African woodlands. *J. Wildl. Manage.*, 33: 40–48.

—— 1969b. Predation as a regulating factor of wild ungulate populations in a Transvaal lowveld nature reserve. *Zool. Afr.*, 4(2): 199–230.

—— 1975. Ungulate-habitat relationships in a South African woodland/savanna ecosystem. *Wildl. Monogr.*, no. 44.

Hitchins, P. M. 1966. Body weights and dressed carcass yields of impala and wildebeest in Hluhluwe Game Reserve. *Lammergeyer*, 6: 20–23.

Honess, R. and K. Winter. 1956. *Diseases of Wildlife in Wyoming*. Cheyenne: Wyoming Game and Fish Comm.

Hoogstraal, H. 1956. *African Ixodiodae*. Part 1, *Ticks of the Sudan/with special reference to Equatoria Province and with preliminary reviews of the genera Boophilus, Margaropus, and Hyalomma*. Washington, D.C.: Bur. Med. Surg., U.S. Navy, 1: 1–1101.

Hopkins, R. A., M. J. Kutilek, and G. L. Shreve. 1986. Density and home range characteristics of mountain lions in the Diablo Range of California. In S. D. Miller and D. D. Everett, eds., *Cats of the World: Biology, Conservation, and Management*, pp. 223–35. Washington, D.C.: National Wildlife Federation.

Hoppe-Dominik, B. 1984. Etude du spectre des proies de la panthere, *Panthera pardus*, dans le Parc National de Tai en Cote d' Ivoire (Prey frequency of the leopard, *Panthera pardus*, in the Tai National Park of the Ivory Coast.) *Mammalia* 48(4): 477–87.

Hornocker, M. G. 1969. Winter territoriality in mountain lions. *J. Wildl. Manage.*, 33 (3): 457–64.

—— 1970. An analysis of mountain lion predation upon mule deer and elk in the Idaho Primitive Area. *Wildl. Monogr.*, no. 21.

Hornocker, M. G. and T. Bailey. 1986. Natural regulation in three species of felids. In S. D. Miller and D. D. Everett, eds., *Cats of the World: Biology, Conservation, and Management*, pp. 211–20. Washington, D.C.: National Wildlife Federation.

Hornocker, M. G. and H. S. Hash. 1981. Ecology of the wolverine in northwestern Montana. *Can. J. Zool.*, 59: 1286–1301.

Huxley, J. 1957. *Religion without Revelation*. New York: Mentor Books.

Ilany, G. 1986. Preliminary observations on the ecology of the leopard *(Panthera pardus jarvisi)* in the Judean Desert. In S. D. Miller and D. D. Everett, eds., *Cats of the World: Biology, Conservation, and Management*, (abstract), p. 199. Washington, D.C.: National Wildlife Federation.

——— 1990. The spotted ambassadors of a vanishing world. *Israel Al,* 31: 16–24.

Isbell, L. A. 1990. A sudden, short-term increase in mortality of vervet monkeys *(Cercopithecus aethiops)* due to leopard predation in Amboseli National Park, Kenya. *Am. J. Primitol.,* 21: 4–52.

Izawa, K. and J. Itani. 1966. Chimpanzees in Kasakati Basin, Tanganyika. *Kyoto Univ. African Studies* 1: 73–156.

Jackson, R. M., and G. G. Ahlborn. 1988a. A radio-telemetry study of the snow leopard *(Panthera uncia)* in West Nepal. *Tigerpaper,* 15(2): 1–14.

——— 1988b. Observations on the ecology of snow leopard in West Nepal. In H. Freeman, ed., *Proc. 5th Intl. Snow Leopard Symp.,* pp. 65–87. Bellevue, Wash.: Snow Leopard Trust.

——— 1989. Snow leopards *(Panthera uncia)* in Nepal—home range and movements. *Natl. Geogr. Res.,* 5(2): 161–75.

Jackson, R. M., G. G. Ahlborn, and K. B. Shah. 1990. Capture and immobilization of wild snow leopards. *Intl. Pedigree Book Snow Leopard,* 6: 93–102.

——— 1991. Snow leopards and other wildlife in the Qomolangma Nature Preserve of Tibet. *Snow Line,* 9(1): 9–12.

Jacobsen, N. H. G. 1974. Distribution, home range, and behaviour patterns of bushbuck in the Lutope and Sengwa Valleys, Rhodesia. *J. South Afr. Wildl. Manage. Assoc.,* 4: 75–93.

Jarman, M. V. 1970. Attachment to home area in impala. *E. Afr. Wildl. J.,* 8: 198–200.

Jarman, M. V. and P. J. Jarman. 1973a. Daily activity of impala. *E. Afr. Wildl. J.,* 11: 75–92.

Jarman, P. J. 1973. The free water intake of impala in relation to the water content of their food. *E. Afr. Agric. For. J.,* 38: 343–51.

——— 1974. The social organization of antelope in relation to their ecology. *Behaviour,* 48: 215–67.

Jarman, P. J. and M. V. Jarman. 1973b. Social behaviour, population structure and reproductive potential in impala. *E. Afr. Wildl. J.,* 11: 329–38.

Jarman, P. J. and A. R. E. Sinclair. 1979. Feeding strategy and the pattern of resource partitioning in ungulates. In A. R. E. Sinclair and M. Norton-Griffiths, eds., *Serengeti: Dynamics of an Ecosystem.,* pp. 130–63. Chicago: University of Chicago Press.

Jonkel, C. J. and I. M. Cowan. 1971. The black bear in the spruce-fir forest. *Wildl. Monogr.,* no. 27.

Joslin, P. 1986. Status of the Caspian tiger in Iran. In S. D. Miller and D. D. Everett, eds., *Cats of the World: Biology, Conservation, and Management*, (abstract), p. 63. Washington, D.C.: National Wildlife Federation.

Joubert, S. C. J. 1986. The Kruger National Park—An introduction. *Koedoe,* 29: 1–11.

Kiley-Worthington, M. 1965. The waterbuck *(Kobus defassa* Rupper 1835 and *K. ellipsiprimnus* Ogilby 1833) in East Africa: Spatial distribution. A study of sexual behavior. *Mammalia,* 29(2): 177–204.

Kingdon, J. 1974. *East African Mammals: An Atlas of Evolution in Africa*. Vol. 1, *Introduction and Primates*. Chicago: University of Chicago Press.

——— 1989. *Island Africa: The Evolution of Africa's Rare Animals and Plants*. Princeton: Princeton University Press.

Kitchener, A. 1991. *The Natural History of the World's Cats*. Ithaca: Cornell University Press.

Kleiman, D. G. and J. F. Eisenberg. 1973. Comparisons of canid and felid social systems from an evolutionary perspective. *Anim. Behav.*, 21: 637–59.

Kolenosky, G. B. and D. H. Johnston. 1967. Radio-tracking timber wolves in Ontario. *Amer. Zool.* 7: 289–303.

Krampitz, H. R., R. Sachs, G. Schaller, and R. Schindler. 1968. Zur Verbreitung von Parasiten der Gattung *Hepatozoon* Miller 1908 (Protozoa, Adeleidae) in ostrafrikanischen Wildsaugetieren. (The extent of parasites of the species *Hepatozoon* Miller 1908 [Protozoa, Adeleidae] in East African wild mammals.) *Z. Parasitenk.*, 31: 203–10.

Kruger National Park. 1974. *Kruger National Park (KNP) Annual Report No. 48 of the National Parks Board of Trustees for the Period 1 April 1973–31 March 1974*.

Kruuk, H. 1972. *The Spotted Hyena: A Study of Predation and Social Behavior*. Chicago: University of Chicago Press.

Kruuk, H. and M. Turner. 1967. Comparative notes on predation by lion, leopard, cheetah and wild dog in the Serengeti area, East Africa. *Mammalia*, 31(11): 1–27.

Kühme, W. 1965. Freilandstudien zur Soziologie des Hyanenhundes (*Lycaon pictus lupinus*. Thomas 1902). (A study of the sociology of free-ranging wild dogs [*Lycaon pictus lupinus*. Thomas 1902]) *Z. Tierpsych.* 22(5): 495–541.

Kutzer, E. 1966. Zur epidemiologie der Sarcoptesräeude. (The epidemiology of sarcoptic mange.) *Agnew. Parasitol.*, 7: 241–48.

Kutzer, E. and K. Onderscheka. 1966. Die Räeude der Gemse und ihre Bekämpfung. (The mange of chamois and their fight against it.) *Z. Jagdwiss.*, 12: 63–84.

Laikre, L. and N. Ryman. 1991. Inbreeding depression in a captive wolf (*Canis lupus*) population. *Consv. Biol.*, 5: 33–40.

Lamprey, H. F. 1963. Ecological separation of the large mammal species in the Tarangire Game Reserve, Tanganyika. *E. Afr. Wildl. J.*, 1: 63–92.

——— 1964. Estimation of the large mammal densities, biomass, and energy exchange in the Tarangire Game Reserve and the Masai steppe in Tanganyika. *E. Afr. Wildl. J.*, 2: 1–46.

Lapage, G. 1968. *Veterinary Parasitology*. Springfield, Ill.: C. C. Thomas.

Laurenson, M. K., T. Caro and M. Borner. 1992. Female cheetah reproduction. *Natl. Geo. Res. and Expl.*. 8(1): 64–75.

Lawick, H. Van and J. Van Lawick-Goodall. 1971. *Innocent Killers*. Boston: Houghton Mifflin.

Lawick-Goodall, J. Van. 1968. The behavior of free-living chimpanzees in the Gombe Stream Reserve. *Animal Behav. Monogr.* 1(3): 161–311.

Ledger, H. P. 1968. Body composition as a basis for comparative study of some East African mammals. In M. A. Crawford, ed., Comparative nutrition of wild animals. *Symp. Zool. Soc. London*, 21: 289–310.

Lee, R. B. 1979. *The !Kung San: Men, Women and Work in a Foraging Society*. New York: Cambridge University Press.

Leopold, B. D. and P. R. Krausman. 1986. Diets of three predators in Big Bend National Park, Texas. *J. Wildl. Manage.*, 50: 290–95.

Le Roux, P. G. and J. D. Skinner. 1989. A note on the ecology of the leopard (*Panthera pardus* Linnaeus) in the Londolozi Game Reserve, South Africa. *Afr. J. Ecol.*, 27(20): 167–71.

Leuthold, W. 1977. *African Ungulates: A Comparative Review of the Ethology and Behavioral Ecology.* New York: Springer Verlag.

Leuthold, W. and B. M. Leuthold. 1976. Density and biomass of ungulates in Tsavo East National Park, Kenya. *E. Afr. Wildl. J.*, 14: 49–58.

Leyhausen, P. 1965. The communal organization of solitary mammals. *Symp. Zool. Soc. London*, 14: 249–63.

———— 1979. *Cat behavior. The predatory and social behavior of domestic and wild cats.* Garland STPM Press, N.Y.

Leyhausen, P. and R. Wolff. 1959. Das Revieu einer Hauskatz. (The territory of the house cat.) *Z. Tierpsychol.* 16: 66–70.

Leyn, G. De. 1962. *Contribution a la Connaissance des Lycaons au Parc National de la Kagera.* (A contribution to the knowledge of wild dogs [Lycaon] at the Kagera National Park.) Brussels: Inst. Parcs. Nat. Congo et du Rwanda.

Lockie, J. D. 1966. Territoriality in small carnivores. *Symp. Zool.Soc. London*, 18: 143–65.

Logan, K. A., L. L. Irwin, and T. Skinner. 1986. Characteristics of a hunted mountain lion population in Wyoming. *J. Wildl. Manage.*, 50: 648–54.

Lord, R. G., Jr. 1957. Estimation of fox populations. Sc.D. diss., John Hopkins University.

Lord, R. G., Jr., F. C. Bellrose, and W. W. Cochran. 1962 Radiotelemetry of the respiration of a flying duck. *Science*, 137: 39–40.

McBride, R. T. 1976. The status and ecology of the mountain lion *Felis concolor stanleyana*, of the Texas-Mexico border. M.Sc. thesis, Sul Ross State University.

McDiarmid, A. 1962. Diseases of free-living wild animals. *F.A.O. Agr. Studies*, no. 57.

McDougal, C. W. 1988. Leopard and tiger interactions at Royal Chitwan National Park, Nepal. *J. Bombay Natl. Hist. Soc.*, 85: 609–11.

McDougal. C. W. 1977. *The Face of the Tiger.* London: Rivington Books and Andre Deutsch.

McLaughlin, R. 1970. Aspects of the biology of cheetahs *Acinonyx jubatus* (Schreber) in Nairobi National Park. M.Sc. thesis. University of Nairobi, Kenya.

Maehr, D. S., R. C. Belden, E. D. Land, and L. Wilkins. 1990. Food habits of panthers in southwest Florida. *J. Wildl. Manage.*, 54: 420–23.

Maglio, V. J. 1978. Patterns of faunal evolution. In V. J. Maglio and H. B. S. Cooke, eds., *Evolution of African Mammals*, pp. 603-19. Cambridge: Harvard University Press.

Makacha, S. and G. Schaller. 1969. Observations on lions in the Lake Manyara National Park, Tanzania. *E. Afr. Wildl. J.* 7: 99–103.

Marais, E. 1939. *My Friend the Baboon.* London: Methuen.

Marshall, A. D. and J. H. Jenkins. 1966. Movements and home ranges of bobcats as determined by radio-tracking in the upper coastal plain of west-central South Carolina. *Proc. S. E. Assoc. Game and Fish Comm.*, 20: 206–14.

Marston, M. A. 1942. Winter relations of bobcats to white-tailed deer in Maine. *J. Wildl. Manage.*, 6: 328–37.

Martin, R. B. and T. De Meulenaer. 1988. *Survey of the Status of the Leopard (Panthera pardus) in Sub-Saharan Africa.* Lausanne, Switzerland: Secretariat on the Convention on Intl. Trade in Endangered Species of Wild Fauna and Flora.

Mason, D. R. 1976. Observations on social organization, behaviour and distribution of impala in the Jack Scott Nature Reserve. *S. Afr. J. Wildl. Res.*, 6(2): 79–87.

Matthiessen. P. 1991. *African Silences.* New York: Random House.

Meadows, R. 1991. How many leopards are there in Africa? In J. D. Seidensticker and S. Lumpkin, eds., *Great Cats: Majestic Creatures of the Wild*, p. 114. Emmaus, Penn.: Rodale Press.

Mech, L. D. 1970. *The Wolf: Ecology and Behavior of an Endangered Species.* New York: Natural History Press.

Meinertzhagen, R. 1938. Some weights and measurements of large mammals. *Proc. Zool. Soc. London*, 108: 433–39.

Mentis, M. T. 1970. Estimates of natural biomass of large herbivores in the Umfolozi game reserve area. *Mammalia*, 34(3): 363–93.

Mentis, M. T. and R. R. Duke. 1976. Carrying capacities of natural veld in Natal for large wild herbivores. *S. Afr. J. Wildl. Res.*, 6(2): 65–74.

Mills, M. G. L. 1984. Prey selection and feeding habits of the large carnivores in the southern Kalahari. *Koedoe Suppl.*, 27: 281–94.

Mitchell, B., J. Shenton and J. Uys. 1965. Predation on large mammals in the Kafue National Park, Zambia. *Zool. Afr.*, 1(2): 297–318.

Miththapala, S. 1991. How to tell a leopard by its spots. In J. D. Seidensticker and S. Lumpkin, eds., *Great Cats: Majestic Creatures of the Wild*, p. 112. Emmaus, Penn: Rodale Press.

Muckenhirn, N. and J. F. Eisenberg. 1973. Home ranges and predation in the Ceylon leopard. In R. L. Eaton, ed., *The World's Cats.* Vol. 1, *Ecology and Conservation*, pp. 142–75. Winston, Ore.: World Wildlife Safari.

Murray, M. 1967. The pathology of some diseases found in wild animals of East Africa. *E. Afr. Wildl. J.*, 5: 37–45.

Murray, M. and A. R. Njogu. 1989, African trypanosomiasis in wild and domestic ungulates: The problem, and its control. In P. A. Jewell and G. M. O. Maloiy, eds., *The Biology of Large African Mammals in Their Environment. Symp. Zool. Soc. London*, 61: 217–40.

Myers, N. 1973. The spotted cats and the fur trade. In R. L. Eaton, ed., *The World's Cats.* Vol. 1, *Ecology and Conservation*, pp. 276–326. Winston, Ore.: World Wildlife Safari.

——— 1974. The leopard hangs tough. *Intl. Wildl.*, 4(6): 3–12.

——— 1975. Silent savannahs. *Intl. Wildl.*, 5: 4–11.

——— 1976. The leopard *Panthera pardus* in Africa. *IUCN Monogr.*, no. 5.

——— 1981. The leopard in Africa: Biological and cultural. *Intl. J. Stud. Anim. Prob.*, 2(1): 5–6.

——— 1986. Conservation of Africa's cats: Problems and opportunities. In S. D. Miller and D. D. Everett, eds., *Cats of the World: Biology, Conservation, and Management*, pp. 437–46. Washington, D.C.: National Wildlife Federation.

Nice, M. M. 1941. The role of territory in bird life. *Am. Midl. Natl.*, 26(3): 441–87.

Nicholls, T. H. and D. W. Warner. 1972. Barred owl habitat use as determined by radiotelemetry. *J. Wildl. Manage.*, 36: 213–24.

Nishida, T. 1968. The social group of wild chimpanzees in the Mahali Mountains. *Primates* 9(3): 167–224.

Norton, B. G. 1987. *Why Preserve Natural Diversity?* Princeton: Princeton University Press.

Norton, P. M. and S. R. Henley. 1987. Home range and movements of male leopards in the Cedarberg Wilderness Area, Cape Province. *S. Afr. J. Wildl. Res.*, 17: 41–48.

Norton, P. M. and A. B. Lawson. 1985. Radio tracking of leopards and caracals in the Stellenbosch area, Cape Province. *S. Afr. J. Wildl. Res.*, 15: 17–24.

Norton, P. M., A. B. Lawson, S. R. Henley, and G. Avery. 1986. Prey of leopards in four mountainous areas of the south-western Cape Province. *S. Afr. J. Wildl. Res.*, 16: 47–52.

Novikov, G. 1961. *Carnivorous mammals of the fauna of the USSR*. Israel Program for Scientific Translations. Springfield, Va.: U.S. Department of Commerce.

O'Brien, S. J., M. E. Roelke, L. Maker, A. Newman, C. A. Winkler, D. Meltzer, L. Colley, J. F. Everman, M. Bush, and D. E. Wildt. 1985. Genetic evidence for species vulnerability in the cheetah. *Science*, 227: 1428–34.

O'Brien, S. J., G. E. Collier, R. E. Benveniste, W. G. Nash, A. K. Newman, J. M. Simonson, M. A. Eichelberger, U. S. Seal, D. Janssen, M. Bush, and D. E. Wildt. 1987. Setting the molecular clock in the Felidae: The great cats, *Panthera*. In R. L. Tilson and U. S. Seal, eds., *Tigers of the World: The Biology, Biopolitics, Management, and Conservation of an Endangered Species*, pp. 10–27. Park Ridge, N.J.: Noyes.

Onderscheka, K., E. Kutzer and H. Richter. 1968. Die Räeude der Gemse und its Bekämpfung. II. Zusammenhaenge Zwischen Ernaehrung und Räeude. (The mange of chamois and their fight against it. Part 2. The relation between nourishment and mange.) *Z. Jagdwiss.* 14: 12–17.

Packer, C. 1986. The ecology of sociality in felids. In D. I. Rubenstein and R. W. Wrangham, eds., *Ecological Aspects of Social Evolution: Birds and Mammals*, pp. 429–51. Princeton: Princeton University Press.

Passmore, J. 1974. *Man's Responsibility for Nature*. New York: Scribner's.

Patterson, W. 1969. Bushbuck solitary but brave. *Africana, E. Afr. Wild. Soc. Rev.*, 3(9): 12–13.

Paynter, D. 1986. *Kruger: Portrait of a National Park*. Johannesburg: Macmillan.

Pellerdy, L. P. 1965. *Coccidia and Coccidiosis*. Budapest: Akademiai Kiado.

Pence, D. B., F. D. Matthews III, and L. A. Windberg. 1982. Notoedric mange in the bobcat, *Felis rufus*, from South Texas. *J. Wildl. Diseases*, 18(1): 47–50.

Penner, L. and W. Parke. 1954. Notoedric mange in the bobcat, *Lynx rufous. J. Mammal.*, 35: 458.

Peterson, R. O., J. D. Woolington, and T. N. Bailey. 1984. Wolves of the Kenai Peninsula, Alaska. *Wildl. Monogr.*, no. 88.

Petrides, G. A. and W. G. Swank. 1965. Population densities and the range-carrying capacity for large mammals in Queen Elizabeth National Park, Uganda. *Zool. Afr.*, 1: 209–25.

Phillips, J. F. V. 1965. Fire—as master or servant, its influence on the bioclimatic regions of trans-Saharan Africa. *Proc. Tall Timbers Fire Ecol. Conf.*, 4: 7–110.

Pienaar, U. De V. 1963. The large mammals of Kruger National Park—their distribution and present day status. *Koedoe*, 6: 1–38.

——— 1968. The ecological significance of roads in a national park. *Koedoe*, 11: 169–74.

——— 1969. Predator-prey relationships amongst the larger mammals of the Kruger National Park. *Koedoe*, 12: 108–76.

——— 1974. Habitat-preference in South African antelope species and its significance in natural and artificial distribution patterns. *Koedoe*, 17: 185–95.

——— 1985. Indications of progressive desiccation of the Transvaal Lowveld over the past 100 years, and implications for the water stabilization program in the Kruger National Park. *Koedoe*, 28: 93–165.

Pienaar, U. De V., P. Van Wyk, and N. Fairall. 1966a. An experimental cropping scheme of hippopotami in the Letaba River of the Kruger National Park. *Koedoe*, 9: 1–33.

——— 1966b. An aerial census of elephant and buffalo in the Kruger National Park and the implications thereof on intended management schemes. *Koedoe*, 9: 40–107.

Pitelka, F. A. 1959. Numbers, breeding schedule and territoriality in Pectoral sandpipers of northern Alaska. *Condor*, 61: 233–64

Powell, A. 1958. *Call of the Tiger*. New York: A. S. Barnes.

Prynn, D. 1980. Tigers and leopards in Russia's Far East. *Oryx*, 15(5): 496–504.

Rabinowitz, A. R. 1989. The density and behavior of large cats in a dry tropical forest mosaic in Huai Kha Khaeng Wildlife Sanctuary, Thailand. *Nat. Hist. Bull. Siam Soc.*, 37(2): 235–51.

——— 1990. Fire, dry dipterocarp forest, and the carnivore community in Huai Kha Khaeng Wildlife Sanctuary, Thailand. *Nat. Hist. Bull. Siam Soc.*, 38(2): 99–115.

——— 1991. Eye of the tiger. *Wildl. Cons.*, 94(6): 48–55.

Rabinowitz, A. R., P. Andau, and P. P. K. Chai. 1987. The clouded leopard in Borneo. *Oryx*, 21: 107–11.

Rabinowitz, A. R. and B. G. Nottingham, Jr. 1986. Ecology and behavior of the jaguar (*Panthera onca*) in Belize, Central America. *J. Zool. Soc. London*, 210: 149–59.

Rabinowitz, A. R. and S. R. Walker. 1991. The carnivore community in a dry tropical forest mosaic in Huai Kha Khaeng Wildlife Sanctuary, Thailand. *J. Trop. Ecol.*, 7(1): 37–47.

Ralls, K., J. D. Ballou, and A. Templeton. 1988. Estimates of lethal equivalents and the cost of inbreeding in mammals. *Consv. Biol.*, 2: 185–93.

Raup, D. M. 1991. *Extinction: Bad Genes or Bad Luck?* New York: W. W. Norton.

Reuther, R. and J. Doherty. 1968. Birth seasons of mammals at San Francisco Zoo. In C. Jarvis, ed., *Intl. Zoo Yearbk.*, 8: 98–101. London: London Zoological Society.

Robinette, W. L. 1963. Weights of some of the larger mammals of N. Rhodesia. *Puku*, 1: 207–15.

Robinette, W. L., J. S. Gashwiler, and O. W. Morris. 1961. Notes on cougar productivity and life history. *J. Mammal.*, 42: 204–17.

Robinette, W. L., C. M. Loveless, and D. A. Jones. 1974. Field tests of strip census methods. *J. Wildl. Manage.*, 38: 81–96.

Robinson, I. H. and M. Delibes. 1988. The distribution of faeces by the Spanish lynx (*Felis pardina*). *J. Zool. London* 216: 577–82.

Robinson, R. 1969. The breeding of spotted and black leopards. *J. Bombay Natl. Hist. Soc.*, 66: 423–29.

Rothman, R. J. and D. L. Mech. 1979. Scent-marking in lone wolves and newly formed pairs. *Anim. Behav.*, 27: 750–60.

Ruggiero, R. G. 1990. Prey selection of the lion (*Panthera leo* L.) in the Manovo-Gounda-St. Floris National Park, Central African Republic. *Mammalia*, 55: 23–33.

Sachs, R. 1967. Live weights and body measurements of Serengeti game animals. *E. Afr. Wildl. J.*, 5: 24–36.

Sachs, R., C. Staak, and C. Groocock. 1968. Sociological investigation of brucellosis in game animals in Tanzania. *Bull. Epizoot. Dis. Afr.*, 16: 93–100.

Sadlier, R. 1966. Notes on reproduction in the larger Felidae. In C. Jarvis, ed., *Intl. Zoo. Yearbook*, 6: 184–87. London: London Zoological Society.

Sankhala, K. 1977. *Tiger! The story of the Indian tiger*. New York: Simon and Schuster.

Santiapillai, C., M. R. Chambers, and N. Ishwaran. 1982. The leopard (*Panthera pardus*) (Meyer 1794) in the Ruhuna National Park, Sri Lanka, and observations relative to its conservation. *Biol. Conserv.*, 23: 5–14.

Saunders, J. K., Jr. 1963. Movements and activities of the lynx in Newfoundland. *J. Wildl. Manage.*, 27(3): 390–400.

Schaller, G. B. 1967. *The Deer and the Tiger*. Chicago: University of Chicago Press.

—— 1972. *The Serengeti Lion*. Chicago: University of Chicago Press.

—— 1977. *Mountain Monarchs: Wild Sheep and Goats of the Himalaya*. Chicago: University of Chicago Press.

Schaller, G. B. and P. G. Crawshaw, Jr. 1980. Movement patterns of jaguar. *Biotropica*, 12: 161–68.

Schenkel, R. 1966. On the sociology and behaviour in impala (*Aepyceros melampus*, Lichtenstein). *E. Afr. Wildl. J.*, 4: 99–114.

Schnabel, Z. E. 1938. Estimation of the total fish population of a lake. *Am. Math. Monthly*, 45: 348–52

Schonewald-Cox, C., R. Azari, and S. Blume. 1991. Scale, variable density, and conservation planning for mammalian carnivores. *Consv. Biol.*, 5: 491–95.

Schutte, I. C. 1986 The general geology of the Kruger National Park. *Koedoe*, 29: 13–37.

Scott, P. P. 1968. The special features of nutrition of cats, with observations on wild felidae nutrition in the London zoo. *Symp. Zool. Soc. London*, 21: 21–36.

Seidensticker, J. C. 1976a. On the ecological separation between tigers and leopards. *Biotropica*, 8: 225–34.

—— 1976b. Ungulate populations in Chitwan Valley, Nepal. *Biol. Conserv.*, 10: 183–209.

—— 1977. Notes on early maternal behavior of the leopard. *Mammalia*, 41: 111–13.

—— 1986. Large carnivores and the consequences of habitat insularization: Ecology and conservation of tigers in Indonesia and Bangladesh. In S. D. Miller and D. D. Everett, eds., *Cats of the World: Biology, Conservation, and Management*, pp. 1–41. Washington, D.C.: National Wildlife Federation.

Seidensticker, J. C., M. G. Hornocker, R. R. Knight, and S. L. Judd. 1970. Equipment and techniques for radiotracking mountain lions and elk. *Forest, Wildl. and Range Exp. Station Bull.*, no. 6.

Seidensticker, J. C., M. G. Hornocker, W. V., Wiles, and J. P. Messick. 1973. Mountain lion social organization in the Idaho Primitive Area. *Wildl. Monogr.*, no. 35.

Seidensticker, J., M. E. Sunquist, and C. W. McDougal. 1990. Leopards living at the edge of the Royal Chitwan National Park Nepal. In J. C. Daniel and J. S. Serro, eds., *Conservation in Developing Countries: Problems and Prospects*, pp. 415–23. Bombay Natural History Society, Bombay, India: Oxford University Press.

Selous, F. C. 1920. *A Hunter's Wanderings in Africa*. London: Macmillan.

Simon, N. M. 1969. Proposals for field investigations of rare and endangered mammals. *Biol. Conserv.*, 1: 280–90.

Simpson, C. D. 1972. An evaluation of seasonal movements in greater kudu populations *Tragelaphus strepiceros* Pallas in three localities in Southern Africa. *Zool. Afr.*, 7: 197–205.

Smith, J. L. D. and C. W. McDougal. 1990. The contribution of variance in lifetime reproduction to effective population size in tigers. *Consv. Biol.* 5(4): 484–90.

Smith, J. L. D., C. W. McDougal, and D. Miquelle. 1989. Scent marking in free-ranging tigers, *Panthera tigris*. *Anim. Behav.*, 37: 1–10.

Smith, J. L. D., C. W. McDougal, and M. E. Sunquist. 1987. Female land tenure system in tigers. In R. L. Tilson and U. S. Seal, eds., *Tigers of the World: The Biology, Biopolitics, Management, and Conservation of an Endangered Species*, pp. 97–109. Park Ridge, N.J.: Noyes.

Smith, R. 1962. Hyena versus leopard. *Afr. Wildl.*, 16: 282–86.

Smith, R. M. 1977. Movement patterns and feeding behaviour of leopard in the Rhodes Matopos National Park, Rhodesia. *Arnoldia*, 13(8): 1–16.

———— 1978. Movement patterns and feeding behaviour of the leopard in the Rhodes Matopos National Park, Rhodesia. *Carnivora*, 1: 58–69.

Smuts, G. L. 1975. Predator-prey relationships in the Central District of the Kruger National Park with emphasis on wildebeest and zebra populations. Memorandum. Dept. of Nature Conservation, Kruger National Park.

———— 1976. Population characteristics and recent history of lions in two parts of the Kruger National Park. *Koedoe*, 19: 153–64.

———— 1978. Interrelationships between predators, prey and their environment. *Bioscience*, 28: 316–20.

———— 1982. *Lion*. Johannesburg: Macmillan South Africa.

Smuts, G. L., B. R. Bryden, V. De Vos, and E. D. Young. 1973. Some practical advantages of CI-581 (Ketamine) for the field immobilization of larger wild felines, with comparative notes on impala and baboon. *Lammergeyer*, 18: 1–14.

Smuts, G. L., J. Hanks, and I. J. Whyte. 1978. Reproduction and social organization of lions from the Kruger National Park. *Carnivore*, 1: 17–28.

Smuts, G. L., I. J. Whyte, and T. W. Dearlove. 1977. Advances in the mass capture of lions *(Panthera leo)*. *Proc. 13th Congress of Game Biologists, Atlanta, Ga.*, 420–31.

Spinage, C. A. 1974. Territoriality and population regulation in the Uganda defassa waterbuck. In V. Geist, and F. Walther, eds., *The Behaviour of Ungulates and Its Relation to Management*, pp. 635–43. Morges, Switzerland: IUCN.

Stevenson-Hamilton, J. 1937. *South African Eden*. London: Cassell.

———— 1947. *Wildlife in South Africa*. London: Cassell.

Storm, G. L. 1965. Movements and activities of foxes as determined by radiotracking. *J. Wildl. Manage.*, 29: 1–13.

Stroganov, S. V. 1969. *Carnivorous mammals of Siberia*. Israel Program for Scientific Translation. Springfield, Va.: U.S. Department of Commerce.

Struhsaker, T. T. 1975. *The Red Colobus Monkey*. Chicago: University of Chicago Press.

Stuart, C. T. 1986. The incidence of surplus killing by *Panthera pardus* and *Felis caracal* in Cape Province, South Africa. *Mammalia*, 50: 556–58.

Stuart, S. N., R. J. Adams, and M. D. Jenkins. 1990. Biodiversity in sub-Saharan Africa and its islands: Conservation, management and sustainable use. *Occasional Papers of the IUCN Species Survival Commission*, no. 6.

Sunquist, M. E. 1981. The social organization of tigers *(Panthera tigris)* in Royal Chitwan National Park, Nepal. *Smithsonian Contrib. Zool.*, 336: 1–98.

———— 1983. Dispersal of three radiotagged leopards. *J. Mammal.*, 64(2): 337–41.

Sweatman, G. 1971. Mites and pentastomes. In J. Davis and R. Anderson, eds., *Parasitic Diseases of Wild Mammals*, pp. 3–64. Ames: Iowa State University Press.

Teer, J. G., and W. G. Swank. 1978. Status of the leopard in Africa south of the Sahara. Unpublished report for the U. S. Fish and Wildlife Service.

Tehsin, R. H. 1980. Do leopards use their whiskers as wind detectors? *J. Bombay Nat. Hist. Soc.*, 77: 128–29.

Theiler, G. 1964. Ecogeographical aspects of tick distribution. In D. H. S. Davis, ed., *Ecological Studies in Southern Africa*, pp. 284–300. The Hague: W. Junk.

Thesen, H. 1972. A leopard sanctuary in the Southern Cape forests. *Afr. Wildl.*, 28(1): 11–13.

Thomson, P. J. 1975. The role of elephants, fire, and other agents in the decline of a *Brachystegia boehmii* woodland. *J. South Afr. Wildl. Manage. Assoc.*, 5: 11–18.

Thomson, W. R. 1974. Tree damage by porcupines in south west Rhodesia. *J. South Afr. Wildl. Manage. Assoc.*, 4: 123–27.

Timberlake, L. 1986. *Africa in Crisis: The Causes, the Cures of Environmental Bankruptcy.* Philadelphia: New Society.

Timoney. T. 1924. The bionomics of the sarcoptic mange parasite of the buffalo, with some observations concerning the relative power of resistance to adverse conditions of the different stages of the acarus and of its egg. *Rep. Proc. 5th Entomol. Meeting:* 180–200.

Tucker, C. J., J. R. G. Townsend, and T. E. Goff. 1985. African land-cover classification using satellite data. *Science,* 227: 369–75.

Turnbull-Kemp, P. 1967. *The Leopard.* Cape Province, So. Afr.: Howard Timmins.

U.S. Department of the Interior, 1982. Fish and Wildlife Service. *Endangered and Threatened Wildlife and Plants.* Code fed. reg. 50, sections 17.11–12.

Van Ballenberghe, V., A. W. Erickson and D. Byman. 1975. Ecology of the timber wolf in Minnesota. *Wildl. Monogr.* no. 43.

Van Hooff, J. 1967. The facial displays of the Catarrhine monkeys and apes. In D. Morris, ed., *Primate Ethology,* pp. 7–68. Chicago: Aldine.

Van Wyk, P. 1971. Veldburning in the Kruger National Park, an interim report on some aspects of research. *Proc. Tall Timbers Fire Ecol. Conf.,* 11: 9–31.

———— 1972. *Trees of the Kruger National Park.* Vol. 1. Cape Town: Purnell and Sons.

Van Wyk, P. and N. Fairall. 1969. The influence of the African elephant on the vegetation of the Kruger National Park. *Koedoe,* 12: 57–89.

Van Wyk, P. and V. A. Wager. 1968. Problems of fire in the Kruger National Park. *Afr. Wildl.,* 22(4): 269–80.

Venter, F. J. and W. P. D. Gertenbach. 1986. A cursory review of the climate and vegetation of the Kruger National Park. *Koedoe,* 29: 139–48.

Walther, F. R. 1969. Flight behavior and avoidance of predators in Thomson's gazelle (*Gazella thomsonii* Guenther 1884). *Behaviour,* 34: 184–221.

Waser, P. 1975. Diurnal and nocturnal strategies of the bushbuck *Tragelaphus scriptus* (Pallas). *E. Afr. Wildl. J.,* 13: 49–63.

Watson, R. and M. Turner. 1965. A count of the larger mammals of the Lake Manyara National Park: Results and discussion. *E. Afr. Wildl. J.,* 3: 95–98.

Western, D. 1975. Water availability and its influence on structure and dynamics of a savannah large mammal community. *E. Afr. Wildl. J.,* 13: 265–86.

———— 1982. Amboseli National Park: Enlisting landowners to conserve migratory wildlife. *Ambio,* 11: 302–8.

Wharton, G. 1960. Water balance in mites. *J. Parasit.,* 46: 6

———— 1963. Equilibrium humidity. In J. Naegele, ed., *Advances in Acarology,* pp. 201–8. Ithaca: Cornell University Press.

White, F. 1983. *The Vegetation of Africa.* Paris: UNESCO.

White, L. J. T. 1992. Here an elephant. *Wildl. Conserv.,* 95 (1): 37–43.

Whyte, I. J. 1976. Aspects of impala (*Aepyceros melampus* Lichtenstein) behaviour and ecology which may effect the epizootiology of foot-and-mouth disease in the Kruger National Park. Cert. in field ecology, thesis, University of Rhodesia.

Wilkie, D. S., J. G. Sidle, and G. C. Boundzanga. 1992. Mechanized logging, market hunting, and a bank loan in Congo. *Consv. Biol.,* 6(4): 570–80.

Wilson, E. O. 1971. Competitive and aggressive behaviour. In J. F. Eisenberg and W. S. Dillon, eds., *Man and Beast: Comparative Social Behavior,* pp. 181–217. Washington, D.C.: Smithsonian Institution Press.

———— 1975. *Sociobiology: The New Synthesis.* Cambridge: Harvard University Press, Belknap Press.

Wilson, V. J. 1965. Observations of the greater kudu (*Tragelaphus strepsiceros* Pallas) from tsetse control hunting schemes in Northern Rhodesia. *E. Afr. Wildl. J.*, 3: 27–37.

────── 1968. Weights of some mammals from eastern Zambia. *Arnoldia*, 3: 1–20.

Wilson, V. J. and G. Child. 1966. Notes of development and behaviour of two captive leopards. *D. Zool. Garten.* (N.F.) 32: 67-70.

Wilson, V. J. and H. Roth. 1967. The effects of tsetse control operations on common duiker in eastern Zambia. *E. Afr. Wildl. J.*, 5: 53–64.

World conservation strategy. 1980. Gland, Switzerland: IUCN.

Wright, B. S. 1960. Predation on big game in East Africa. *J. Wildl. Manage.*, 24: 1–15.

Wright, S. 1977. *Evolution and the Genetics of Populations.* Vol. 3. Chicago: University of Chicago Press.

Yanez, J. L., J. C. Cardenas, P. Gezelle, and F. M. Jaksic. 1986. Food habits of the southernmost mountain lions *(Felis concolor)* in South America: Natural versus live-stocked ranges. *J. Mammal.*, 67: 604–6.

Young, E. 1972. Observations on the movement patterns and daily home range size of impala, *Aepyceros melampus* (Lichtenstein), in the Kruger National Park. *Zool. Afr.*, 7: 187–95.

Young, E. and L. J. J. Wagener. 1963. The impala as a source of food and by-products. *J. South Afr. Vet. Med. Assoc.*, 39: 81–86.

Young, E., F. Zumpt, and I. J. Whyte. 1972a. *Notoedres cati* (Hering, 1838) infestation of the cheetah: Preliminary report. *J. South Afr. Vet. Med. Assoc.*, 43: 205.

────── 1972b. Sarcoptic mange in free-living lions. *J. South Afr. Vet. Med. Assoc.*, 43: 226.

Zong-Yi, W. and W. Sung. 1986. Distribution and recent status of the Felidae in China. In S. D. Miller and D. D. Everett, eds., *Cats of the World: Biology, Conservation, and Management*, pp. 201–9. Washington, D.C.: National Wildlife Federation.

Zuckerman, S. 1953. The breeding season of mammals in captivity. *Proc. Zool. Soc. London*, 122: 827–950.

Index